Conserving Forest Diversity through Ecosystem Management

Conserving Forest Diversity through Ecosystem Management

Trees, Forests, Environment, Habitat, and Wildlife

David R. Patton
Northern Arizona University

Bruce E. Fox
Northern Arizona University

John D. Bailey
Oregon State University

WAVELAND PRESS, INC.
Long Grove, Illinois

For information about this book, contact:
Waveland Press, Inc.
4180 IL Route 83, Suite 101
Long Grove, IL 60047-9580
(847) 634-0081
info@waveland.com
www.waveland.com

13-digit ISBN 978-1-4786-3785-1
10-digit ISBN 1-4786-3785-4

Printed in the United States of America

7 6 5 4 3 2 1

In Memoriam

Elbert L. Little, Jr. (1907–2004)
Chief Dendrologist
USDA Forest Service

At some time in their career, forestry and wildlife professionals will see the name Elbert Little on a publication that they use in their research or management activities. Elbert was an authority on trees as evidenced by the 25 books about trees published in his career of 70 years, mostly when he was an employee of the USDA Forest Service. He had a BA and BS in botany and zoology from the University of Oklahoma and an MS and PhD in botany from the University of Chicago. For eight years he was a forest ecologist with the USDA Forest Service based in Tucson, Arizona. His research emphasis was on poisonous range plants, watershed management, and pinyon-juniper woodland in New Mexico and Arizona. In 1942 he moved to Washington, DC where he was a dendrologist with the Forest Service for 34 years. Some of his most influential books include a five-volume atlas series on trees of the United States and a field guide to North American Trees, eastern and western regions (*Atlas of United States Trees*, 6 vols., USDA 1971–1981).

In his career Elbert received the Superior Service Award and the Distinguished Service Award from the Forest Service and a Distinguished Service Award from the American Forestry Association. Because he was fluent in Spanish, the United Nations used his expertise in the Food and Agriculture Organization in Mexico, Ecuador, and Nicaragua. He was a member and a fellow of the Society of American Foresters, and a member of both the American Association for the Advancement of Science and the International Society of Plant Taxonomists. Although he retired from the Forest Service in 1975, Elbert continued to work as a research associate with the Department of Botany, U.S. National Museum of Natural History in the Smithsonian Institute.

During his time with the Forest Service Elbert visited the locations of the research stations and one of his favorite places was the Forest Hydrology Lab on the Arizona State University campus. As I was the director's representative at the lab, he always visited with me to be brought up to date on forestry research in Arizona and in particular to hear about ongoing studies in the pinion-juniper woodland. During some of these visits, he traveled with scientists to see their study sites and the collection of data. These trips were beneficial to him and the scientists because of the exchange of ideas that took place. When I see the word *tree* or *dendrology* I am reminded of Elbert Little, his trips to Arizona, and the enormous contribution he made to botany and our forestry and wildlife professions.

David R. Patton
Professor and Dean Emeritus
Northern Arizona University

About the Authors

David R. Patton (drpwild@msn.com) is professor and dean emeritus, School of Forestry, Northern Arizona University. He earned a BS degree in forestry from West Virginia University, an MS degree in wildlife management from Virginia Tech, a PhD in watershed management from the University of Arizona, and is a Certified Wildlife Biologist. His career includes assignments on the Cleveland National Forest, the Santa Fe National Forest, and research biologist and project leader for the Rocky Mountain Forest and Range Experiment Station, Fort Collins, Colorado, and wildlife biologist for the U.S. Fish and Wildlife Service in Decatur, Alabama. He has lived and worked in wildlife and forestry positions in Zambia, Kenya, and Australia. David is a member of the Society of American Foresters and The Wildlife Society and in 2017 he was presented with a Lifetime Achievement Award by the Arizona Chapter of The Wildlife Society.

Bruce E. Fox (bruce.fox@nau.edu) is professor of Forest Management, School of Forestry, Northern Arizona University. He earned his BS in forestry and MF degrees from the University of California, Berkeley, and a PhD from the University of Michigan. He worked in private industry before coming to Northern Arizona University. He served as coordinator of academic affairs and as chair for the School of Forestry, director of the Honors Program, and director of the Master of Administration Program at Northern Arizona University. He teaches forest operations and management in the school's integrated undergraduate forestry program. He also continues to work as a consultant for the forest and aerospace industries. He is a Registered Professional Forester in the State of California.

John D. Bailey (john.bailey@oregonstate.edu) is professor of Silviculture and Fire Management, College of Forestry, Oregon State University. He earned his BS and MS degrees in forestry and forest biology from Virginia Tech, and his PhD in forest science and silviculture from Oregon State University. His career includes research with the Environmental Protection Agency laboratory in Corvallis, Oregon, and nine years on the faculty at Northern Arizona University including work with the Rocky Mountain Research Station. His current research and teaching focuses on forest restoration and fuels treatments in dry, fire-prone forest types, and the implications of silvicultural treatments on fire behavior, biomass availability and growth, wildlife habitat, and other associated ecosystem services. In 2019, John was awarded Fellow, Society of American Foresters.

Contents

Foreword

I was recently asked to describe this book in a few words, and I said "animals, people, and trees." It is exactly the sort of good book I would expect from these authors. This book is, to my knowledge, the first modern attempt at bringing together in one place the basic current ideas about forestry and wildlife management for beginning students. In that sense it is both a remedy for the disciplinary separation that has been reinforced by increasingly (perhaps until recently) narrow academic views of what makes a successful professional program, what makes a successful professor teaching in that program, and for the intellectual segregation in the respective research communities.

The historical figures that are most often cited as founders of the United States versions of professional forest and wildlife managers expressed views that probably should have produced integrated, ecosystem-focused programs earlier. Aldo Leopold, who laid the foundation for the study of wildlife management, was proud to call himself a "consulting forester." Gifford Pinchot, the forestry pioneer, said that the amount of timber to be harvested for wood products was that amount that could be "spared" from the support of other forest functions and benefits like wildlife habitat and watershed protection.

As one studies forests and their contents using this book it is worthwhile to consider at the outset two questions: the first is "to what actual degree are forests 'integrated systems' and not collections of individual organisms?" and the second is "how similar are the major categories of organisms, that is trees and wildlife?" Both questions go to the root of why students of trees and students of wildlife have often gone different ways. I will not give my answers here, but I challenge users of this book to use its content to answer them for themselves.

I consider this book the best on its subject I have run across, and I think many, including me, who are no longer formal students will find it both intriguing reading and a constantly valuable reference. My own approach to books is somewhat idiosyncratic: I rarely read from start to finish, unless it is a work of fiction or a court summons. For this book, I would recommend starting with Chapter 3, "Big Issues." This approach will prime the reader to put the huge helping of relevant information and knowledge in the book in an active and practical "so what" context.

In sum, I have waited a long time to see a book like this, and now, in my 80th year I am delighted to have done so.

John C. Gordon Pinchot
Professor Emeritus and Former Dean
Yale School of Forestry and Environmental Studies
jgordon@iforest.com

Preface

Many people have contributed to the forestry and wildlife professions but three, Aldo Leopold, Gifford Pinchot, and Charles Elton, are ecological pioneers who have had major influences on how we think about trees, forests, environment, habitats, and wildlife. Although several books published by these authors are more than 50 years old, their influences are still evident today in the forestry and wildlife literature and in decisions made by professionals in state, local, and federal agencies.

Aldo Leopold, educated as a forester, is considered the founder of the wildlife profession and authored the first authoritative book on game management (*Game Management* 1933). His first employment was as a forester on the Apache National Forest in Arizona. His book *A Sand County Almanac* (1949) discusses environmental ethics and the conservation of nature and wildlife. He advocated that science be the foundation for management instead of relying on game refuges and hunting regulations to protect game animals. Leopold stated his land ethic in *A Sand County Almanac* as: "A thing is right when it tends to preserve the integrity, stability and beauty of the biotic community and it is wrong when it tends otherwise." Leopold was the first person in the United States to be appointed to a university position as a professor in game management at the University of Wisconsin.

Gifford Pinchot was educated as a forester at the French National School of Forestry in Nancy, France. He was the first chief of the forest service and established the Society of American Foresters. Pinchot was from a wealthy family and endowed funds to Yale University to support the first forestry school where Leopold started as a graduate student. Pinchot's book *Breaking New Grounds* (1947) stated, "***The greatest good for the greatest number over the longest time,***" that became the foundational philosophy upon which the Forest Service's conservation policy is based. Pinchot's contributions to forestry were to help in setting aside millions of acres of land for national forests, provide leadership to elevate management to a scientific basis and to have forests on a use and renewal path.

Charles Elton was educated at Liverpool College and Oxford University in England. His thoughts presented in *Animal Ecology* (1927) are from the view of a field zoologist and an observer of nature. In this book he focuses on life history of animals, succession, food chains, size of food items, ecological niche, animal communities, and the concept of a pyramid of numbers that are major topics in scientific journals and wildlife literature. Leopold recommended *Animal Ecology* to students of game as "a good general reading." This book is more than general reading for it indicates the topics that were being discussed before wildlife and forestry had professional societies. In *The Ecology of Invasions by Animals and Plants* (1958), Elton discussed conservation and the need to maintain species diversity, which is now a common issue in conservation biology and land management activities. He was one of the first ecologists to suggest that different kinds of habitats have a characteristic group of animals.

The Focus of This Book

This book was written for an undergraduate course with students who want a basic background in forestry and wildlife management principles and practices as part of their educational program. For the working professional the content serves as an updated review of subjects related to their current position or for a continuing education and certificate program. It is important to acknowledge that over 50 percent of the forest lands in the U.S. are in private ownership, so chapters include topics that will be of interest to these landowners for planning and managing forests and wildlife jointly on their property.

Acknowledgments

In a textbook project the work of the authors is only one part of the overall process. We appreciate the assistance of Don Rosso, Carrie Campbell, Katy Murphy, and Deborah Underwood of Waveland Press for their contributions in getting this book in print.

Bruce Fox recognizes the late Edward C. Stone for his guidance in preparing him for a career in forestry and a framework of understanding ecosystems; his students in the spring 2019 forest planning course at Northern Arizona University who gave significant feedback that improved the subject matter in this book; and he thanks his wife Kathrin and children Leah and Ethan for their wholehearted support by giving up his family time so he could be an author of this textbook.

John Bailey is grateful for the quality educational and work experiences that span his career: Virginia Tech, Northern Arizona University, and Oregon State University. There are numerous mentors and colleagues from these institutions who helped form his understanding of silviculture and forestry, but none more important than David W. Smith and Richard Vasey. He also acknowledges the collective wisdom of the Society of American Foresters D2 Silviculture instructors. None of this work is possible, however, without support from his wife Bobbi and the family.

We appreciate the efforts of a group of Forest Service and state biologists who worked with the senior author during the time of the Resource Planning Act Assessment to develop a relational database for field application. We hope these biologists and foresters will recognize in many parts of this book their contributions that are still relevant in this twenty-first century.

The authors thank Doris Patton for fact checking tables, figures, and references by reading each chapter line by line to insure continuity of subject matter; and we recognize the contributions made to the education of forestry and wildlife management students by the late Dr. Steven Dewhurst, School of Forestry at Northern Arizona University. Royalties from the three authors will be contributed to the Northern Arizona University Foundation to support the Dewhurst Memorial Scholarship Fund established in the School of Forestry.

In the Beginning 1

Since the beginning of time, trees have been a part of the human environment. The Christian creation story identifies two types of trees: the tree of life and the tree of the knowledge of good and evil. The Buddhist sacred scripture, the Tipitaka, references over 200 species of trees, and the Buddha developed a great appreciation for and knowledge of trees. The Islamic tradition identifies the responsibility of Muslims to act as stewards of the environment, which includes treating trees with respect and describes planting trees is an act of charity that brings blessings to the planter. Native Americans and First Nations also have rich spiritual connections to trees, from the origin accounts of the Iroquois to the inclusion of trees into Pueblo clan names. A love of trees and communing with nature in forests is a basic tenet of the Druid belief system.

In modern times, trees have been classified by plant taxonomists as gymnosperms or angiosperms (Hardin et al. 2001). Gymnosperm ("naked seed") do not produce flowers and their seeds are directly exposed to the air. Angiosperms have seeds inside a fruit and are therefore "hidden." In general terms, gymnosperms are often called "conifers" and angiosperms are called "hardwoods." Within these two classes there is variation so the two terms are not mutually exclusive. Two other terms, softwoods (conifers) and broad leaved (hardwoods), are also commonly used to classify trees.

Anyone who has walked through a forest can provide a description of a tree that would be easily understood because a tree is a plant life-form recognizable as different from other plants. Trees are not a taxonomic classification but for management purposes they are included in a life-form group with recognizable physiognomic characteristics.

1-1 A Biotic View of Land

Based on a recent archeological discovery, evidence suggests that humans (*Homo*) and animals have been linked for over 130,000 years in North America (Holen et al. 2017). It is only recently that humans have come to appreciate the importance of wildlife, largely because we have learned that their existence is inextricably tied to our own. This change of view was noted by Aldo Leopold who published *A Biotic View of Land* (1939) in which he explained mankind's relationship to the environment. Leopold proposed that land is not just soil but energy flowing through a circuit of plants and animals and that an ecosystem is much more than the sum of its parts. He also emphasized that wildlife, which was once restricted to game ani-

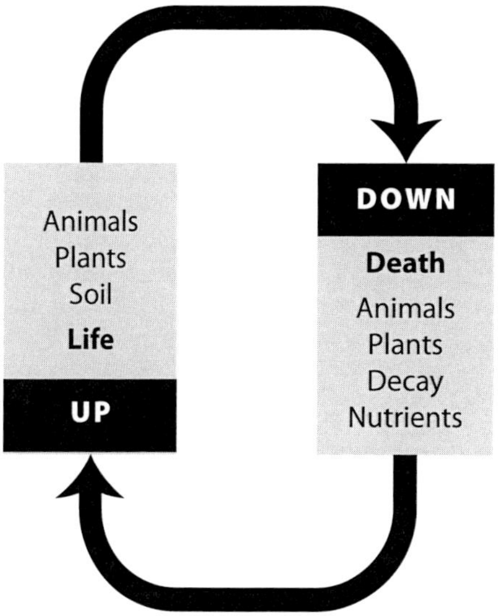

Figure 1-1 Energy flowing through an ecosystem has an Up-circuit and a Down-circuit.

Source: Leopold 1939.

mals, now included predators, songbirds, and all wild things. In Leopold's view, the biotic pyramid has an **Up-circuit** for energy that starts with the soil and progresses upward in food chains; the **Down-circuit** returns nutrients to the soil from the death and decay of living organisms (Figure 1-1).

Management of game animals alone was emphasized from the early 1930s to the 1960s largely due to the influence of hunters who were paying the cost through the purchase of hunting and fishing licenses and special taxes on ammunition and equipment. Starting in 1960, several laws were passed that set a new direction for management of federal lands. This shift culminated in 1973 with an event that marked the change from game management to wildlife management. In that year the Wildlife Management Institute reissued "The American Game Policy of 1930" as the "North American Wildlife Policy" (Allen 1973).

Additional emphasis was given to conserving all forms of wildlife when the World Wildlife Fund sponsored two workshops resulting in the publication of "New Principles for the Conservation of Wild Living Resources" (Holt and Talbot 1978). The year 1973 marked another seminal event in wildlife management: the passage of the Endangered Species Act (ESA) (U.S. Congress 1973). The ESA created mechanisms for the identification and protection of both plant and animal species listed as either "Threatened" or "Endangered." This piece of legislation has had major impacts on the management of wildlife and wildlife habitat on both public and private lands.

The period from 1960 to 1980 was a time of environmental awareness devoted to interpreting the laws and developing a broader awareness of the environment and how humans are affecting it. In many ways the health and sustainability of wildlife populations became a surrogate for protecting forest health. By building on past knowledge and developing new theories and techniques we have transitioned into the twenty-first century with a renewed framework for a better understanding of natural resource management as one part of ecosystem functioning. This framework encompasses the interactions between all plant and animal species, including humans. The ideas of biodiversity (Wilson 1976) and conservation biology (Soule 1986) have become a fundamental part of the new emphasis on ecosystem management.

Working Definitions

The scientific foundation of the wildlife and forestry professions rests on five very important words and the interactions between them: **trees, forests, environment,**

habitat, and **wildlife**. Because of the amount of literature available in professional journals, state and federal publications, and college textbooks, variation exists in the definitions of these terms. We encourage readers to consult other documents for definitions appropriate for management agencies depending on the laws and regulations under which they operate. For clarity we use the following working definitions in the narrative sections of this book:

- *Tree:* A woody perennial plant, typically large with a well-defined stem or stems carrying a more or less definite crown and a mature height of at least 16 feet.
- *Forest:* An ecosystem characterized by a more or less dense and extensive tree cover, often consisting of stands varying in characteristics such as species composition, structure, and associated processes. In contrast, a woodland is a plant community in which the mature trees are often small, characteristically short-boled (stem/trunk) relative to their crown depth, and frequently forming only an open canopy with the intervening area being occupied by lower vegetation, commonly in grass.
- *Environment:* All the abiotic and biotic factors that affect an animal's or plant's ability to survive, reproduce, and thereby maintain a population over time.
- *Habitat:* All the factors in the environment (food, cover, water, space, climate, vegetation structure, topography, etc.) that affect an animal's ability to survive and reproduce in a specific place (see Figure 1-2).

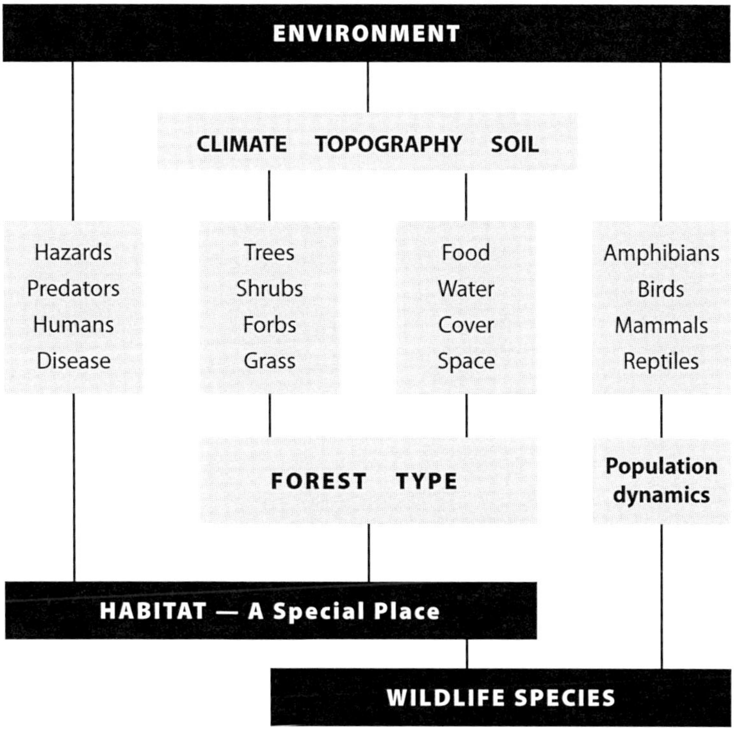

Figure 1-2 Conceptual model for defining habitat.

The phrase "Habitat—A Special Place" was featured in a public information film produced by the USDA Forest Service in 1979. The film presented examples of the use and importance of maintaining habitat for humans and wildlife in different setting across the United States.

- *Wildlife* (general definition): All vertebrate species that occur in the wild, undomesticated and free-roaming in a natural environment. Some definitions include plants as wildlife. Other definitions do not include fish because they are often managed as a separate resource.
- *Wildlife* (limited definition): Wildlife, as used in forest type checklists, is limited to birds and mammals.
- *Wildlife* (endangered species): Defined under the Endangered Species Act (U.S. Congress 1973) and listed by the United States Fish and Wildlife Service.

Ecological Principles

We present the following ten fundamental principles based on accumulated literature and years of knowledge passed on by generations of practicing managers. This is rooted in a body of evidence that, taken collectively, can be synthesized and easily translated into ecological principles or axioms (Patton 1992). For use in wildlife conservation and management, these principles are formulated to guide the thinking and actions of managers in decision-making, as well as for their use in the planning and assessment process.

Principle 1. Plant communities, under the influence of the environmental factors of soil, climate, topography, hydrology, etc., will change over time as a result of natural disturbances or human activities.

Principle 2. Animal species are directly or indirectly dependent on plants for their energy to survive and reproduce and, as such, will change as a plant community changes in structure, size, and species composition.

Principle 3. The primary factors that wildlife need to survive, reproduce, and maintain their population over time are: food, cover, water, and space.

Principle 4. Factors that inhibit wildlife from maintaining a population over time are primarily predators, disease, accidents, starvation, humans, and changes in habitat due to changes in plant communities.

Principle 5. Wildlife with large home ranges are less likely to be affected by changes in plant communities than those with small home ranges.

Principle 6. Wildlife species are integral parts of ecosystem functioning, providing pathways for nutrient cycling and energy flow through food chains and webs.

Principle 7. No ecosystem is completely stable, and change (e.g., succession and disturbance) is the rule rather than the exception.

Principle 8. Humans are ecological dominants with the ability to make environmental changes that speed up, slow down, and alter plant succession and disturbance processes.

Principle 9. In the minds of most humans, not all wildlife has equal value; some are favored at one time and others at another time.

Principle 10. Diversity is a natural phenomenon of life with different levels of recognition (genetic, species, ecosystem, landscape) that can be observed in its present form but can change over time and space.

Understanding Diversity

Diversity is all around us in our everyday life, evident in physical and social differences in the environments where we work and live. The word diversity is often used to describe something that exists or is perceived but is often difficult to explain or quantify. Diversity, as stated in Principle 10, has its roots in the works of Charles Darwin (Burkhardt 1996) and Carl Linnaeus (1758) in their early efforts to understand how the natural worlds of plants and animals are connected and function. The term diversity has been in use in indirect ways for many years. The definition of biological diversity evolved over time to include the variation of life at all levels of biological organization (Gaston and Spicer 2004).

Douglas Pimlott (1969) stated that the strongest case for diversity is the intangible values relating to life (e.g., the presence of wolves, coyotes, and goshawks as predators in ecosystems, and the value of understanding complex environmental interrelationships through bird watching). Raymond Dasmann (1981) introduced the idea of "declining diversity" in his wildlife biology textbook. The term biodiversity (hereinafter called "diversity") was first used by Thomas Lovejoy in 1980 when he was a conservation biologist working for the World Wildlife Fund.

The National Forest Management Act (U.S. Congress 1976, par. 219.3) defines diversity as, "the distribution and abundance of different plant and animal communities and species within the area covered by a land or resource management plan." Federal regulations emphasize diversity to ensure species are part of the forest planning process and in efforts to restore or maintain the integrity of terrestrial and aquatic ecosystems. Using the best available science is a requirement in this process. Detailed definitions by John Helms (1998), David Lindenmayer and Jerry Franklin (2002), and Richard Primack (2006) provide a variety of inherent factors. Stephen Boyce and Noel Cost (1978) suggest that an increase in the diversity of habitats increases the potential livelihood for different kinds of organisms. Creating a broad mixture of stand conditions over time, and their associated disturbance regimes, ensures the broadest diversity of species.

Quantifying Diversity. Quantifying diversity requires the measurement of abundance of all component species, which might not be physically practical or economically feasible, especially for animal species. Pimlott (1969, pg. 374) stated, "the message is clear that diversity of habitat is the life blood of the majority of species and the ramifications extend from the subsistence of an individual to the viability of a population and to the survival of species."

An accounting of species coupled with a variety of natural biological communities in an area is believed to indicate a degree of ecological diversity. The concept of diversity has recently expanded to imply that the more complex a community, the greater its stability and the more readily it can adjust to perturbations from within without major effects on the structure of the community as a whole. The many food chains and webs and energy pathways in complex communities create buffers that minimize the effects of the destructive forces of nature and humankind on the community. Intuitively, diversity is thought of as "good" because complex interactions are harder to destroy than simple ones. However, the view that diversity promotes stability has been challenged (May 1973).

Regardless, the number of species in an area goes a long way toward characterizing a community (May 1975).

Habitat diversity—to increase **species richness**—has been used in a practical way to manage forests in the eastern region of the Forest Service (Siderits and Radtke 1977). Species richness (SR) is the number of species occurring in a defined area and is often referred to as "**alpha diversity**" (MacArthur 1965). The term is somewhat misleading because it could include those species that are rare and not common to the forest type considered. SR gives as much weight to species with a few individuals as those with a high density. An example of SR in a major forest type is the ponderosa pine in Arizona (Patton et al. 2014) (Table 1-1). The type has 125 plant species in 36 families and 628 animal species in 147 families for a SR value of 753.

Table 1-1 Variety of species and life-forms in Arizona's ponderosa pine (pp) forest type.

Life-forms	Families in pp	Species in pp	Percent of total species
		PLANTS	
Forbs	11	27	4
Grasses	1	30	4
Shrubs	14	37	5
Trees	6	24	3
Other*	4	7	1
Sub-Total	36	125	17
		ANIMALS	
Amphibians	5	10	1
Anthropods	80	422	56
Birds	3	111	15
Mammals	19	48	6
Reptiles	10	37	5
Sub-Total	147	628	83
TOTAL	183	753	100

*Includes cacti, fern, agave, vine, sedge.

The number of species identified shows the difficulty of accounting for diversity in a forest type. It also indicates that there is a complex set of wildlife habitat relationships at work; to benefit one plant or animal species will have a ripple effect on a number of other species. Species richness by itself does not account for the relative abundance of individuals but is a snapshot of their presence in time. Having a number associated with factors such as the number of species of trees or animals is one requirement for an inventory to be used in the planning process.

In this textbook there are several hundred trees, birds, and mammals whose common names are used in narrative sections and tables. Common names represent everyday usage but may vary by geographic locality. A Latinized scientific name is the authority added to a common name so that there is no confusion in identification of a species. Scientific and common names of birds, mammals, and tree species are listed in Appendix A (available on the book's web page at waveland.com).

Levels of Diversity. Because diversity is a complex subject, it is useful to consider the different levels from the narrowest to the broadest perspective (USDA 1990). Four general categories (genetic, species, ecosystem, and landscape) illustrate the range of complexities:

> *Genetic diversity* refers to inherited characteristics found among individual representatives of a species that are critical to the survival of the species. A variety of genes provide resilience under environmental stress, which allows a species to adapt to changing conditions.
>
> *Species diversity* is the level of biodiversity that receives the most attention. It is the variety of living organisms found in a particular place—for example, the hundreds of different species inhabiting a ponderosa pine forest—including all life-forms of plants and animals. To address species diversity requires considering how the species changes from place to place as well as over time in the same place.
>
> *Ecosystem diversity* includes the variety of species and ecological processes (nutrient cycling and energy flow) that occurs in different physical settings, such as in an old-growth forest, a riparian zone, or a desert habitat.
>
> *Landscape diversity* deals with the dispersion of ecosystems, such as grasslands, wetlands, ponds, streams, and forests, in a geographical setting. A broad expanse of forests or grasslands is not as diverse as the same expanse containing a mixture of ecosystems.

Conservation Biology. Conservation biology is a discipline that has resulted from understanding the four biodiversity categories with the emphasis on the protection of species, their habitats, and ecosystems (Soule 1986). A somewhat broader definition is: A science that conducts research on biological diversity, identifies threats to biological diversity, and plays an active role in the preservation of biological diversity. Implicit in these definitions are rates of extinction, which ultimately encompass population demographics, viable populations, and threatened and endangered species (Primack 2006). Biodiversity and conservation biology were presented in a different way as related to forests by Lindenmayer and Franklin (2002). Their approach was to examine the "matrix" outside the "set-aside" or reserved lands and how "management practices on these lands will largely determine how successful human society is at conserving forest biodiversity and maintaining forest health" (pg. 5).

One of the disciplines supporting biodiversity and conservation biology is landscape ecology (Turner et al. 2001). As humans, we see large areas in our immediate field of vision and call it a "landscape," but landscapes, like ecosystems, have no definite space limitation. Of significance is the arrangement of landscape components into recognizable patterns, how these patterns are distributed, and what causes these patterns to change over time. The need for a landscape view has resulted from having to manage large areas following natural disturbances, such as fire, floods, and hurricanes, or human-induced changes, such as timber harvesting. All of these factors start a process that results in different plant and animal communities. Where these events occur, there may be need for a program to rehabilitate the land to a previous or different vegetative state (Walker and del Moral 2003).

COMMON TERMINOLOGY. With any new discipline, also comes new terminology—such as **matrix**, **patch**, and **metapopulation**—and these terms are now common in ecology journals and textbooks (Bissonette 1997; Lindenmayer and Franklin 2002; Smith and Smith 2009; Millspaugh and Thompson 2009). Techniques important for collecting landscape data include remote sensing, aerial photography, and geographic information systems, all of which require some level of on-the-ground verification. Aided by powerful computer storage and retrieval software, landscape analysis is becoming a necessary management tool for natural resource managers. As with any new way of doing business, software is often developed that facilitates the use of large amounts of data for collation and analysis. This has happened in landscape ecology, with many ways to collect information on forest components, including wildlife populations (Morrison 2002; Skalski and Robson 1992), along with the earlier developments for using spatial statistics (Ripley 1981) and models for management planning for large landscapes (Millspaugh and Thompson 2009).

With all the new technology and systems available for processing information that is now available in the literature, natural resource managers may be facing the paradox of too many choices in selecting options to address desired outcomes. For a detailed discussion of this topic we recommend *The Paradox of Choice: Why More Is Less* (Schwartz 2004). This book suggests that "more is less" and "good enough" may be the best that can be expected or attained. In making choices managers are concerned with maintaining diversity in stands and forests in horizontal and vertical space.

Horizontal diversity results from the arrangement of plant life-forms, successional stages, and varying amounts of forest components in **space** along a horizontal plane. Vertical diversity is the degree of divergence in height space between **layers** or vegetation stages in a stand. It portrays the effect of layering and separation of plant life-forms. A vertical line from above the tree tops to below the soil surface, at any point along a forest floor, could pass through several layers from tree canopy to underground material. Along the horizontal forest floor, the vertical strata can change from stands to openings with the contents in the space occupied by the vertical physical components, which often create an edge effect where the different strata are overlapping (Figure 1-3).

1-2 All About Trees

President Theodore Roosevelt (American Forestry Foundation 2016) said, "To exist as a nation, to prosper as a state, and live as a people, we must have trees." Trees have a life history that includes reproducing and maintaining a population

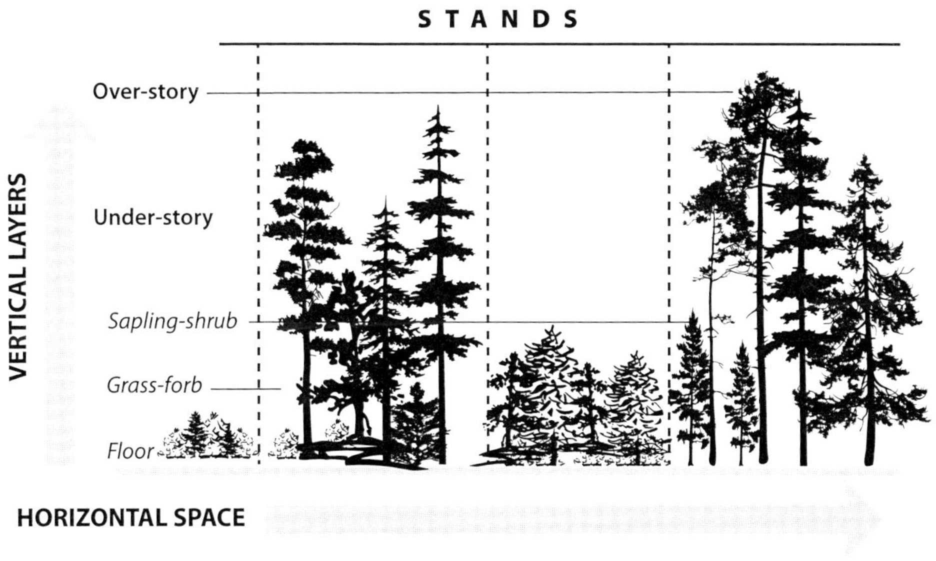

Figure 1-3 A conceptual model of layers in stands and stages occupying space and creating vertical and horizontal diversity.

defined by genetics and site characteristics. Given all the conditions that affect the life of a tree, it is remarkable that a giant sequoia in California that has a seed of less than 0.1 inch width, can grow to a diameter over 240 inches, a height of 250 feet, and attain an age of over 2,000 years.

Stand—The Basic Unit

A forest **stand** is the basic unit for inventory and management of trees and has slightly different applications across professions. A stand is a three-dimension entity with vertical layers and horizontal components. The reference in forestry is to a distinguishable land management unit that shares characteristics typically associated with the distribution of tree age, size, and composition (pure or mixed). Stands include shrubs, forbs, and grasses as well as trees, which also reflect stages in forest succession. The ecological definition of a stand is a contiguous group of similar plants (Helms 1998).

In a *Journal of Forestry* article, Kevin O'Hara and Linda Nagel (2013) emphasized that defining stands should be flexible for different uses. A stand can be any size and composition or structure that makes sense for on-the-ground management. Stands contain **structural stages** based on the distribution of tree diameter or age, such as opening, seedling, sapling, young, pole or intermediate, mature, and old growth (these stages are addressed in more depth in Chapter 2). Over time, stands move relatively predictably through stages of stand dynamics:

stand initiation, when trees and associated vegetation are first getting established and competition for growing space is severe;

stem exclusion (self-thin) is a stage after trees reach crown closure and begin to reduce light to their neighbors that are less tolerant of shade and competition;

understory reinitiation, when regular and irregular disturbances create sufficient growing space in the understory for small, isolated pockets of regeneration to survive and grow into the main canopy; and

old growth, structural conditions after many decades of successful understory reinitiation that establish a complex, multi-aged, and often multi-species stand.

Stand density, the number of trees per acre, and **stand basal area** relate well to percent crown closure (open, thin, thick, and closed canopy). Density can be expressed by several indices constructed from both number of stems and mean tree diameter (Tappeiner et al. 2007). The terms **even-aged** and **uneven-aged/multi-aged** are commonly used to describe stand structure (Figure 1-4). Even-aged stands are composed of a single age class that does not vary in age by more than 20 percent of the rotation age (the planned age of harvest); plantations represent classic even-aged stands and the stand initiation and stem exclusion stages of stand dynamics. Two-aged stands can develop when a uniform crop of regeneration, typically of a shade-tolerant species, establishes under an older main canopy. Multi-aged stands reflect three or more regeneration cycles and often of multiple species, until a relatively smooth age/size distribution exists from the youngest and smallest individuals to the oldest and largest, all in one stand.

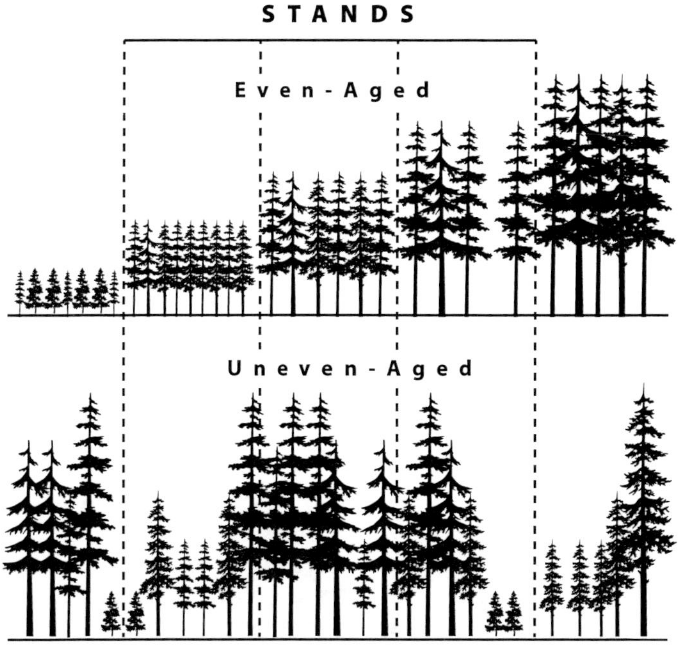

Figure 1-4 Visual representation of even-aged and uneven-aged stands.

Standing Live Trees

Three morphological components define a tree: **crown**, **trunk**, and **roots**. The crown is composed of branches with secondary branching by twigs with lateral attachments of green leaves containing chlorophyll. Leaves can have different shapes such as needle-like, scale-like, round, oblong, triangular, and other variations (Figure 1-5).

Figure 1-5 Diversity of leaf shapes.

In forest management tree crowns in even-aged stands are often grouped into vigor classes of dominant, codominant, intermediate, and suppressed or overtopped based on the amount of sunlight reaching the crown. The commonly used four crown classes are defined as:

Dominant trees have crowns consistently above the level of the main canopy surrounding them and receive full sunlight to most of their crown;

Codominant trees form the general canopy cover of the stand and receive direct sunlight from above, but little from the sides;

Intermediate trees, which are shorter than dominant and codominant trees, have crowns in the lower part of the main canopy layer; and

Suppressed or overtopped trees that receive no direct sunlight from above are positioned below the general canopy level.

Wolf tree is a free growing tree with a broad and open crown (Figure 1-6) and can be an important habitat component for some wildlife species (Figure 1-7).

Tree crowns have a width, length, surface area, and volume. These metrics are an indication of the ability of a tree to grow and is related to the photosynthetic

Figure 1-6 Visual characteristics of a ponderosa pine wolf tree that is a turkey roost tree.

Figure 1-7 Aluminum plate attached to a tree, such as a wolf tree, that is to be saved from harvesting.

process of the leaves. When the crown is projected to the ground it provides a measure of canopy cover. Crowns also contain the primary growth points of the tree: buds that contain the next year's foliage and branches. Buds are formed (and protected) to ensure future growth and survival, and a good growing season in one year promotes good growth in subsequent years.

The totality of leaf surface in a crown is referred to as **foliage**. Each leaf in the crown is a chemical factory that via photosynthesis absorbs carbon dioxide and other gasses, produces carbohydrates for the plant to survive and grow, and emits oxygen back to the air. Specialized cells called "**stomata**" are present on the top and bottom of the leaf to control water loss from transpiration. The outer surface of the leaf is the **epidermis** and has a protective coating of waxy cuticle. In the fall season deciduous trees have a leaf loss by **abscission** preceded by a dramatic color change in the foliage. The colors change to yellow, orange, or red based on the loss of chlorophyll production as the leaf dies. In many parts of the U.S. the color change in leaves is an event for nature photographers and a tourist attraction.

The trunk of a tree is the woody section that provides strength to hold the crown loaded with twigs and leaves and to provide them with water. Two types of cells dominate the function of tree trunks. **Xylem** cells bring water and minerals up from the roots to the leaves and **phloem** cells move the carbohydrate rich sap produced out of the leaves. Tree trunk in forest management literature is often referred to as a "bole." There are several layers of bark on the trunk that insulate it from moisture, heat, and cold. An outer bark is for protection; an inner bark is the phloem that transports the sap. Inside that is a layer, the **cambium**, which is the secondary growing part of the tree. New growth each year moves the tree diameter outward producing phloem to the outside and xylem to the inside.

The process of creating new xylem creates an identifiable ring that can be used to determine tree age because early-season growth produces larger cells and lighter wood relative to late-season growth. This annual ring can also show periods of growth resulting from a management activity (e.g., thinning) or environmental factors (e.g., fire, abnormal rainfall). The outer xylem of a tree is called "**sapwood**," which moves water from the roots to the leaves. As new sapwood is produced over the years the inner most xylem cells lose their ability to transport water, resulting in filled/clogged cells known as **heartwood** in the center of the tree (typically darker and more decay resistant). See Figure 1-8.

Roots are the underground part of a tree that anchor a trunk to the ground. Roots can grow straight down as a taproot or spread out in a lateral arrangement to absorb water. Any damage to the root system from surface disturbance can affect the ability of the root to transmit water upward. Root hairs on the fine branches of the root system increase the absorption area in contact with the soil and the tree's ability to transmit nutrients and water. Indeed, most tree species form mutualistic association with fungi or bacteria to further increase the water and nutrient absorption capacity or to fix nitrogen that is particularly important to trees.

Most forest management decisions ultimately involve consideration of the size of an individual tree as expressed in its **diameter**. Diameter (D) is conventionally measured at breast height (dbh) or 4.5 feet from the ground surface. Tree diameter

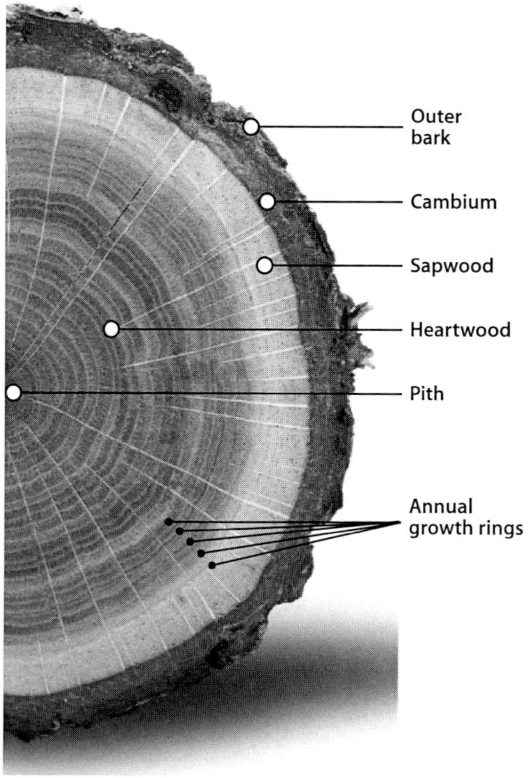

Figure 1-8 Cross section of a tree showing annual rings and growing tissue.

is important because it indicates a capacity for potential forest products, aesthetics, and/or a habitat component for a particular wildlife species. The common way to determine diameter is with a tape that converts circumference in inches to diameter. Every 3.1416 (π) inches on the tape is 1 inch of diameter to the nearest 0.1 inch. Dbh is used to compute the cross-sectional area of the trunk as basal area (BA) using the formula $A = \pi r^2$. For example, the BA of a 10-inch diameter tree is:

$$3.1416 \times (10/2 \text{ in.})^2 = (3.1416 \times 25) = 78.54 \text{ square inches.}$$

For management purposes, BA is expressed in square feet, therefore:

$$78.54 \text{ square inches} / 144 \text{ square inches per square foot} = 0.5454 \text{ square feet.}$$

A collection of trees, either as a plot or a stand, will have an average size, and average size has meaning for even-aged populations (i.e., plantations and other bell-shaped distributions centered on some average size). However, foresters typically use **a quadratic mean diameter** (QMD) rather than a simple arithmetic mean because of the above BA calculation. QMD is calculated by squaring each diameter, taking the average, and then taking the square root of the resultant value.

The arithmetic mean and QMD are identical for a collection of trees with the same diameter, and they are very similar for a collection with very similar diameters. But QMD can be significantly larger than the arithmetic mean for broad distributions of sizes, bimodal distributions, and collections of trees that include exceptionally large trees due to the power of squaring. Such stands don't really have an "average" size, though, so why calculate either. Diameters are generally grouped into size classes (stand tables) for management purposes, such as in tree structure and frequency distribution in a forest stand.

There is no set range of diameters in a class but a 2- and 4-inch spread in sizes are common to balance the detail needed in stock (summaries of volume per size class per unit of area) and stand (summaries of trees per size class per unit of area) tables with field work and presentation efficiency (Table 1-2). A distribution using 4-inch size classes can be expanded depending on the need for analysis.

Table 1-2 Tree distribution based on a 4-inch diameter distribution.

Dbh Class	Dbh in Inches	Range
Seedling < 4.5 ft in height	none	none
Sapling	2	0.1–3.9
Small tree (pole)	6	4.0–7.9
Medium tree	10	8.0–11.9
Small saw log	14	12.0–15.9
Large saw log	18	16.0–19.9
Extra large	20	20+

These groupings are particularly useful in describing structural stages as well as size distribution. Too many size classes lead to an overabundance of detail that may make it more difficult to associate structural stages as a wildlife habitat component. Tree diameter and height are used to compute **volume**. However, it is not that simple. The volume of interest is the content of wood in the trunk and trees are not perfect cylinders but have a taper from bottom to top, particularly in the crown.

There are two volumes of interest in a tree: the bole (trunk) volume and the crown volume. Volumes for boles of trees are used as a guide of the amount of wood available for forest products, such as lumber measured in board feet (1 in. thick × 12 in. wide × 12 in. long) or chips measured in cubic volume (or even weight). Estimates are made based on different lengths of logs contained in a tree. Because there is taper in the trunk from base to top, a form factor is used to account for the taper in each length of log. For example, tables are available that give the volume in board feet using diameter at each 16-foot section (log).

Crown volume is calculated from average crown diameter and length, which can also be used for crown surface area. For our purposes, crown volume or surface area is of interest because they contain features that can be used by different species of wildlife for both food and cover. An assumption is that the tree is a geometric solid such as a cone. For an approximation, the formula for cubic feet is:

$$\text{Vol} = (\pi r^2) \times (h/3)\text{: r is the crown base radius and h is crown length.}$$

$$\text{Example: } r = 3 \text{ ft., } h = 18 \text{ ft.}$$

$$\text{Vol} = (3.1416 \times 9) \times (18/3) = 170 \text{ cubic ft.}$$

The **age** of a standing tree can be determined by counting the number of annual rings that extends from the center (pith) to the outer ring of xylem just inside the bark (Figure 1-8). This is accomplished by using an increment borer to remove a sample core about the size of a soda straw. Annual rings can also provide insight about how fast a tree is growing, the stage of stand development, and the effects of management practices, such as thinning. They are also used to construct growth and yield charts and tables for managers to project options for the future.

Tree height is measured for the total length from ground level to top of the crown. Small trees can be measured directly using telescoping poles. The traditional method for measuring larger individuals is with a manual clinometer at a fixed distance from a tree such as 50, 66, or 100 feet. Tangents of angles from an observer's eye to the tree top and another to the bottom are multiplied by the horizontal distance and added to get the height. Modern clinometers use either sound waves or lasers to measure distance to the tree, which is then internally and electronically calculated using the angles to the top and bottom of the stem in order to get total tree height. **Height** is an important metric because it is used to compute a **site index**, which is an indicator of site productivity (i.e., how much bole and crown volume can be produced over time).

Base ages are used to compare height growth potential. Usually the base age is 100 for slower growing forest types in the West (Burkhart et al. 2019). For example, if the height is 70 feet, the site index is 70 and is more productive than the site with a tree height of 60 at the same age. To determine a site index requires knowing the age of a tree in the dominant and codominant crown classes as defined earlier.

One of the key factors that affects a tree's ability to survive and grow relative to its neighbors is its **tolerance** to shade. A relative scale is used to indicate tolerance as very intolerant, intolerant, intermediate, tolerant, and very tolerant (Appendix B). Some examples are in Table 1-3. Managers use these scales in making decisions on which regeneration method to use on a specific site. For example, a clearcutting method would be used to promote reproduction for a very intolerant shade species such as aspen.

Table 1-3 Examples of tree species according to tolerance to shade.

Tolerance Level	Species
1. Very intolerant	longleaf pine, aspen
2. Intolerant	ponderosa pine, loblolly pine
3. Intermediate	Douglas-fir, eastern white pine
4. Tolerant	red maple, redwood
5. Very tolerant	balsam fir, hemlock

The replacement of shade intolerant species by a more tolerant species leads to the development of multistory, older forests and, eventually, the climax stage of succession (see Chapter 2). Competition in the overstory of a tolerant species leads to the climax of the same species or potentially another tolerant species. However, sometimes disturbances make the progression of stands through time difficult to project. Regular disturbances (e.g., trees being overtopped or dying of old age) are more predictable than irregular or chaotic disturbances like lightning, landslides, and wildland fire.

Dead Trees and Snags

Forests are more than live, straight trees used for timber production. As a by-product of live trees, dead trees become **snags** through a decay process from solid trunk with brown needles or leaves, to the final stage during which most of the bark and branches are gone and the trunk has become soft and easy to penetrate due to decay. A snag may break apart and/or fall at any point in that progression, becoming a solid or partially decayed fallen log. After that point, it is continually reduced to organic matter, and its nutrients are returned to the soil to be used again.

The process from a seed to live tree and later to a snag is an ecological trail demonstrating the importance of understanding ecological systems in managing forests for wildlife. A live tree progressively provides food, homes, and special-use sites for many species of wildlife while it grows and matures, but use continues well after the tree is partially or completely dead. Indeed, dead trees are as important to ecological functioning in a forest as live trees, and this has to be taken into account (Franklin and Forman 1987). Animals use dead wood in trees and snags in many ways, including those exhibited in Table 1-4.

Table 1-4 Examples of animal species that use dead trees and snags for life necessities.

Types of Use	Species
Open nests	osprey, bald eagle, and heron
Cavity nests	goldeneye, woodpeckers, owls, and chickadees
Hunting perches	hawks and flycatchers
Food storage	squirrels and jays
Food and shelter	chickadees and bats
Resting and roosting	swallows, hawks, nuthatches, turkey, bats

Richard Jefferies (1848) wrote about a tree in a state of decay that should be saved from the axe. His narrative below describes the relationships that exist in a single tree and is an example of the functioning of a small ecosystem. What Jefferies says is from personal observation and not a scientific study. However, his very ecological description could have been written today.

> Those hollow trees, according to woodcraft, ought to come down by the axe without further loss of time. Yet it is fortunate that we are not all of us, even in this prosaic age, imbued with the stern utilitarian spirit; for a decaying tree is perhaps more interesting than one in full vigour of growth. The starlings make their nests in the upper knot-holes; or, lower down, the owl feeds her young. . . . The woodpecker comes for the insects that flourish on the dying giant; so does the curious little tree-climber, running up the trunk like a mouse; and in winter, when insect-life is scarce, it is amusing to watch there the busy tomtit. . . . For the acorns the old oak still yields some rooks, pigeons, and stately pheasants, with their glossy feathers shining in the autumn sun. Thrushes carry wild hedge-fruit up on the broad platform formed by the trunk where the great

> limbs divide, and, pecking it to pieces, leave the seeds. . . . Sometimes wild bees take up their residence in the hollow, slowly filling it with comb, buzzing busily to and fro; and then it is not to be approached so carelessly, though so ready are all creatures to acknowledge kindness that ere now I have even made friends with the inhabitants of a wasp's nest. . . . In the evening the goat-sucker or nightjar comes with a whirling phantom-like flight, wheeling round and round: a strange bird, which will roost all day on a rail, blinking or sleeping in the day-light, and seeming to prefer a rail or branch without leaves to one that affords cover. Here also the smaller bats flit in the twilight, and, if you stand still, will pursue their prey close to your head, wheeling about it so that you may knock them down with your hand if you wish. . . . All the year through the hollow tree is haunted by every kind of living creature, and therefore let us hope it may yet be permitted to linger awhile safe from the axe. (pgs. 79–81)

Excessive amounts of dead and decaying woody material, however, can be serious fire and safety hazards to property, forest animals, and human life. As a result, there is a trade-off in how much material should be retained versus the risks of retaining it, especially in a wildland-urban interface and actively used areas (e.g., job sites and recreational areas).

Coarse Woody Debris

A by-product of dead or dying trees that accumulates on the forest floor is coarse woody debris (CWD). CWD is defined as any piece of dead woody material, e.g., dead boles, limbs, and large root masses, on the ground in forest stands or in streams (Helms 1998). CWD comes from disease, natural disturbances of fire, wind, tree mortality caused by insects, and management activities such as timber harvesting. Nutrients from the decay process are consumed by fungi, bacteria, and insects that in turn are eaten by other life-forms. By providing food and microhabitats to many species, such as ants and termites, CWD helps maintain the functioning of a forest ecosystem. In some instances, a fallen tree becomes a decaying log covered with mosses and lichens where tree seeds germinate and grow in a straight line along the top and, in these conditions, becomes a **nurse log**.

1-3 Where Are the Trees?

Trees are most often discussed as being in a forest type, but their contribution to a forest is more than just being a member of a stand. Trees are ecological dominants in that their dimensions (height, width, volume) have an influence on what plants and animals occupy space in their general vicinity, particularly under the crown and on the forest floor. In general, the presence—or absence—of trees reflects the conditions and combinations of environmental factors that exist on a given site. Temperature, radiation, moisture, soil, elevation, and physiographic features determine what species are present over time. Trees and their attributes can be either a food and/or cover component of habitat for many wildlife species.

Tree Species in States

Two volumes of silvics of tree species published by the USDA Forest Service have distribution maps and life history information for hardwood and coniferous species in the continental states and Alaska (Burns and Honkala 1990). An interpretation of the maps in these two volumes yields the following lists:

- Number of tree species that have a native range in 49 states (Table 1-5).
- The natural range of hardwoods by state (Appendix C).
- The natural range of conifers by state (Appendix D).

A comparison of the states with the five highest number of tree species by geographic region is more or less indicative of the difference in environmental conditions that exist from east to west. The figures for the western trees are much lower than the eastern trees because of the influence of elevation, temperature

Table 1-5 States in the continental U.S. and Alaska listed by number of hardwood and conifer tree species in their natural range.

State	#Species	Hardwoods	Conifers	State	#Species	Hardwoods	Conifers
Upper 1/3 = 16 states							
Virginia	92	76	16	Pennsylvania	72	59	13
North Carolina	90	74	6	Ohio	72	64	8
Tennessee	87	75	12	Illinois	69	65	4
Alabama	84	2	12	Arkansas	68	65	3
South Carolina	83	69	14	Maryland	67	56	11
Georgia	78	62	16	New York	67	54	13
West Virginia	77	66	11	Indiana	66	60	6
Mississippi	75	66	9	Missouri	66	63	3
Middle 1/3 = 20 states							
Florida	65	53	12	Delaware	53	43	10
Texas	65	57	8	New Hampshire	53	41	12
Kentucky	62	55	7	California	51	14	37
Louisiana	62	54	8	Wisconsin	51	41	10
New Jersey	61	50	11	Maine	47	35	12
Michigan	60	50	10	Rhode Island	45	39	6
Oklahoma	59	55	4	Minnesota	43	33	10
Connecticut	57	46	11	Oregon	42	14	28
Massachusetts	56	45	11	Iowa	37	34	3
Vermont	54	43	11	Kansas	35	34	1
Lower 1/3 = 13 states							
Washington	28	8	20	Colorado	15	5	10
Nebraska	27	24	3	Utah	15	4	11
Montana	25	7	18	Nevada	14	3	11
South Dakota	25	4	4	Wyoming	14	5	9
Idaho	24	4	20	New Mexico	13	3	10
North Dakota	21	19	2	Alaska	10	1	9
Arizona	19	5	14				

Table 1-6 The highest number of trees for five states in the western and eastern geographic area.

Number of Tree Species			
West	**No.**	**East**	**No.**
California	51	Virginia	92
Oregon	42	N. Carolina	90
Washington	28	Tennessee	87
Montana	25	Alabama	84
Idaho	25	S. Carolina	83

ranges, and rainfall patterns (Table 1-6). These figures are only for trees that are considered in their native range. Some interesting facts about trees are:

- Virginia has the most tree species: 92
- Virginia has the most hardwood species: 76
- California has the most conifer species: 37
- Quercus (oak) is the most common genera of hardwood species: 35
- Pinus (pine) is the most common genera of conifer species: 28
- Box elder is the most widely distributed hardwood: 39 states
- Eastern red cedar is the most widely distributed softwood: 35 states

The Value of Trees

Sometimes it is easy to overlook the total value of a tree in discussing its place in the ecological realm of diversity. Every part of a tree—leaves, bark, twigs, roots, and trunk—is found in some product used on a daily basis by humans (Figure 1-9). Probably the most ecological value of trees is in the fact that they produce oxygen and sequester carbon thereby reducing carbon dioxide emission that contributes to the greenhouse effect. While trees have many uses that add to our quality of life, they also are used as symbols to represent the general ambience of the people and government of a state.

All fifty states have a designated species as their state tree (Table 1-7 on pg. 22). Nine tree species are listed as a state tree in more than one state:

- sugar maple (NY, VT, WI, WV),
- eastern cottonwood (WY, NE, KS),
- white oak (IL, CT, MD),
- tulip tree (IN, KY, TN),
- American elm (ND, MA),
- eastern white pine (ME, MI),
- flowering dogwood (MO, VA),
- longleaf pine (NC, AL), and
- sabal palm (SC, FL).

In addition to being designated a state tree in four states, sugar maple occurs in 23 other states. Sugar maple is a major component of seven forest types and has commercial value for saw logs and pulpwood and is often used as an ornamental. It also has special value as the principal source of maple sugar. Most state trees occur in more than one forest type.

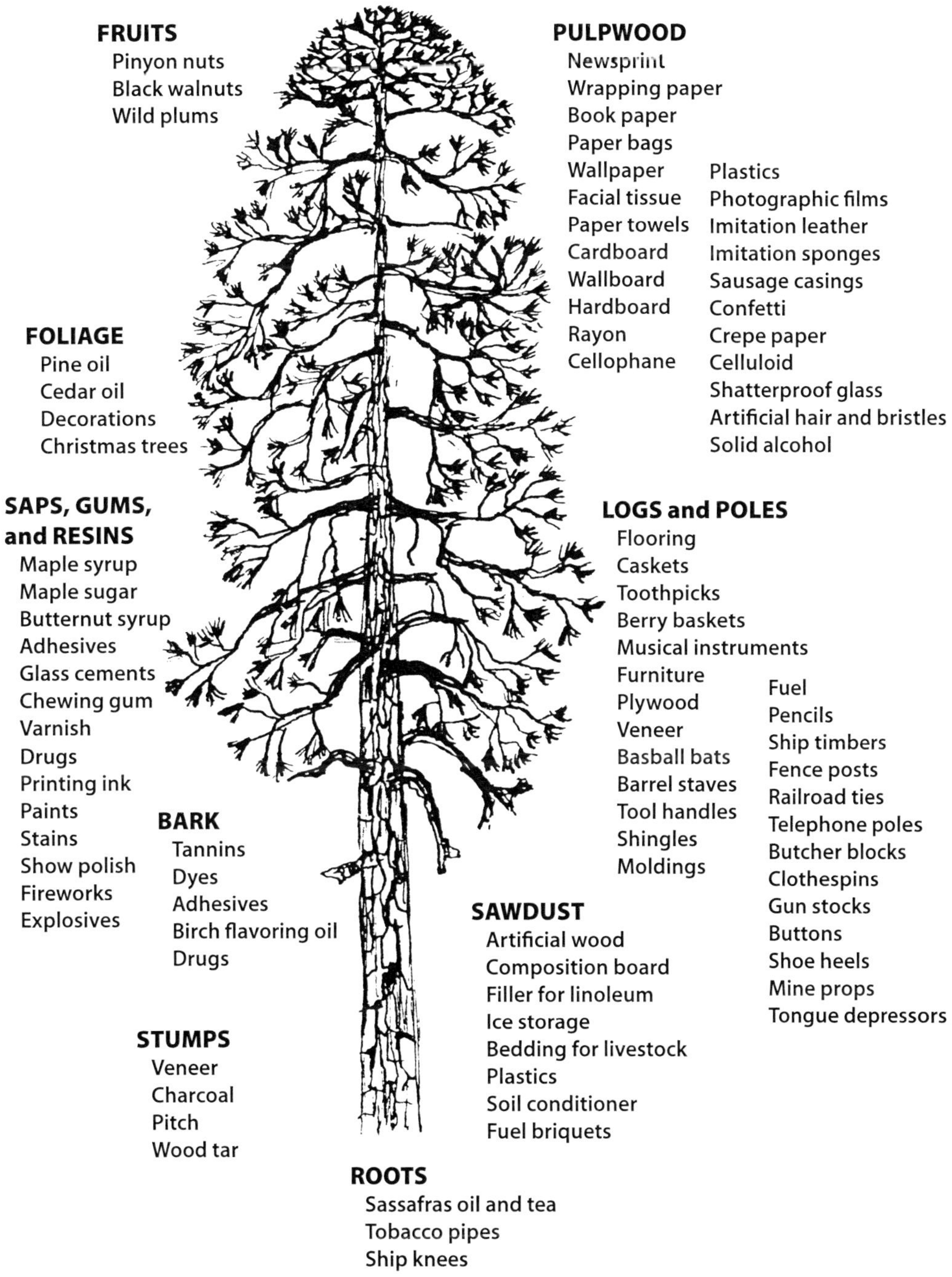

Figure 1-9 Products derived from trees.

Table 1-7 State trees.

State	Common Name	Scientific Name	Date Established
Alabama	Longleaf pine	*Pinus palustris*	1997
Alaska	Sitka spruce	*Pica sitchensis*	1962
Arizona	Blue palo verde	*Parkinsonia florida*	1954
Arkansas	Loblolly pine	*Pinus taeda*	1939
California	Coast redwood	*Sequoia sempervirens*	1937
Colorado	Blue spruce	*Picea pungens*	1939
Connecticut	White oak	*Quercus alba*	1947
Delaware	American holly	*Ilex opaca*	1939
Florida	Sabal palm	*Sabal palmetto*	1953
Georgia	Southern live oak	*Quercus virginiana*	1937
Idaho	Western white pine	*Pinus monticola*	1935
Illinois	White oak	*Quercus alba*	1973
Indiana	Tulip tree	*Liriodendron tulipifera*	1931
Iowa	Bur oak	*Quercus macrocarpa*	1961
Kansas	Eastern cottonwood	*Populus deltoids*	1937
Kentucky	Tulip tree	*Liriodendron tulipifera*	1994
Louisiana	Baldcypress	*Taxodium distichum*	1963
Maine	Eastern white pine	*Pinus strobus*	1945
Maryland	White oak	*Quercus alba*	1941
Massachusetts	American elm	*Ulmus Americana*	1941
Michigan	Eastern white pine	*Pinus strobus*	1955
Minnesota	Red pine	*Pinus resinosa*	1945
Mississippi	Southern magnolia	*Magnolia grandflora*	1938
Missouri	Flowering dogwood	*Cornus florida*	1955
Montana	Ponderosa pine	*Pinus ponderosa*	1949
Nebraska	Eastern cottonwood	*Populus deltoids*	1972
Nevada	Single-leaf pinyon	*Pinus monophilia*	1953
New Hampshire	American white birch	*Betula papyrifera*	1947
New Jersey	Northern red oak	*Quercus rubra*	1950
New Mexico	Pinyon pine	*Pinus edulis*	1956
North Carolina	Longleaf pine	*Pinus palustris*	1963
North Dakota	American elm	*Ulmus Americana*	1947
Ohio	Ohio buckeye	*Aesculus glabra*	1953
Oklahoma	Eastern redbud	*Cercis canadensis*	1937
Oregon	Douglas-fir	*Pseudotsuga menziesii*	1939
Pennsylvania	Eastern hemlock	*Tsuga canadensis*	1931
Rhode Island	Red maple	*Acer rubrum*	1964
South Carolina	Sabal palm	*Sabal palmetto*	1939
South Dakota	Black Hills spruce	*Picea glauca*	1947
Tennessee	Tulip tree	*Lirodendron tulipifera*	1947
Texas	Pecan	*Carya illinoinensis*	1919
Utah	Quaking aspen	*Populus tremuloides*	1933
Vermont	Sugar maple	*Acer saccharum*	1949
Virginia	Flowering dogwood	*Cornus florida*	1956
Washington	Western hemlock	*Tsuga heterophylia*	1947
West Virginia	Sugar maple	*Acer saccharum*	1949
Wisconsin	Eastern cottonwood	*Populus deltoids*	1947
Wyoming	Eastern cottonwood	*Populus deltoids*	1947

Common and scientific name authority is the National Arboretum, U.S. Department of Agriculture, Washington, DC.

Many of the trees selected as a state tree have a commercial value such as Douglas fir in Oregon, loblolly pine in Arkansas, and white oak in Kentucky. Some species are trees but are not part of a forest type nor do they have commercial value, but are part of the landscape; for example, palo verde in Arizona is not part of a forest type and does not have commercial value, but it is a large part of the semi-desert landscape. On the other end of the spectrum, quaking aspen is the state tree in Utah; the species occurs in 30 states, has a commercial value, is used as an ornamental, and is a valuable food source for browsing wildlife such as deer and elk.

Conversion Factors. In measurements used to account for trees and their products there are two systems used in the United States: metric and English. While metric units are for scientific research and application, English units have been in place for a length of time that for land management agencies it is still the preferred system. In the following chapters, English units are presented without a side-by-side metric equivalent. When there is a need to convert from English to metric then the factors in Table 1-8 on the following page can be used to make a conversion.

1-4 Summary

Every living individual has a beginning and an end. In the middle is a series of events that takes place in time where energy is part of a cycle of "ups" and "downs" that defines the working of ecosystems. Leopold's biotic pyramid provides the basic components for understanding the workings of a complex system. For forestry and wildlife professionals the systems of concern include trees, forests, environment, habitats, and wildlife. Ten ecological principles are presented for use in forest wildlife conservation and management that can guide the thinking and actions of managers in decision-making. Forests are composed of trees with morphological components of crown, trunk, and roots. Trees form stands with structural stages that form the basic management unit.

One of the major factors influencing decisions is the concept of diversity. Diversity has evolved over time to include the variation of life at all levels of biological organization and is a natural phenomenon of life that exists now and can be observed in its present form but constantly changes in time and space. Diversity has four levels of complexity: genetic, species, ecosystem, and landscape. Federal regulations emphasize that diversity is a way to ensure wildlife species are part of the forest planning process. Creating a broad mixture of stand conditions over time, and their associated disturbance regimes, ensures the broadest diversity of species. A variety of species coupled with a variety of biological communities in an area is believed to indicate a high degree of ecological diversity.

Table 1-8 Conversion factors for metric and English units.

MULTIPLY	BY	TO CONVERT INTO
LENGTH		
feet (ft)	0.30480	meters
inches (in)	2.5400	centimeters (cm)
kilometers (km)	0.62137	miles
meters (m)	3.28083	feet
meters	0.04971	chains
millimeters	0.03937	inches
miles (mi)	1.60934	kilometers
yards (yd)	0.91440	meters
AREA		
acres (ac)	0.40468	hectares
acres	43,560.00	square feet
hectares (ha)	2.47104	acres
hectares	10,000.0	square meters
square feet (ft^2)	0.09290	square meters
square kilometers (km^2)	0.38610	square miles
square kilometers	100.00	hectares
square kilometers	247.104	acres
square meters (m^2)	10.76387	square feet
square miles (mi^2)	2.58999	square kilometers
square miles	640.00	acres
BASAL AREA		
square feet per/ac	0.2296	square meters/ha
square meters per/ha	4.3560	square feet/ac
WEIGHT		
kilograms (kg)	2.20462	pounds
kilograms/hectare (kg/ha)	0.8922	pounds/ac
pounds (lb)	453.592	grams (g)
pounds/acre (lb/ac)	1.1208	kilograms/ha
VOLUME		
board ft	0.00566	cubic meters
board ft/acre (bd ft/ac)	0.01399	cubic meters/ha
cubic feet (ft^3)	0.02831	cubic meters
cubic feet/acre (ft^3/ac)	0.06997	cubic meters/ha
cubic meters (m^3)	35.3145	cubic feet
cubic meters	176.57	board feet
cubic meters/hectare (m^3/ha)	71.4571	board feet/ac
cubic meters/hectare	14.291	cubic feet/ac
gallons (gal)	3.785	liters
liters (l)	0.26417	gallons
TEMPERATURE		
Centigrade	0.5556 (F - 32)	Fahrenheit
Fahrenheit	32 + (1.80 C)	Centigrade

Source: Munns et al. 1935.

REFERENCES

Allen, D.L. (chair). 1973. *North American wildlife policy.* Wildl. Manage. Inst., Washington, DC.
American Forestry Foundation. 2016. *We must have trees.* General Publication, Washington, DC.
Bissonette, J.A. (ed.). 1997. *Wildlife and landscape ecology: Effects of pattern and scale.* Springer-Verlag, New York.
Boyce, S.G., and N.D Cost. 1978. Forest diversity: New concepts and applications. Res. Paper SE-194. USDA Forest Service, Asheville, NC.
Burkhardt, F.D. (ed.). 1996. *Charles Darwin's letters: A selection 1825–1859.* Cambridge University Press, Cambridge, UK.
Burkhart, H.E., T.E. Avery, and B.P. Bullock. 2019. *Forest measurements,* Sixth edition. Waveland Press, Long Grove, IL.
Burns, R.M., and B.H. Honkala. 1990. *Silvics of trees of North America.* Vol. 1, conifers, Vol. 2, hardwoods. Agric. Hdbk. 654. USDA Forest Service, Washington, DC.
Dasmann, R.F. 1981. *Wildlife biology.* John Wiley and Sons, New York.
Franklin, J.F., and R.T. Forman. 1987. Creating landscape patterns by forest cutting: Ecological consequences and principles. *Landscape Ecology* 1:5–18.
Gaston, K.J., and J.I. Spicer. 2004. *Biodiversity: An introduction.* Blackwell Scientific Publishing, Oxford, UK.
Hardin, J.W., D. Leopold, and F. White. 2001. *Textbook of dendrology.* McGraw-Hill, New York.
Helms, J. A. (ed.). 1998. *The dictionary of forestry.* Society of American Foresters, Bethesda, MD.
Holen, S.R., T.A. Demere, and D.C. Fisher. 2017. A 130,000 year-old archaeological site in Southern California, U.S.A. *Nature* 544(7651):479–483.
Holt, S.J., and L.M. Talbott. 1978. New principles for the conservation of wild living resources. *Wildl. Mono.* 59. The Wildlife Society, Washington, DC.
Jefferies, R. 1848. *The gamekeeper at home: The amateur poacher.* Oxford University Press, UK.
Leopold, A. 1939. A biotic view of land. *J. For.* 37(9):727–730.
Lindenmayer, D.B., and J.F. Franklin. 2002. *Conserving forest biodiversity: A comprehensive multiscaled approach.* Island Press, Washington, DC.
Linnaeus, C. 1758. *Systema Naturae.* International Commission on Zoological Nomenclature, London, England, UK.
MacArthur, R.H. 1965. Patterns of species diversity. *Biol. Rev.* 40:510–533.
May, R.M. 1973. *Stability and complexity in model ecosystems.* Princeton University Press, Princeton, NJ.
May, R.M. 1975. Patterns of species abundance and diversity. P. 81–120 in *Ecology and environmental communities,* M.L. Cody and J.M. Diamond (eds.). Belknap Press, Cambridge, MA.
Millspaugh, J.J., and F.R. Thompson, III. 2009. *Models for planning wildlife conservation in large landscapes.* Academic Press, San Diego.
Morrison, M.L. 2002. *Wildlife restoration: Techniques for habitat analysis and animal monitoring.* Soc. for Ecol. Restoration. Island Press, Washington, DC.
Munns, E.N., V.A. Clements, and T.G. Hoerner. 1935. Converting factors and tables of equivalents used in forestry. USDA Forest Service Misc. Pub. No. 225.
O'Hara, K.L., and L.M. Nagel. 2013. The stand: Revisiting a central concept in forestry. *J. For.,* 111(5):335–340.
Patton, D.R. 1992. *Wildlife habitat relationships in forested ecosystems.* Timber Press, Portland, OR.
Patton, D.R., R.W. Hofstetter, J.D. Bailey, and M.A. Benoit. 2014. *Species richness and variety of life in Arizona's ponderosa pine forest type.* Gen. Tech. Rep. RMRS-GTR-332. Rocky Mountain Research Station, USDA Forest Service, Fort Collins, CO.
Pimlott, D.H. 1969. The value of diversity. *Trans. N. Amer. Wildl. and Nat. Resour. Conf.* 34:265–280.

Primack, R.B. 2006. *Essentials of conservation biology.* Sinauer Associates, Inc., Sunderland, MA.
Ripley, B.D. 1981. *Spatial statistics.* John Wiley and Sons, Inc., New York.
Schwartz, B. 2004. *The paradox of choice–why more is less. How the culture of abundance robs us of satisfaction.* Harper Collins Publishers, New York.
Siderits, K., and R.E. Radtke. 1977. Enhancing forest wildlife habitat through diversity. *Trans. N. Am. Wildl. Nat. Resour. Conf.* 42:425–433.
Skalski, J.R., and D.S. Robson. 1992. *Techniques for wildlife investigations: Design and analysis of capture data.* Academic Press, New York.
Smith, R.L., and T.M. Smith. 2009. *Elements of ecology.* Benjamin Cummings, New York.
Soule, M.E. (ed.). 1986. *Conservation biology: The science of scarcity and diversity.* Sinauer and Associates, Sunderland, MS.
Tappeiner, J.C., D.A. Maguire, and T.B. Harrington. 2007. *Silviculture and ecology of western U.S. forests.* Oregon State University Press, Corvallis, OR.
Turner, M.G., R.H. Gardner, and R.O. O'Neill. 2001. *Landscape ecology in theory and practice: Pattern and process.* Springer Verlag, New York.
U.S. Congress. 1973. Endangered Species Act. Public Law 93-205. Government Printing Office, Washington, DC.
U.S. Congress. 1976. National Forest Management Act. Public Law 94-988. Government Printing Office, Washington, DC.
USDA. 1990. *Conserving our heritage, America's biodiversity.* USDA Forest Service, Washington, DC.
Walker, R.W., and R. del Moral. 2003. *Primary succession and ecosystem rehabilitation.* Cambridge University Press, UK.
Wilson, E.O. (ed.). 1976. *Biodiversity.* The National Academy of Science, National Academy Press, Washington, DC.

The Forest for the Trees 2

There is an old expression "you can't see the forest for the trees" that is often used to describe someone or an entity that is so focused on minute details that they fail to recognize a larger problem or see the whole picture. During the discussions when the National Forest Management Act (U.S. Congress 1976) was being proposed to prevent national forests from being managed only for the production of trees, Senator Hubert S. Humphrey from Minnesota stated:

> The days have ended when the forest may be viewed only as trees and trees viewed only as timber. The soil and the water, the grasses and the shrubs, the fish and the wildlife, and the beauty of the forest must become integral parts of the resource manager's thinking and actions. (cited in United States History, par. 1)

The National Forest Management Act directs the Secretary of Agriculture to protect and improve the future productivity of all renewable resources on forest land, including improvements of maintenance and construction, reforestation, and wildlife habitat management on timber sales. After several years of implementing national forest plans with public review and selection of alternatives, in 1992 the chief of the Forest Service, Dale Robertson, issued a policy letter to regional foresters and station directors and in part it states: "an ecological approach to managing National Forests and Grasslands is the right way to go because forests are dynamic and complex ecosystems" (par. 3).

Ecosystems bring together the phenomenon of life (**Principle 10,** Chapter 1) in a way that includes all forms of life unified in a system that considers not just forest and trees but also life-forms of shrubs, grass, birds, mammals, etc., and presents the elements for a unified approach to forest management. Before the National Forest Management Act was passed, the USDA Forest Service initiated a two-year study (1970–1972) on the nation's forest and range resources. One result of this study was a system of classifying land area of the 48 contiguous states into units called "ecosystems" based on broad geographic forest types. Our descriptions of the **forest types** are synonymous with the **ecosystems** listed in Garrison et al. (1977).

2-1 Forests Are Ecosystems

The term ecosystem was defined early on as the whole complex of physical factors forming the environment (Tansley 1935). Eugene Odum's (1971) definition clarified the concept: "Any unit that includes all the organisms in a given area interacting with the physical environment so that a flow of energy leads to a clearly defined trophic structure, biotic diversity, and material cycles is an ecological system or ecosystem" (pg. 8).

One way to understand the workings of an ecosystem is to progressively add to the complexity, starting with a single individual as the first level of integration or resolution (Figure 2-1). **Individuals** are distinct and cannot be subdivided into units that can live as single entities. An individual is uniform in its genetic composition but there are differences between individuals in a population. As more organisms of the same species come together in an interbreeding group, the result is a population. A population of one species when added to a population of another becomes a **community** and the various combinations of plant and animal communities create a degree of ecological diversity. We speak of several populations of different plant species as a plant community or of animals as an animal community.

Structure and Function

An ecosystem is an energy-nutrient processing system. Ecosystems have physical **structure**, for example, soil, plants, and animals, as well as **processes** of energy flow and nutrient cycling involving biotic and abiotic components (defined below). In addition, ecosystems have time and space attributes. While the term ecosystem has been used in the scientific literature since the 1930s, there is often reluctance by managers to embrace the concept because they do not directly consider energy and nutrients in their decision-making process.

An ecosystem describes an area and the animals and plants (**biotic components**) within it as well as their physical environment (**abiotic components**). In all cases the ecosystem functions as a unit through nutrient cycling and energy flow through plants and animals. Nutrient cycling differs from energy flow (defined below) as nutrients eventually return to the system through the death and decay of plants and animals in which they are temporarily bound. Functional structure develops from trophic levels (often called "biotic pyramids") when the nutrients cycle from a large number of producers (green plants) to a smaller number of primary and secondary consumers (animals) and ultimately to decomposers.

All elements go through a cycling process from their reservoirs of living organisms, air, water, and soil. Three types of cycles exist: **biological**, **geological**, and **chemical**. Collectively these single cycles are often referred to as "**biogeochemical** cycles." While about half of the basic elements are essential for life, four (carbon, nitrogen, oxygen, and hydrogen) constitute about 97% percent of the weight of protoplasm (Emmel 1973) and cycles most easily. Carbon is a basic building block of the molecules necessary for life so that the cycling of carbon through the use of carbon dioxide (CO_2) by plants and respiration by animals is a critical life-sustaining process (Figure 2-2 on p. 30).

At death, microbes (decomposers) break down the organic matter in a decaying process that releases carbon bound in animals to the soil, air, and water reservoirs. There is a growing concern that carbon dioxide is increasing because of increased human activities. Living animals take in oxygen (O_2) and release carbon dioxide from respiration. Carbon dioxide is contributing to the greenhouse effect

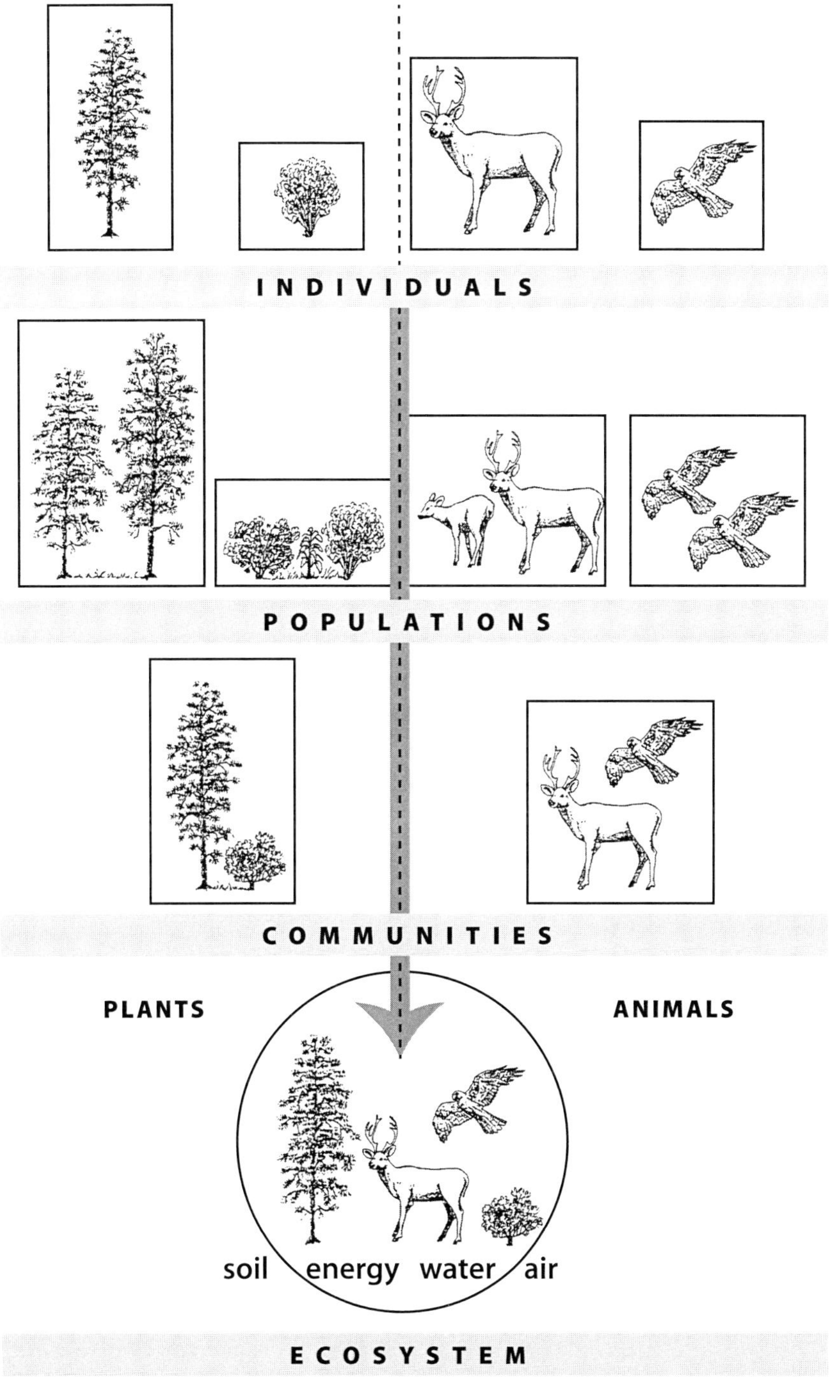

Figure 2-1 The hierarchical levels of an ecosystem.

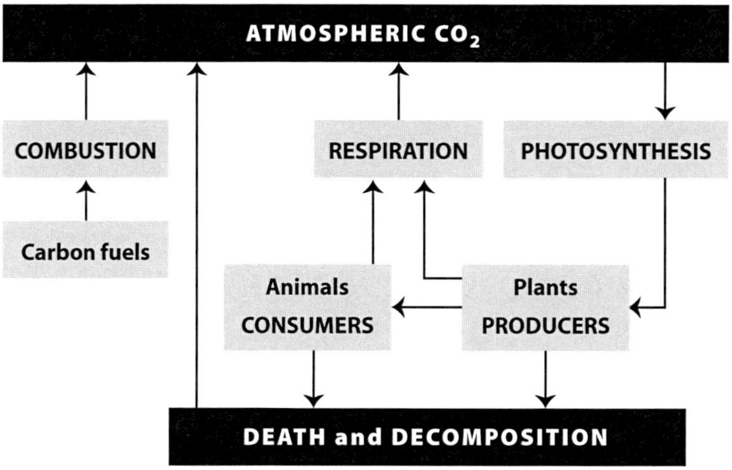

Figure 2-2 Carbon cycle in an ecosystem.

and the theory is that this is causing an increase in global warming. The clearing of forests for agriculture can decrease carbon held in the trees as much as 90% (Houghton 1995).

About 67% of the sun's total energy warms the atmosphere and land surface at noon on a clear day (Gates 1965). It takes approximately eight minutes for light energy to arrive on the earth's surface and less than 2% is trapped by plants and stored as chemical energy (Emmel 1973). The primal source of energy for all of earth's biological systems is the sun. The air we breathe is composed of 78% nitrogen (N), 21% oxygen, and about 0.03% carbon dioxide. The chlorophyll in the leaves of green plants absorbs a small portion of the sun's energy, which is then used to combine carbon dioxide from the air and water (H_2O) from the soil into starches and sugars that are contained in the plants eaten by animals. Thus plants (autotrophic) depend on the sun's energy, chlorophyll, and photosynthesis to produce their own food; in contrast, **animals do not produce their own food**. It is important to remember this fact when reviewing theories about habitat, wildlife populations, and factors regulating populations (Figure 2-3).

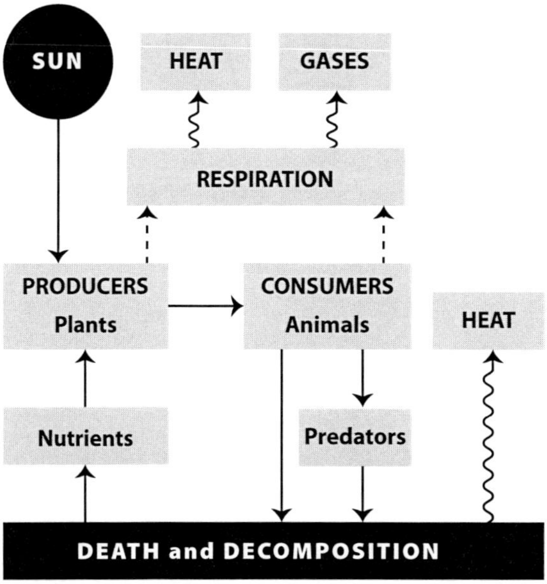

Figure 2-3 Energy flow in an ecosystem.

Energy flow is a one-way transfer from plants to animals in an herbivore relationship or from animals to animals in a carnivore relationship. Energy has been changed in the transfer, but none has been created or destroyed; it has merely changed in form (**first law of thermodynamics**). During the transfer process a loss of energy as heat through respiration always occurs, so to keep the ecosystem functioning there must be a constant flow of energy from the sun through the system (**second law of thermodynamics**).

Biological Engines

Energy flow and nutrient cycling in an ecosystem requires pathways. These pathways are found in animal **food chains** and **food webs** (Figure 2-4). In simple ecosystems, as in the arctic, the pathways are short and few in number, but in many ecosystems, such as the tropics, the pathways are long, numerous, and hence complex. Whatever the length and complexity, every food chain must start with plants. Energy from plants is transferred through a series of animal species each of which eats one or more species preceding it in the chain.

The shorter the food chain the greater the energy available for converting food items into living matter and metabolic activity. Food chains and food webs are the **biological engines** of ecosystems and without them there would be no sustainable life. Anytime a chain is broken a link in the system is eliminated. The question is: How many chains can be broken before energy and nutrient paths are degraded to a point where life cannot be sustained for an individual species? What would happen if the two plant producers, grass and forbs, were eliminated in the food web in Figure 2-4?

2-2 Classification Schemes

Scientists, in trying to describe and understand the complicated world in which we live, are constantly developing new ways to classify the enormous number of diverse plant and animal species into recognizable categories of like things for

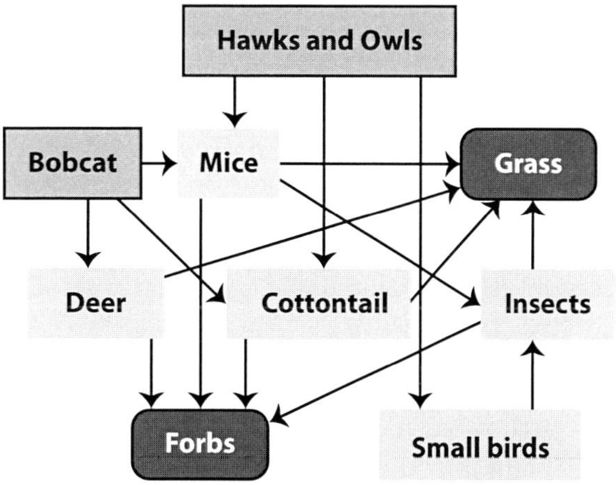

Figure 2-4 Food chains and food webs are the biological engines of ecosystems.

practical understanding, awareness, and management. Classification schemes are a product of thinking in an organized manner, which tends to be hierarchical. Hierarchies are a convenient method of abstraction allowing us to reduce complex systems to simpler units that we can understand and study (O'Neill et al. 1986); information at each level of the hierarchy can be aggregated upward. This same procedure can be used regardless of the type of data contained in each distinct level. Hierarchical descriptions also provide a way for managers and researchers to communicate. Thinking in hierarchical terms allows us to discover how individual levels are integrated into more complex systems.

The UNESCO System

In the 1970s, there was an effort to develop a classification system that would bring some commonality to naming vegetation and plant communities so ecologists could speak the same language. A major effort by the United Nations Education, Scientific, and Cultural Organization resulted in a worldwide vegetation classification scheme (UNESCO 1973). The system is open-ended and hierarchical, using plant physiognomic characteristics for descriptors. Under the UNESCO system, there are five mutually exclusive vegetation classes in the first level of the hierarchy: forest land, woodland, shrubland, dwarf shrubland, and herbaceous vegetation.

In ecological literature, different terms (i.e., biome, biotic community, life zone, plant community, formation, association, ecosystem, cover type, vegetation type, and habitat type) are commonly used to describe vegetation at various levels of integration. In the past, there has been considerable effort by ecologists to delineate vegetation over continental or large regional areas, for example:

- biomes (Shelford 1963);
- biotic provinces (Dice 1943);
- ecological regions of North America (CEC 1997);
- ecoregions (Bailey 1998);
- forest and range ecosystems (Garrison et al. 1977);
- habitat type (Daubenmire 1968);
- life zones (Merriam, 1898);
- North American biotic communities (Brown 1998);
- North American forests (Walker 1999); and
- North American terrestrial vegetation (Barbour and Billings 2000).

Michael Barbour and William Dwight Billings (2000) list 17 **formations** for North America. The names of the formations consist of combinations of life-forms, climate, geography, and plants. A chapter for each formation describes its ecological characteristics in detail and is an excellent reference for students and researchers. A subunit of a formation can be a vegetation type named by its dominant species, such as ponderosa pine, oak-hickory, beech-birch-maple, or redwoods.

The effect of climate and change in elevation creates a distinct and identifiable vegetation zone. Such a zone was recognized by Merriam during his travels in the western United States as a member of the U.S. Biological Survey (1898). His **life**

zones were identified with increasing elevation as Sonoran, Transition, Canadian, Hudsonian, and Alpine. These life zones are useful in the West in mountainous country, but the concept is difficult to apply to lower elevations across the United States. Merriam's life zone classification was indicative of vegetation descriptions plant ecologists were trying to formulate to create a common understanding of classification that would allow discussion about areas where vegetation exists, why, and what causes it to change.

Homer Shantz and Raphiel Zon (1924) published *Forest Vegetation of the United States in Natural Vegetation: Atlas of American Agriculture*. The Shantz-Zon map was interpreted and reformatted by the U.S. Forest Service to create Forest Regions of the United States (USDA 1948). Later, August Kuchler (1966) developed a map of potential natural vegetation. His vegetation map is widely used because it includes shrub and grassland, not just forest land and forest types. A later version of the map was produced by the U.S. Forest Service (USDA 1967) to be included in the **National Atlas** and has continued to be used to depict the distribution of forest vegetation. The 1967 map, along with Kuchler's 1966 map, were interpreted for a map, "Forest and Range Ecosystems of the United States" (USDA 1974).

One of the first maps to depict forest vegetation in the United States was published in a geography textbook (Tarbell 1899). Basically, the map delineated the country into two forest regions: coniferous and deciduous. The northern part of the United States extending along the border with Canada was coniferous, the middle part from Ohio south to Tennessee was deciduous, and the southern portion from Louisiana to Florida was coniferous. All of the forested area west of the Great Plains was shown as coniferous except for the middle and southern part of California, which was designated as deciduous.

2-3 Vegetation Succession

The change in plant and animal species composition over time on the same piece of land results in a natural cycle of both in the phenomenon of diversity (Principle 10, Chapter 1). Succession as an ecological process was a major topic of discussion that began after the publication of an analysis of the development of vegetation by Frederic Clement in 1916. The thesis declared that when one plant community is established, it slowly creates conditions in seral stages that allow new species from other communities to invade, grow, and reproduce. In this new community the plant species composition, abundance, and structure are different from what occupied the same space in the past. The end result of succession is a self-perpetuating climax stage. There is no disagreement by ecologists that succession occurs but the idea of a final self-perpetuating stage, the **climax**, has generated much discussion in science literature (Watt 1947; Horn 1976; Noble and Slatyer 1977; Tansley 1935; Knapp 1974; Gleason 1939; Whittaker 1953; Curtis 1959).

Agents of Change

Succession can be caused by plants and animals from within the ecosystem (**autogenic**) or from outside (**allogenic**) influences, such as wind or fire. In addition to the agents of change, the starting point of succession can be either **primary** or **secondary**. Primary (starting from bare soil) successional processes have a beginning but the end may be difficult to determine. One of the common disturbances in stands is small gaps created by the loss of a single or a few trees that result in a micro-succession, which proceeds as primary succession from bare ground (Watt 1947). These small spaces (**nooks and crannies**) are important in creating niches for a variety of animal species.

There comes a time in an undisturbed plant community when a balance between all the factors causing change is achieved. Under equilibrium conditions the wastes from living matter are recycled so that the ecosystem is self-perpetuating until a catastrophic event occurs; then the process begins again progressing toward climax. The factors influencing succession can be expressed in a conceptual equation (Jenny 1958) as:

$$Ve = f\ (So + Or + To + Cl + Ti)$$

Where Ve = vegetation
f = function of
So = soil
Or = organisms
To = topography
Cl = climate
Ti = time

In reality changes attributable to external factors, such as human intervention, commonly interrupt the climax or seral stages; therefore, ecosystems are never stable in the true sense of the word, but are **resilient**.

Seral Stages

Vegetation in forested ecosystems advances through general succession seral stages (Figure 2-5) often identified for forestry and wildlife management purposes as:

- VSS-1 Opening (bare, grass, forb)
- VSS-2 Seedling
- VSS-3 Sapling
- VSS-4 Young
- VSS-5 Intermediate (pole)
- VSS-6 Mature
- VSS-7 Old growth

Stages 3 to 7 will commonly be associated with a diameter class (Chapter 1). Structure class terminology used by resource agencies may change according to regional standards and vegetation type. Succession is a powerful force that shapes vegetation structural stages, and a thorough understanding of the processes involved is critical to maintaining habitats for different wildlife species. Disturbances in a successional process often occur and may initiate a new stand devel-

Figure 2-5 Conceptual view of succession progressing from seedlings to old forest conditions over time.

Regression to a former stage can occur from disturbances caused by natural events of fire, wind, or from a regeneration cut.

opment phase with different characteristics. Chadwick Oliver and Bruce Larson (1990) suggest that these development phases follow a certain pattern as follows:

1. Stand initiation
2. Stem exclusion
3. Understory reinitiation
4. Old growth

Rate of change between stages is variable and dependent on local conditions. The ability of humans to modify succession using the axe, plow, and cow is well-documented, but we now have other and more precise means gained by the accumulated knowledge of not only the development of plant communities but also the invention and use of tools to redirect succession over large areas in a short time. Ecological knowledge together with the tools of the chain saw, bulldozer, and fire have truly made humans **ecological dominants**: organisms which by their presence can radically alter the environment of other species.

THE EDGE EFFECT. When two distinct vegetative life-forms (trees, shrubs, grass) or structures within life-forms (mature trees, saplings) meet, a boundary or edge is created between the two (Figure 2-6). The natural change of plants from one

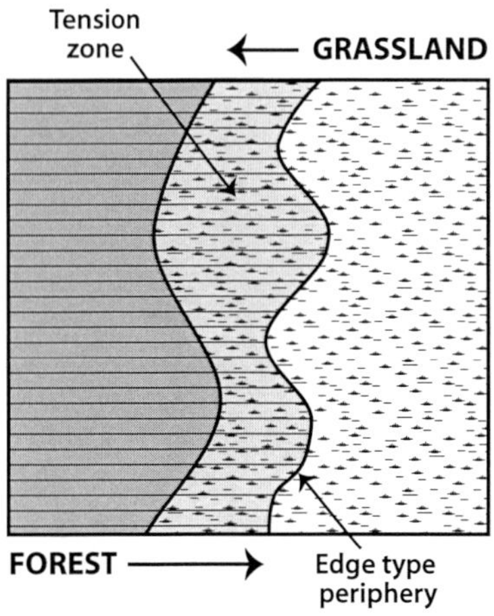

Figure 2-6 An edge is created when two vegetation types, forest types, or stand structural stages meet.

stage to another is generally not abrupt but is a gradual replacement process. This change intrinsically produces an increase in variety or diversity of habitats. Leopold (1933) referred to this change or increase as a **phenomenon of an edge** from which he developed his **law of dispersion**: "The potential density of game of low radius requiring two or more types is, within ordinary limits, proportional to the sum of type peripheries" (pg. 132).

Edge can be of two types: inherent or induced (Thomas 1979). An **inherent edge** is the natural outcome of the evolution of successional and structural stages of plant communities. **Tension zones** arise in cases where the position of the naturally evolved inherent edge is constantly in flux. **Induced edges** result from management practices such as fire, timber harvesting, grazing, or seeding and planting that are direct but short-term activities undertaken to manipulate succession. Both types of edges can be either abrupt or meandering depending upon the nature of the vegetation types or successional stages.

Leopold's law is based on the premise that home range is fixed for a species as an innate characteristic and determines how far an animal will go in search of food and cover. While a rabbit may be restricted by instinct to remain within a small area (acres), a deer is not. Yet, both rabbits and deer use edges to meet their life needs (food, cover, water, space), but at different scales. Although Leopold's law of dispersion was developed in relation to small game and nongame species, his edge principle applies in varying degrees to other wildlife. Larry Harris (1988) has suggested that while many species such as the white-tailed deer benefit from edge, species needing large blocks of land with interior habitats often do not. Wildlife managers have to keep in mind that there is a point where creating more edge than naturally occurs may result in fragmentation of habitat that is detrimental to other wildlife species.

Ecological Amplitude

Coupled with a plant community, successional stages have associated changes in animal species. These changes occur because animal species, like plant species, have different life requirements and therefore different tolerances (ecological amplitude) to the existing environmental conditions. As plant composition changes through succession there is in turn a change in animal species. Habitat change over a 20-year period was recorded on the Connecticut Hill Game Management Area. Evidence is presented that as changes occurred in vegetation composition, so did occupancy by the ruffed grouse (Bump 1950). In other cases, David Johnston and Eugene Odum (1956) found the same type of changes in bird populations in Georgia (Table 2-1) and changes in mammal populations were recorded by Catherine Ream and George Gruell (1980) (Table 2-2) in succession following clearcutting and burning.

Because of the differences in ecological amplitudes of species, many will have overlapping habitat requirements during the changes in successional stages and even after climax is attained. The degree of overlap is a function of ecological amplitude and niche separation. In many instances, animals that man most desires to perpetuate, such as deer and rabbits, are members of early seral rather than climax stages. Thus, some wildlife species and many of the tree species (e.g., aspen) that are valuable to man thrive best in temporary communities.

Old Growth

One of the major problems with old growth issues is definition. Foresters usually consider old growth as over-mature or decadent stands exceeding rotation age based on the criterion of culmination of mean annual increment (MAI). Environ-

Table 2-1 Changes in bird species association with vegetation types as a result of succession on abandoned farmland in Georgia.

Species	Grassland	Shrubs	Pine forest	Hardwood forest
Grasshopper sparrow	***********			
Meadowlark	************			
Yellow-breasted chat		**********		
Cardinal		*********************************		
Pine warbler			**************	
Carolina chickadee			*******************************	
Crested flycatcher			******************	
Acadian flycatcher				*************

Source: Johnston and Odum 1956.

Table 2-2 Changes in mammal species association as a result of succession retarded by clearcutting and burning.

Species	Clearcut and burn	Herbaceous growth	Brush, Seedlings, Saplings	Mature trees
Deer mouse	***			
Microtines		**************************************		
Ground squirrel		**************************************		
Chipmunks		**********************		
Rabbits		**********************		
Red squirrel			******************	
Flying squirrel			******************	

Source: Ream and Gruell 1980.

mental groups conceive old growth stands as possessing pristine or virgin conditions. As a result of the issues that developed in the Pacific Northwest and because no common understanding of old growth existed, the Society of American Foresters (SAF) appointed a task force to formulate an ecological definition of an average old growth acre (Heinrichs 1983).

The definition applies specifically to Douglas-fir forests of the Pacific Northwest but many of the characteristics cited are applicable to other vegetation types. The items included relate not just to growth of a single tree but to stands and their structure and composition as well. The SAF task force definition states that an old growth forest:

- consists of two or more tree species of a wide range of sizes and ages; often a long-lived seral dominant such as Douglas-fir is associated with a shade-tolerant species such as western hemlock;
- contains a deep, multilayered crown canopy;
- contains more than 10 trees at least 200 years old or 40 inches diameter at breast height (dbh);
- contains 10 snags over 20 feet tall, and more than 20 tons of down logs; and
- contains at least four large snags and an equal number of logs 25 inches diameter at breast height and 50 feet long.

Another definition of old growth is broader and based on the premise that old growth forests are relatively old and undisturbed by humans (Hunter 1989). Clearly old growth is an ecological entity, and humans have placed a high value on preserving part or all of it for its ecological significance, wildlife habitat, or recreational and aesthetic values. However, several issues pertaining to old growth remain and need to be addressed by state and federal agencies for landscape level management. Old growth questions are: How many acres should be conserved? Can silvicultural techniques duplicate existing conditions for future stands? At what point in time should old growth be regenerated?

The first question will take several years to answer since it involves hotly debated social and political issues. The second question cannot be addressed easily and it may take several generations before an answer is found. The third question will also take time because it deals with the problem of when does old growth cease to be old growth and become a decadent stand, which may lose the value for which it was protected in the first place. The third question raises another question: Is old growth a renewable resource?

James Kimmins (1997) makes the point that the notion of a renewable resource should not be based on whether it is living or dead, but whether it can be restored to a point of reuse after a period of time. The period of time must either be within our current economic or social planning scale, or it must be renewed at a rate that renders investment in its renewal economically attractive. Resources not meeting these criteria are classed as nonrenewable. Kimmins further observes that with the general antipathy of the public towards cutting 1,000-year-old redwoods in California or a 300-year-old oak tree in France, these trees are nonrenewable on the basis of his conditions: once cut the social value the trees provided is gone, never to be renewed.

There are some characteristics of old growth forests that seem to be consistent no matter what the forest type. Old trees are only one of the defining characteristics of old growth and there is no point in time that stands obtain old growth status, but there is a slow long term process of maturation that may last several hundred years. The term old growth does not adequately describe the total ecological meaning. Old forest conditions may represent the best descriptive term. Old forest conditions contain large trees that are past rotation age, are ecological dominants that influence site conditions, contain dead and dying trees, and have a ground layer that has scattered logs and stumps in a process of advanced decay.

Old forest conditions will have small openings with shrubs, forbs, and grasses creating an impression of being biologically diverse with a large amount of carbon tied up in the volume of large trees and understory components. Old forest descriptive conditions will vary across the continent and forest type and there is no one definition that fits all situations. The description of the maturation of a mature stand to old forest conditions parallels the description by Richard Jefferies (1848) of a hollow tree destined to go before the axe (Chapter 1). A more detailed definition of old growth can be found in *The Dictionary of Forestry* (Helms 1998). Foresters and wildlife managers can manipulate **succession** in a way that will maintain the desired stand conditions so as to have a sufficient number of areas continually coming into the desired seral stages. This idea is the basic premise of **silviculture** and forest management. The change in plant and animal species composition over time on the same piece of land results in a natural cycle of both. Populations in an area may not change in total numbers of individuals but their location within that area will change. This concept is important to wildlife management because it introduces the element of **ecological time** (seedling to old forest conditions), which is beyond the span of the "here and now" (5–10 years) typically incorporated into management plans.

2-4 Forests and Woodlands

Our focus in this book is on forests and woodlands because they contain trees as the dominant plant life-form. The distinction between the two classes is the degree of canopy closure. The woodland class has a more open canopy, in the range of 25–60%. Vegetation over large landscapes is not too difficult to identify to a low level of continuity or commonality. In mountain areas of the western United States, even the most casual observer can see that as elevation increases from 1,000 to several thousand feet, vegetation changes, and some species of plants will become common and dominate the scene. Observers most likely will attribute this change in plant species to temperature and moisture differences because they also can physically experience the same elements. In addition, observers will notice that vegetation along the route changes gradually, and at close range a boundary is not distinct. They will notice that vegetation is not entirely homogeneous, but there are locations where plants form areas containing different species that seem to be influenced by local conditions, including soil and exposure to sun on a slope. While pinyon-juniper woodland is shown as a forest type on the National Atlas it is also considered as one of 15 rangeland types.

Forest Types

Within the continental limits of the United States and Alaska, vegetation dominated by trees can be categorized into forest types (ecosystems) containing native tree species (Table 2-3). Forest type has been a classification system in use by the USDA Forest Service since 1929 and is widely accepted by land managers, plant ecologists, and wildlife biologists as a way to group trees for management purposes (Eyre 1980). Tree diversity in the types results from the interactions of the range of soil, topography, and climatic conditions found within the geographical areas for each type.

Table 2-3 Forests types (Continental U.S.) and the number of states where they occur.

Eastern Forests	Number of States	Western Forests	Number of States
10—White-red-jack pine	13	20—Douglas-fir	9
11—Spruce fir	8	21—Hemlock-Sitka spruce	2
12—Longleaf-slash pine	8	22—Ponderosa pine	14
13—Loblolly-shortleaf pine	17	23—Western white pine	2
14—Oak-pine	17	24—Lodgepole pine	7
15—Oak-hickory	32	25—Larch	4
16—Oak, gum-cypress	17	26—Fir-spruce	8
17—Elm-ash-cottonwood	17	27—Redwood	2
18—Maple-beech-birch	13	28—Pinyon-juniper	9
19—Aspen-birch	3	29—Western hardwoods	10
Total	**145**	**Total**	**68**

Identified by Garrison et. al (1977) and on the National Atlas map.

Cover is both a forestry and a wildlife term, but with different meanings. Cover or cover type in forestry and plant ecology is the vegetation that overlays soil material. In wildlife, cover is used to designate areas for protection and shelter of animals and can be something other than vegetation. In this textbook, the term **cover type** is changed to **forest type** or **ecosystem** to avoid confusion between the two definitions. The criteria used to describe forest types are as follows:

- The dominant trees must cover at least 25% of the area.
- The type must occupy a fairly large area but not necessarily in continuous stands.
- The type must be based entirely on biological considerations.

NATIONAL ATLAS. The most recent work on forest types was produced by advanced very-high-resolution radiometer (AVHRR) composite images during the 1991 growing season (Zhu and Evans 1994) and is the basis for a National Atlas Map (USDA 2000). All of the forest types are essentially the same as when they were first described by Homer Shantz and Raphiel Zon (1924). The Atlas map validates the excellent fieldwork to document type location and ecological descriptions by early plant ecologists. **Caveat**: A type is not continuous and homogeneous but will have gaps and intrusions of patches of other types. Types occur in different areas because some trees have a wider ecological tolerance than others. Types are named according to the plurality of the dominant tree species. Other trees can be associates of the dominant species therefore creating a biological community that includes many life-forms of plants and animals.

2-5 Where Are the Forests?

The geographical distribution of forest types is identified on the National Atlas map of the continental United States and Alaska. The two most dominant types (based on size of area occupied) in each state are associated with other types within a state (Table 2-4 on the following page). Because of the number, size, and scale of complexity of forest types it is difficult to show the detail of a specific type on small scale maps (1:7,500,000). Statistical information is often reported by a defined geographical region as done in the Resource Planning Act Assessment (USDA 1981). The following sections outline forest types by region.

Eastern U.S.

White-Red-Jack Pine—10 . This type is found on smooth to irregular plains and tablelands of the northern lake states and provinces and parts of New York, New England, and Canadian Maritime provinces with an extension southward through the Appalachians. Eastern white pine, red pine, and jack pine are the major components. Associates include eastern hemlock, aspens, northern white cedar, maples, and birches. Precipitation averages 25 to 45 inches, distributed evenly throughout the year. More than 50 percent of the area is gently sloping. Timber harvesting and recreation are the main land use activities.

Occurs in: 13 States

Dominant in: ME, NH, VT

Present in: CT, MA, MI, MN, NC, NY, RI, TN, VA, WI

Spruce-Fir—11. Flat plains and tablelands in the lake states and provinces, New England states, the Maritime provinces, and at high elevations in the Appalachian Mountains are occupied by the spruce-fir type. Red spruce, white spruce, and balsam fir are associated with northern white cedar, eastern hemlock, eastern white pine, and tamarack. Hardwoods include maples, birches, and aspens. Normal precipitation is 30 to 40 inches. The type occurs at high elevations in the Appalachian Mountains as far south as West Virginia. Change in elevations is less than 500 feet and often below 100 feet. Timber harvesting and recreation are the main land use activities.

Occurs in: 8 States

Dominant in: MN, VT

Present in: ME, NH, NY, WI, WV, MI

Table 2-4 Dominant and associated forest types for each state.

State		Dominant Types	Associated Types
AK	21, 20	Western hemlock-Sitka spruce	—
AL	13, 14	Loblolly-shortleaf pine/Oak-pine	12, 15, 16
AR	13, 15	Loblolly-shortleaf pine/ Oak-hickory	14, 16
AZ	28, 22	Pinyon-juniper/Ponderosa pine	26
CA	22, 29	Ponderosa pine/Hardwoods	26, 24, 21, 20, 27
CO	26, 22	Fir-spruce/Ponderosa pine	28, 20, 24, 29
CT	15, 17	Oak-hickory/Elm-ash-cottonwood	18, 10
DE	13, 15	Loblolly-shortleaf pine/Oak-hickory	14
FL	12, 15	Longleaf-slash pine/Oak-hickory	16, 14, 13
GA	12, 13	Longleaf-slash pine/Loblolly-shortleaf	14, 16, 15
IA	15, 17	Oak-hickory/Elm-ash-cottonwood	16
ID	20, 23	Douglas-fir/White-pine	24, 25, 26, 28, 29
IL	15, 17	Oak-hickory/Elm-ash-cottonwood	16
IN	15, 18	Oak-hickory/Maple-beech-birch	16, 17
KS	15, 17	Oak-hickory/Elm-ash-cottonwood	—
KY	15, 14	Oak-hickory/Oak-pine	16, 17
LA	16, 13	Oak-gum-cypress/Loblolly-shortleaf pine	12, 14
MA	15, 18	Oak-hickory/Maple-beech-birch	17, 13, 10
MD	15, 13	Oak-hickory/Loblolly-shortleaf pine	14
ME	11, 18	Spruce-fir/Maple-beech-birch	10
MI	19, 18	Aspen-birch/ Maple-beech-birch	15, 10, 11, 17
MN	19, 11	Aspen-birch /Spruce-fir	10, 15, 18, 17
MO	15, 14	Oak-hickory/ Oak-pine	17, 16
MS	13, 14	Loblolly-shortleaf pine/Oak-pine	15, 12, 16
MT	20, 22	Douglas-fir/Ponderosa pine	24, 25, 29, 28
NC	13, 15	Loblolly-shortleaf pine/Oak-hickory	16, 12, 10, 14
ND	29, 17	Hardwoods/Elm-ash-cottonwood	15, 22
NE	17, 22	Elm-ash-cottonwood/Ponderosa pine	—
NH	18, 10	Maple-beech-birch/White-red-jack pine	11
NJ	13, 15	Loblolly-shortleaf pine/Oak-hickory	17
NM	28, 22	Pinyon-juniper/Ponderosa pine	20
NV	28, 22	Pinyon-juniper/Ponderosa pine	—
NY	18, 10	Maple-beech-birch/White-red-jack pine	17, 11, 15
OH	15, 18	Oak-hickory/Maple-beech-birch	17
OK	15, 14	Oak-hickory/Oak-pine	13
OR	20, 22	Douglas-fir/Ponderosa pine	26, 24, 28, 25, 27
PA	15, 18	Oak-hickory/Maple-beech-birch	14
RI	15, 17	Oak-hickory/Elm-ash-cottonwood	10
SC	13, 12	Loblolly-shortleaf pine/Longleaf-slash pine	16, 14, 15
SD	22, 17	Ponderosa pine/Elm-ash-cottonwood	29
TN	15, 14	Oak-hickory/Oak-pine	16, 13, 10
TX	15, 14	Oak-hickory/Oak-pine	28, 16, 12
UT	28, 29	Pinyon-juniper/Hardwoods	26, 22, 20
VA	13, 15	Loblolly-shortleaf pine/Oak-hickory	10, 14, 16
VT	18, 11	Maple-beech-birch/Spruce-fir	10
WA	20, 26	Douglas-fir/Fir-spruce	21, 22, 23, 25, 24
WI	19, 18	Aspen-birch/Maple-beech-birch	15, 10, 11
WV	15, 18	Oak-hickory/Maple-beech-birch	13, 11
WY	24, 26	Lodgepole pine/Fir-spruce	22, 20, 29

Forest type names are from Eyre (1980).

Longleaf-Slash Pine—12. This type is restricted to flat and irregular southern Gulf Coastal Plains of the U.S. where local relief is less than 300 feet. Major species include longleaf and slash pine. Associated species include shortleaf, loblolly, and Virginia pine with a mixture of oaks and hickories. Annual precipitation ranges from 40 to 60 inches.

Occurs in: 8 States

Dominant in: SC, FL, GA

Present in: NC, TX, AL, LA, MS

Loblolly-Shortleaf Pine—13. The loblolly-shortleaf type generally occurs on irregular Gulf Coastal Plains and the eastern Piedmont region where local relief is 300 to 1,000 feet. This type is of the greatest extent in the South and Southeast. The major species are loblolly and shortleaf pine in association with oaks, hickories, and maples. Annual precipitation is 40 to 60 inches, evenly distributed with a midsummer peak. About 60 to 70 percent of the land use is in forestry, mostly farm woodlots.

Occurs in: 17 States

Dominant in: MD, DE, NJ, VA, NC, SC, GA, AL, AR, LA, MS

Present in: MA, WV, FL, TX, TN, OK

Oak-Pine—14. The oak-pine type occurs on diverse land forms from the southern-most ridges and valleys of the Appalachians westward across the coastal plains and north into the Ozark Plateaus. White oak, northern red oak, scarlet oak, black oak, chestnut oak, with loblolly pine, shortleaf, and Virginia pine are the major species. Precipitation is relatively high averaging 40 to 55 inches.

Occurs in: 17 States

Dominant in: KY, TN, TX, OK, AL, MS

Present in: MD, PA, DE, MO, VA, NC, SC, FL, GA, AR, LA

Oak-Hickory—15. Oak-hickory is the most extensive type in the United States and occurs in the east where conditions are moister than the surrounding areas. White oak, northern red and black oak, with associates of pignut, shagbark, mockernut, and bitternut hickory are the major species in the type. The forest type is best developed in the Ozarks. Annual precipitation in the type averages between 35 to 45 inches. Oak-hickory varies from open to closed woods with a strong to weak understory of shrubs, vines, and herbaceous plants.

Occurs in: 32 States

Dominant in: KS, NY, CT, MA, MD, OH, WV, PA, RI, DE, NJ, MI, IA, IL, IN, MO, VA, NC, FL, KY, TN, TX, OK

Present in: ND, NH, MN, WI, SC, GA, AL, AR, MS

Oak-Gum-Cypress—16. This type is characterized by the vegetation of the Mississippi Valley and other bottom lands in every southern state, the cypress savanna west of the Everglades in Florida, the mangrove swamps south of the Everglades, and the east coast of Florida, Georgia, and the Carolinas. Blackgum, sweetgum, baldcypress, Nuttall, shumard, overcup, and cherrybark oak are the major species in the type. Other hardwoods such as willows, maples, sycamore, cottonwoods, and beech are present depending on location. Annual precipitation varies from 60 inches in Florida to 35 inches in the northern extremity of the type in Indiana.

Occurs in: 17 States

Dominant in: AR, LA

Present in: IL, IN, MO, VA, NC, SC, FL, GA, KY, TN, TX, OK, AL, MS, IA

ELM-ASH-COTTONWOOD—17. The elm-ash-cottonwood type occurs in narrow belts along major streams or scattered areas of dry swamps largely on the lower terraces and flood plains of the Mississippi, Missouri, Platte, Kansas, and Ohio Rivers from the Dakotas, Minnesota, and Ohio south through Kansas and Missouri. Cottonwood is the dominant species in the type with associates of green ash, white ash, American elm, and black willow. Precipitation varies from 10 inches near the foothills of the Rocky Mountains to 50 inches in the southern and northeastern areas. Vegetation is of low to tall broadleaved deciduous trees, varying from open to dense and often accompanied by vines.

Occurs in: 17 States

Dominant in: KS, NE, SD, ND, RI, IA, IL, MO

Present in: NY, MA, OH, NJ, MI, MN, IN, KY, CT

MAPLE-BEECH-BIRCH—18. The maple-beech-birch type is best developed in the New England states and Maritime provinces. Sugar maple, American beech, and yellow birch are the dominant tree species. Associates are red maple, eastern hemlock, white ash, black cherry, sweet birch, and northern red oak. Typically, the type occurs on open high hills and low mountains. Annual precipitation ranges from 40 to 48 inches.

Occurs in: 13 States

Dominant in: ME, VT, NY, NH, CT, MA, OH, WV, PA, MI, WI, IN

Present in: MN

ASPEN-BIRCH—19. The aspen-birch type is found within the Great Lakes region. Quaking aspen, big tooth aspen, balsam poplar, paper birch, and gray birch are the dominant species. Balsam fir and maples are also common. The type is closely associated with moraines and outwash plains of recent glacial origin. Annual precipitation for the type is between 30 to 35 inches.

Occurs in: 3 States

Dominant in: MN, WI

Present in: MI

Western U.S.

DOUGLAS-FIR—20. Just south and west of the Hemlock-Spruce forest and inland from the coast of British Columbia, Washington, and Oregon extending south to northern California is the Douglas-fir forest. This forest contains at least 50 percent Douglas-fir with varying amounts of western hemlock, western redcedar, and redwood. Pacific yew and bigleaf maple are common understory species. The type constitutes one of the largest blocks of timber in the west. In the Rocky Mountains the type is developed best in Idaho, Montana, British Columbia, and Alberta but its range extends southward to Utah, Colorado, New Mexico, and Arizona. Associated species are lodgepole pine, western larch, grand fir, and quaking aspen. The Douglas-fir forest is the most widespread type in the northern Rocky Mountains, but southward it is scattered. Precipitation is 40 to 80 inches in the extreme west to 20 to 30 inches in the interior.

Occurs in: 10 States

Dominant in: OR, WA, ID, MT

Present in: CA, WY, NM, UT, CO, AK

WESTERN HEMLOCK-SITKA SPRUCE—21. Southward along the coast of British Columbia, Washington, Oregon, and northern California, western hemlock and Sitka spruce are the dominant species but Douglas-fir and western redcedar are common associates. Red alder and black cottonwood are also present. Sitka spruce generally exceeds 200 feet in height and can attain ages of more than 1,000 years. Big leaf maple is the only major hardwood component. Precipitation ranges from 60 to 115 inches annually. This type probably has the most productive understory of any coniferous forest in the United States.

Occurs in: 5 States

Dominant in: WA, OR

Present in: CA, ID, AK

PONDEROSA PINE—22. Extending south through central British Columbia, Washington, Oregon, and into California is the narrow ponderosa pine type. Associated with ponderosa pine in the northern areas are Douglas-fir, lodgepole pine, grand fir, and western larch. South into California common associates are white fir, Jeffrey pine, and incense cedar. Ponderosa pine also occupies warm dry sites in the Rockies with precipitation from 15 to 20 inches. Its range extends from eastern British Columbia, Washington, and Oregon into Idaho, Montana, Wyoming, South Dakota, and south through Colorado, New Mexico, and Arizona. At higher elevations Douglas-fir is associated with ponderosa pine. In the northern part of its range lodgepole pine, blue spruce, and quaking aspen are common species.

Occurs in: 14 States

Dominant in: OR, CA, CO, MT, NM, AZ, NE, SD

Present in: WA, ID, WY, NV, UT, ND

WESTERN WHITE PINE—23. The western white pine type occurs in the high mountains of the northern Rocky Mountains of western Alberta, Montana, and northern Idaho. It is also present in scattered areas in the Cascade Mountains of Oregon and Washington. White pine is the dominant species in combination with sugar pine. Western redcedar is an associated species. Annual rainfall in the type is from 20 to 30 inches.

Occurs in: 2 States

Dominant in: ID

Present in: WA

LODGEPOLE PINE—24. East of the ponderosa pine type in California is the lodgepole pine type. It extends northward as a disjunct population primarily in Oregon. The major associated species are red fir and mountain hemlock. The type occurs in pure stands in British Columbia, Idaho, and Montana but it ranges from Oregon and Wyoming south into Colorado. Common associated species are Douglas-fir, Engelmann spruce, subalpine fir, and aspens. Lodgepole pine as a pioneer species gets established following fire but it can also form a climax forest. Annual rainfall is from 20 to 50 inches.

Occurs in: 7 States

Dominant in: WY

Present in: OR, WA, ID, MT, CO, CA

LARCH—25. Located in the high mountains of Alberta, eastern Oregon, northern Idaho, and western Montana is the larch forest type. Western larch is the dominant species in association with white pine. Also Douglas-fir and grand fir may be present. Annual precipitation is 20 to 30 inches in eastern Oregon and 20 to 50 inches in the northern Rockies.

Occurs in: 4 States

Dominant in: —

Present in: OR, WA, ID, MT

FIR-SPRUCE—26. Adjacent to the ponderosa pine but at higher elevations is another long, narrow band of vegetation comprised of silver fir, subalpine fir, red fir, white fir, and Engelmann spruce and sometimes in combination with mountain hemlock. In the Southwest the type is developed best in Colorado but it also occurs in northwest Wyoming and in disjunct areas in western Montana and eastern Idaho. Fir-spruce occurs in small areas at high elevation in Arizona and New Mexico. Species in the type are subalpine and white fir with Engelmann spruce, blue spruce, and aspens. Annual precipitation ranges from 22 inches in the Rocky Mountains to 75 inches in the Sierras.

Occurs in: 8 States

Dominant in: WA, CO, WY

Present in: OR, CA, ID, UT, AZ

REDWOOD—27. The Redwood type occurs only in the low coastal mountains of northern California and the southwestern portion of Oregon. Redwoods are the tallest trees in the world, often reaching 300 feet. Douglas-fir, western hemlock, and Sitka spruce are common associates. Annual precipitation is approximately

60 inches. Hardwood species found in the Redwood type are bigleaf maple, red alder, and willows.

Occurs in: 2 States

Dominant in: CA

Present in: OR

Pinyon-Juniper Woodland—28. The pinyon-juniper type occurs in the Basin and Range province of Utah, Nevada, southern Idaho, and southeast Oregon. Large areas of pinyon-juniper occur in Arizona and New Mexico. Single leaf and Colorado pinyon with Utah, alligator, and one seed juniper are the main species. The name "pygmy forest" characterizes the pinyon and juniper woodland. The trees occur as dense to open woodland and savannah woodland. Annual precipitation is about 10 inches.

Occurs in: 9 States

Dominant in: NM, AZ, NV, UT

Present in: OR, ID, MT, CO, TX

Western Hardwoods—29. This type occurs in California and Oregon and in the Rocky Mountain regions with rainy winters and dry summers. Precipitation varies from 40 to 100 inches. Vegetation consists mostly of Oregon white oak, Coulter and digger pine, blue oak, canyon live oak, and interior live oak. In some areas aspen and other conifers are present with an understory of shrubs, grasses, and forbs.

Occurs in: 10 States

Dominant in: CA, UT

Present in: OR, ID, WY, SD, MT, CO, NV, ND

Alaska (Region I)

Spruce-Hardwoods—30. This type is restricted to the interior region of Alaska containing white spruce and where hardwood species exist in pure or mixed stands of black cottonwood, balsam popular, and paper birch. Boreal conditions of warm summers and cold winters restrict the composition of forest cover types where rainfall varies from 60 inches in the southern part to six inches in the Arctic region of Alaska. Because of a simple composition of tree species the Alaska interior region is considered an area where the effects of climate change might be observed.

Western Hemlock-Sitka Spruce—31. Found along the southern coast of Alaska and borders British Columbia to the Pacific Northwest and the state of Washington.

2-6 Where Are the Habitats and Wildlife?

The seven forest regions in Figure 2-7 combine the commonality of types into different regions with an associated checklist of wildlife species. A regional aggrega-

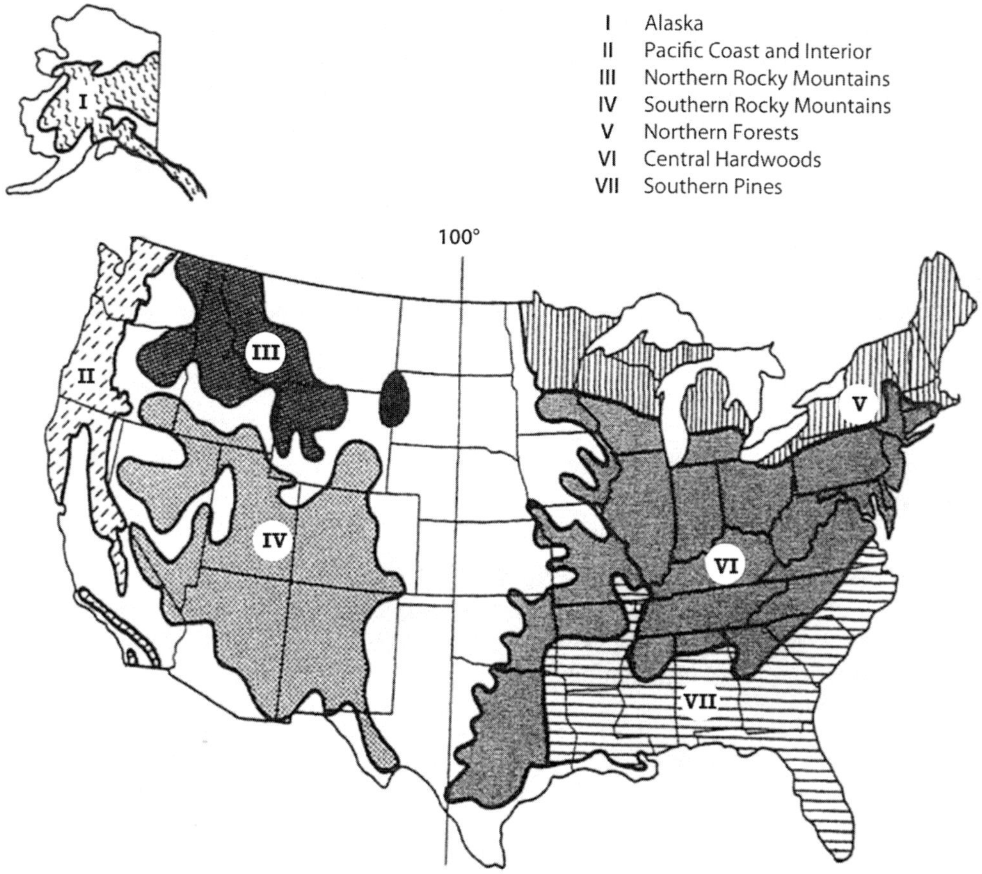

Figure 2-7 Forest Regions across the continental U.S. and Alaska.

tion of types has been in use by the U.S. Forest Service since the publication of *Forest Regions of the United States* (USDA 1948) but newer ways to consolidate data and describe geographic associations for planning and management purposes are becoming more prominent (Bailey 1998).

All Things Considered

A region is not habitat in its entirety but only where the environmental factors (both direct and indirect) fulfill the needs for a species to survive and reproduce in a specific forest type. Identifying wildlife species to a forest type in a state creates a large data set that is beyond the scope of this textbook. However, a checklist for species in forest regions (Appendix E) is a place to start and provides the first level for a forest wildlife habitat relationships data model. Charles Elton in his textbook on Animal Ecology (1927) stated, "One of the first things with which an ecologist has to deal is the fact that each different kind of habitat contains a characteristic set of animals" (pg. 5). A checklist in our context is an inventory of associated wildlife at the landscape level and does not indicate a degree of abundance or density. The checklist for each region is from a database created by the USDA Forest Service to meet the reporting requirements of the first Resource Planning Act Assessment (U.S. Congress 1974). Species are listed if they could be present either during summer, winter, or yearlong. The list should not be considered definitive but is an approximation of occurrences based on an educated guess from literature.

The number of regions where an animal occurs is one way to quantify ecological amplitude. The hierarchical scheme of forest region with forest types can be used to synthesize information. Forest regions form the basis for identifying vegetation with plant associations dominated by trees. Forest types in a region have a characteristic set of wildlife consisting of different life-forms (i.e., birds and mammals) that adds to species richness as one component of diversity.

Functional Groups

To give meaning to a checklist, we can examine the natural characteristics of individual species to develop a perception of how diversity is created in different forest ecosystems (Chapter 6). The **phenomenon of life** creates functional groups that facilitate energy flow through the system. The distribution of wildlife species in each region indicates a level of complexity because each species has a different set of environmental conditions that defines their habitat. Functional groups are a coarse level in a hierarchy that with refinement could identify factors that indicate how a species adapts to its environment, e.g., the way cover, water, food, or space is used (the niche of a species).

I. Alaska: Western hemlock-Sitka spruce, Douglas-fir, Spruce-hardwoods.

Functional Group-Birds		Functional Group-Mammals	
Birds of Prey	7	Carnivores, Large	3
Birds of Prey, secondary cavity users	2	Carnivores, Small	7
Birds, Primary cavity creators	4	Insectivores	1
Birds, Secondary cavity users	10	Herbivores, Large	2
Birds, Ground nesting	3	Rodents, Small	4
Birds, Perching	37	Rodents, Large	2
		Rodents, Tree squirrels	2
Total	63	Total	21

II. Pacific Coast and Interior: Douglas-fir, Ponderosa pine, Redwood, Western hemlock-Sitka spruce, Lodgepole pine, Larch, Fir-spruce, Pinyon-juniper, Western hardwoods.

Functional Group-Birds		Functional Group-Mammals	
Birds of Prey	9	Carnivores, Large	3
Birds of Prey, secondary cavity users	6	Carnivores, Small	14
Birds, Primary cavity creators	11	Insectivores	2
Birds, Secondary cavity users	14	Herbivores, Large	4

Birds, Ground nesting	7	Rodents, Small	16
Birds, Perching	62	Rodents, Large	2
		Rodents, Tree squirrels	4
		Omnivores	1
Total	109	Total	46

III. Northern Rocky Mountains: Lodgepole pine, Douglas-fir, Larch, Ponderosa pine, Fir-spruce, Western white pine, Pinyon juniper, Western hardwoods.

Functional Group-Birds		**Functional Group-Mammals**	
Birds of Prey	9	Carnivores, Large	3
Birds of Prey, secondary cavity users	5	Carnivores, Small	13
Birds, Primary cavity creators	9	Insectivores	3
Birds, Secondary cavity users	13	Herbivores, Large	5
Birds, Ground nesting	4	Rodents, Small	12
Birds, Perching	59	Rodents, Large	2
		Rodents, Tree squirrels	2
Total	99	Total	40

IV. Southern Rocky Mountains: Pinyon juniper, Ponderosa pine, Fir spruce, Lodgepole pine, Douglas-fir, Western hardwoods.

Functional Group-Birds		**Functional Group-Mammals**	
Birds of Prey	8	Carnivores, Large	2
Birds of Prey, secondary cavity users	6	Carnivores, Small	14
Birds, Primary cavity creators	10	Insectivores	2
Birds, Secondary cavity users	14	Herbivores, Large	5
Birds, Ground nesting	4	Rodents, Small	14
Birds, Perching	57	Rodents, Large	2
		Rodents, Tree squirrels	3
		Herbivores, medium	1
		Omnivores	1
Total	99	Total	44

V. Northern Forests: Aspen-birch, Maple-beach-birch, Spruce-fir, White-red-jack pine, Oak-hickory, Elm-ash-cottonwood.

Functional Group-Birds		**Functional Group-Mammals**	
Birds of Prey	10	Carnivores, Large	2
Birds of Prey, secondary cavity users	3	Carnivores, Small	13
Birds, Primary cavity creators	8	Insectivores	3
Birds, Secondary cavity users	11	Herbivores, Large	2
Birds, Ground nesting	3	Rodents, Small	9
Birds, Perching	61	Rodents, Large	2
		Rodents, Tree squirrels	4
		Omnivore	1
Total	96	Total	36

VI. Central Hardwoods—Great Plains: Oak-hickory, Oak-pine, Maple-beach-birch, Oak-gum, Elm-ash-cottonwood (Great Plains extension), Ponderosa pine, Western hardwoods.

Functional Group-Birds		Functional Group-Mammals	
Birds of Prey	8	Carnivores, Large	1
Birds of Prey, secondary cavity users	3	Carnivores, Small	11
Birds, Primary cavity creators	8	Insectivores	3
Birds, Secondary cavity users	11	Herbivores, Large	1
Birds, Ground nesting	3	Rodents, Small	11
Birds, Perching	61	Rodents, Large	1
		Rodents, Tree squirrels	4
		Omnivore	2
Total	94	Total	34

VII. Southern Pines: Loblolly-shortleaf pine, Longleaf-slash pine, Oak-pine, Oak-hickory, White-red-jack pine, Oak-gum-cypress, Pinyon juniper.

Functional Group-Birds		Functional Group-Mammals	
Birds of Prey	10	Carnivores, Large	2
Birds of Prey, secondary cavity users	3	Carnivores, Small	11
Birds, Primary cavity creators	8	Insectivores	2
Birds, Secondary cavity users	10	Herbivores, Large	1
Birds, Ground nesting	2	Rodents, Small	6
Birds, Perching	49	Rodents, Large	1
		Rodents, Tree squirrels	3
		Omnivore	2
Total	82	Total	28

The accounting of birds and mammals in each region indicates the complexity that managers have to consider in developing a forest plan. A summary of the seven forest regions (Table 2-5) indicates there are large differences in numbers of birds and mammals from western to eastern landscapes. This difference is probably related to the effects of climate in mountain ranges from low to high elevation with rough topography. These changes result in differing physical structures and vege-

Table 2-5 Birds and mammals associated with different forest regions.

Western U.S.	Birds	Mammals
I Alaska	63	21
II Pacific Coast and Interior	109	46
III Northern Rocky Mountains	99	40
IV Southern Rocky Mountains	99	44
Total	370	151
Eastern U.S.	**Birds**	**Mammals**
V Northern Forests	96	36
VI Central Hardwoods	94	34
VII Southern Pines	82	28
Total	272	98
TOTAL	**642**	**249**

tation composition that create more habitats occupied by different species. All birds and mammals in a region contribute to an ecosystem where energy flows from the birds of prey and large carnivores to rodents and perching birds in a prey base (Figure 2-8).

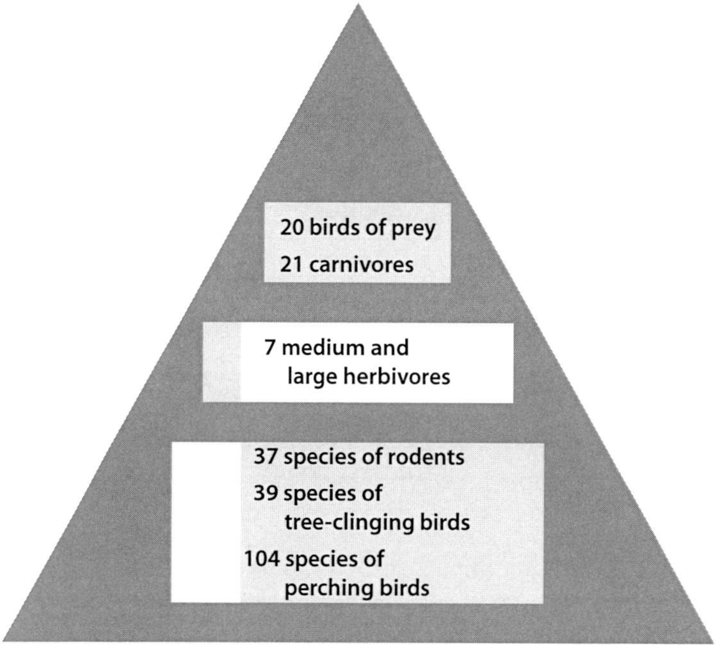

Figure 2-8 Energy flowing through forest ecosystems from predators to prey in food chains and webs across the seven forest regions.

An accounting of where animals occur provides interesting facts about how well some wildlife species have adapted to various forest types in forest regions while others are more restricted (Table 2-5). For example:

19 species occur in all seven forest regions: American robin, Cedar waxwing, Chipping sparrow, Dark-eyed junco, Fox sparrow, Golden-crowned kinglet, Hermit thrush, Northern oriole, Pine siskin, Song sparrow, White-crowned sparrow, Yellow-rumped warbler, Coyote, Mink, Red fox, River otter, Black bear, American beaver, American robin.

30 species occur only in one region: American swallow-tailed kite, Mississippi kite, Nuttall's woodpecker, Strickland's woodpecker, Bridled titmouse, Dusky-capped flycatcher, Prothonotary warbler, California quail, Anhinga, Bay-breasted warbler, Cape May warbler, Elegant trogon, Grace's warbler, Hermit warbler, Kirkland's warbler, Red-faced warbler, Tennessee warbler, Virginia's warbler, White-nosed coati, Collared peccary, Wild boar, Brush rabbit, Gray-collared chipmunk, Mexican woodrat, Red tree vole, Sewell, Townsend's chipmunk, Abert's squirrel, Douglas's squirrel, Western gray squirrel.

There are many ways to present information on the distribution and association of wildlife with their habitats and it all starts by having access to an inventory. One of the problems in inventorying resources is that individual plants are stationary in time but animals are mobile and very difficult to census. The U.S. Forest Service maintains the Forest Inventory and Analysis Program in cooperation with the State and Private Forestry and National Forest Systems but there is no comparable ongoing program to inventory the nation's wildlife resources.

Life History

There is one factor that all plant and animal species have in common—a story of their life events from birth to death. It includes specific information on reproduction as well as food and cover requirements that are needed to survive and reproduce. Life history is often separated into two categories, reproduction and habitat requirements, but can include detailed information on biology, depending on the intended use. A complete life history has distribution data as well as a taxonomic classification. State and federal agencies publish life history information in various forms, such as books and brochures, for use by the public or in an educational setting. An understanding of wildlife habitat relationships (by associating birds and mammals with habitats) is enhanced when one also has a basic knowledge of their life history. A synopsis of the life history of animals used in association with forest types is provided in Appendix F.

2-7 Summary

Forest types across the continental U.S. are ecosystems that have both structure and function. Ecosystems describe an area containing biotic components (plants and animals) and abiotic components (physical factors). The functioning process involves nutrient cycling through biological, chemical, and geological processes. Carbon is a major building block of the molecules through the use of CO_2 by plants and respiration by animals in a life-sustaining process involving energy flow and nutrient cycling. The cycling process requires pathways facilitated by food chains and food webs that are an ecosystem's biological engines and without them there would be no sustainable life. Forests are part of a classification system that categorizes the grouping of tree species into like categories for practical understanding, awareness, and management. Forests develop in time from bare ground to mature forests through a succession process.

The factors influencing succession can be expressed in a conceptual way in that vegetation (a forest of trees) is a function of soil, organisms, climate, time, and topography. The change in plant and animal species over time on the same piece of land results in a natural cycle of both as a phenomenon of diversity. The geographical distribution of forest types is identified for the continental U.S. and Alaska on a National Atlas map. Ten types and their tree species associations are described for the western and ten for the eastern states. Two types are found in Alaska. These twenty two forest types (ecosystems) contain habitats for a variety of wildlife based on the stand conditions present at a specific location within the type. Another way to look at forest ecosystems is by combining the commonality of types into different regions, which results in a convenient way to present forest types with an associated checklist of wildlife species.

References

Bailey, R.G. 1998. Description of the ecoregions of the United States. USDA Forest Service Misc. Pub. No. 1391.

Barbour, M.G., and W.D. Billings. 2000. *North American terrestrial vegetation,* Second edition. Cambridge University Press, Cambridge, UK.

Brown, D.E. (ed.). 1998. *A classification of North American biotic communities.* The University of Utah Press, Salt Lake City.

Bump, G. 1950. *Wildlife habitat changes in the Connecticut Hill game management area: Their effect on certain species of wildlife over a twenty-year period.* Cornell University Agric. Exp. Stn., Ithaca, NY.

CEC. 1997. *Ecological regions of North America: Toward a common perspective.* Commission for Environmental Cooperation, Quebec, Canada.

Curtis, J.P. 1959. *The vegetation of Wisconsin.* University of Wisconsin Press, Madison.

Daubenmire, R.F. 1968. *Plant communities.* Harper and Row, New York.

Dice, L.R. 1943. *The biotic provinces of North America.* University of Michigan Press, Ann Arbor.

Elton, C. 1927. *Animal ecology.* Sedgwick and Jackson, London.

Emmel, T.C. 1973. *An introduction to ecology and population biology.* W.W. Norton and Company, Inc., New York.

Eyre, F.H. (ed.). 1980. *Forest cover types of the United States and Canada.* Society of American Foresters, Washington, DC.

Garrison, G.A, A.J. Bjustad, and D.A. Duncan. 1977. Vegetation and environmental features of forest and range ecosystems. Agric. Hdbk. 475. USDA Forest Service, Washington, DC.

Gates, D.M. 1965. Radiant energy, its receipt and disposal. *Metero. Mono.* 6:1–26.

Gleason, H.A. 1939. The individualistic concept of plant associations. *Amer. Mid. Nat.* 21:92–10.

Harris, L.D. 1988. Reconsideration of the habitat concept. *Trans. N. Amer. Wildl. Nat. Resour. Conf.* 53:137–144.

Heinrichs, J. 1983. Old growth comes of age. *J. For.* 81:776–779.

Helms, J.A. (ed.). 1998. *The dictionary of forestry.* Society of American Foresters, Bethesda, MD.

Horn, H.S. 1976. Succession. P. 18–204 in *Theoretical ecology principles and application,* R.M. May (ed.). Blackwell Scientific Publications, Oxford, UK.

Houghton, R.A. 1995. Land-use changes and the carbon cycle. *Global Change Biology* 1(4):275–287.

Hunter, M.L., Jr. 1989. What constitutes and old-growth stand? *J. For.* 87:33–35.

Jefferies, R. 1848. *The gamekeeper at home: The amateur poacher.* Oxford University Press, UK.

Jenny, H. 1958. Role of the plant factor in the pedogenic functions. *Ecology* 39:6–16.

Johnston, D.W. and E.P. Odum. 1956. Breeding bird populations in relation to plant succession on the Piedmont of Georgia. *Ecology* 37:50–62.

Kimmins, J.P. 1997. *Forest ecology: A foundation for sustainable management.* Macmillan, New York.

Knapp, R. 1974. Cyclic successions and ecosystem approaches in vegetation dynamics. P. 91–100 in *Handbook of Vegetation Science.* W. Junk Publishers. The Hague, Netherlands.

Kuchler, A.W. 1966. Potential natural vegetation (map). University of Kansas. USDA Forest Service, Washington, DC.

Leopold, A. 1933. *Game management.* Charles Scribner's Sons, New York.

Merriam, C.H. 1898. Life zones and crop zones of the United States. *U.S. Biological Survey Bull.* 10:1–79. Government Printing Office, Washington, DC.

Noble, I.R., and R.O. Slatyer. 1977. Post-fire succession of plants in mediterranean ecosystems. P. 27–63 in *Environmental consequences of fire and fuel management in Mediterranean ecosystems,* H.A. Money and C.E. Conrad (eds.). Gen. Tech. Rep. WO-3, USDA Forest Service, Washington, DC.

Odum, E.P. 1971. *Fundamentals of ecology.* W. B. Sanders Co., Philadelphia, PA.

Oliver, C.D. and B.C. Larson. 1990. *Forest stand dynamics. Biological resource management series.* McGraw-Hill, Inc., New York.

O'Neill, R.V., D.L. DeAngelis, and J. B. Waide et al. 1986. *A hierarchical concept of ecosystems. Mono. Population Biology 23.* Princeton University Press, Princeton, NJ.

Ream, C.H., and G.E. Gruell. 1980. Influences of harvesting and residue treatments on small mammals and implication for forest management. P. 455–467 in *Environmental consequences for timber harvesting in Rocky Mt. coniferous forests.* Gen. Tech. Rep. INT-90. USDA Forest Service, Missoula, MT.

Robertson, F.D. 1992. Ecosystem management on the National Forests and Grasslands. Letter to regional foresters and station directors, 1330–1331. USDA Forest Service, Washington, DC.

Shantz, H.L., and R. Zon. 1924. Forest vegetation of the United States in natural vegetation: Atlas of American agriculture. Agric. Hdbk. 271. U.S. Department of Agriculture, Washington, DC.

Shelford, V.E. 1963. *The ecology of North America.* University of Illinois Press, Urbana.

Tansley, A.G. 1935. The use and abuse of vegetational concepts and terms. *Ecology* 16:284–307.

Tarbell, H.S. 1899. *The complete geography.* American Book Company, New York.

Thomas, J.W. 1979. Wildlife habitats in managed forests: The Blue Mountains of Oregon and Washington. Agric. Hdbk. 53. USDA Forest Service, Washington, DC.

UNESCO. 1973. International classification and mapping of vegetation. *Ecol. and Cons. Ser.* 6, United Nations Educational, Scientific, and Cultural Organization, Paris.

United States History. (n.d.). *National Forest Management Act.* Available at https://www.u-s-history.com/pages/h1604.html

U.S. Congress. 1974. Forest and Rangeland Renewable Resources Planning Act. Public Law 93 378. Government Printing Office, Washington, DC.

U.S. Congress. 1976. National Forest Management Act. Public Law 94-988. Government Printing Office, Washington, DC.

USDA. 1948. Forest vegetation of the United States (map). Agric. Hdbk. 41. USDA Forest Service, Washington, DC.

USDA. 1967. Major forest types (map). National Atlas. USDA Forest Service, Washington, DC.

USDA. 1974. Forest and range ecosystems of the United States (map). USDA Forest Service, Washington, DC.

USDA. 1981. *An assessment of the forest and rangeland situation in the U.S. Department of Agriculture.* USDA Forest Service, Washington, DC.

USDA. 2000. The national atlas of the United States. Forest cover types. USDA Forest Service and USDI Geological Survey, Denver, CO.

Walker, L.C. 1999. *The North American forests: geography, ecology, and silviculture.* CRC Press, Taylor and Francis Group, New York.

Watt, A.S. 1947. Pattern and process in the plant community. *J. Ecol.* 35:1–22.

Whittaker, R.H. 1953. A consideration of climax theory: The climax as a population and pattern. *Ecol. Mono.* 23:41–78.

Zhu, Z., and Evans, D.L. 1994. U.S. forest types and predicted percent forest cover from AVHRR data. USDA Forest Service and USDI Geological Survey, Washington, DC.

The Big Issues 3

Over time, in what is now the United States, various major social-economic-ecological-political forces have shaped the landscape. Native Americans managed forests to enhance game habitat and to keep clear for agriculture (Bonnicksen 2000). When European settlers first arrived, agricultural lands were occupied, towns constructed, and timber/cattle grazing in the forests were early and lucrative cash crops. Until the early twentieth century, forests were cleared indiscriminately for wood products and often abandoned; and the expansion of farmland resulted in the cutting and burning of millions of acres of forests. Massive market hunting efforts along with human-caused habitat destruction drove the passenger pigeon to extinction. Hunting for meat—sometimes only the tongue—and hides resulted in the near extinction of the American bison. And fire, once an essential land management tool for maintaining resistant and resilient ecosystems, was largely excluded from the landscape. The list of ecological impacts, unfortunately, continues to grow to this day. In this chapter we discuss key social-economic-ecological-political forces that have shaped—and continue to shape—forest management in the United States.

3-1 Land Tenure

In 1630 the estimated forest land was 46% of the land base—just over 1 billion acres. By 1910, forest land had declined by 26% to 754 million acres, some 34% of the land area (USDA 2014). Conversion to agriculture (deforestation) accounted for most of the loss in forest land. Wildlife species, especially large herbivores and predators, birds, and bats, were not controlled by administrative land boundaries, however. In the absence of barriers, such as very strong and durable fences, many species easily move across vast territories having multiple human landowners. For example, coyotes crossed the Golden Gate Bridge into San Francisco and have also expanded their range all the way to the East Coast (Sacks et al. 2006). Thus, to manage land for forests and wildlife requires knowledge of **who owns what and where**.

Until the Revolutionary War, British policy for land east of the Appalachian Mountains encouraged the development of agriculture (which Native Americans had practiced for thousands of years) that not only decreased forested land but moved this land into private holdings. After the war, the federal government took over the claims that many states had on land west of the Appalachian Mountains in exchange for paying off the war debts of the states. The new government of the United States also encouraged development and the conversion of forest land to agriculture.

The United States continued with a dual strategy. It acquired millions of acres of land through a combination of annexation (e.g., Texas), war (e.g., most of the Southwest, Nevada, and California), purchase (the most famous being the Louisiana Purchase and Alaska), and diplomacy (e.g., Oregon Territory) (Figure 3-1). Land acquisition occurred without the willing consent of Native American populations. At the same time, the explicit goal was to move this land from public to private hands through a variety of mechanisms, most notably by the various Homestead Acts and grants to the railroads.

The federal government began to reserve some of its acquired land as early as 1862 (the transfer of Yosemite Valley to the state of California, and the reservation of what became Yellowstone National Park in 1876), but not until the late 1800s did the federal government begin a major policy shift from disposition to retention of land. By then, most of the land east of the Rocky Mountains had moved into private ownership, as had most of the high-quality agricultural land west of the Rockies (i.e., valley bottoms with water). By the mid-1920s, the major federal land dispositions had ended.

During the Great Depression of the 1930s, significant forest and agricultural land came back into a state and local ownership because of foreclosures. The fed-

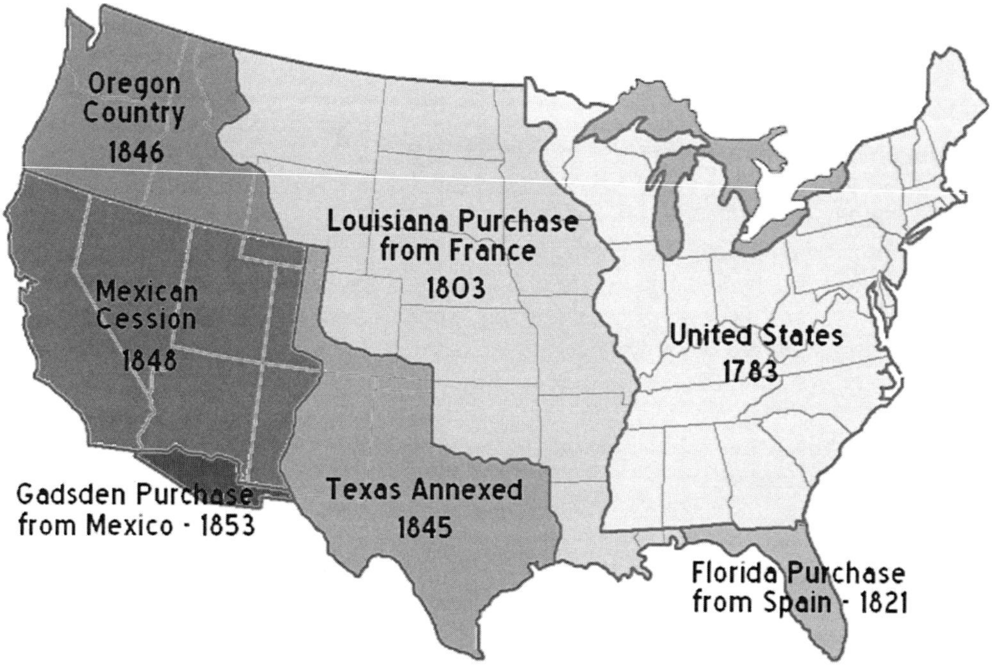

Figure 3-1 History of land acquisition of the United States.

From Hamilton 2016.

eral government also purchased some of these lands east of the Mississippi River. Although some changes in the ownership has occurred, the current forest land ownership has stayed relatively stable since then. These policies, and the resulting land ownership patterns, occurred with little or no consultation or acquiescence of the Native Americans who had called these lands home. In many instances, Native Americans were forcibly removed from lands where they lived to make these lands available for Anglo settlers.

For resource planning purposes, the U.S. Forest Service has divided the United States into four regions with eight subdivisions (Figure 3-2) by states (Table 3-1) (Oswalt et al. 2014). These planning regions do not correspond to the forest regions described in Chapter 2, Figure 2-7.

Total forest land is unevenly distributed across the country geographically by four major ownership classes: public, private corporate, private non-corporate, and tribal holdings. As of 2012, the United States had a total of about 766,234,000 acres of forest land—just over one-third of the country (Table 3-2 on pg. 61). The southern region has both the highest number of acres of forest land (244,716,000) and the highest percentage of forest land (46%). In addition, over 52 million acres

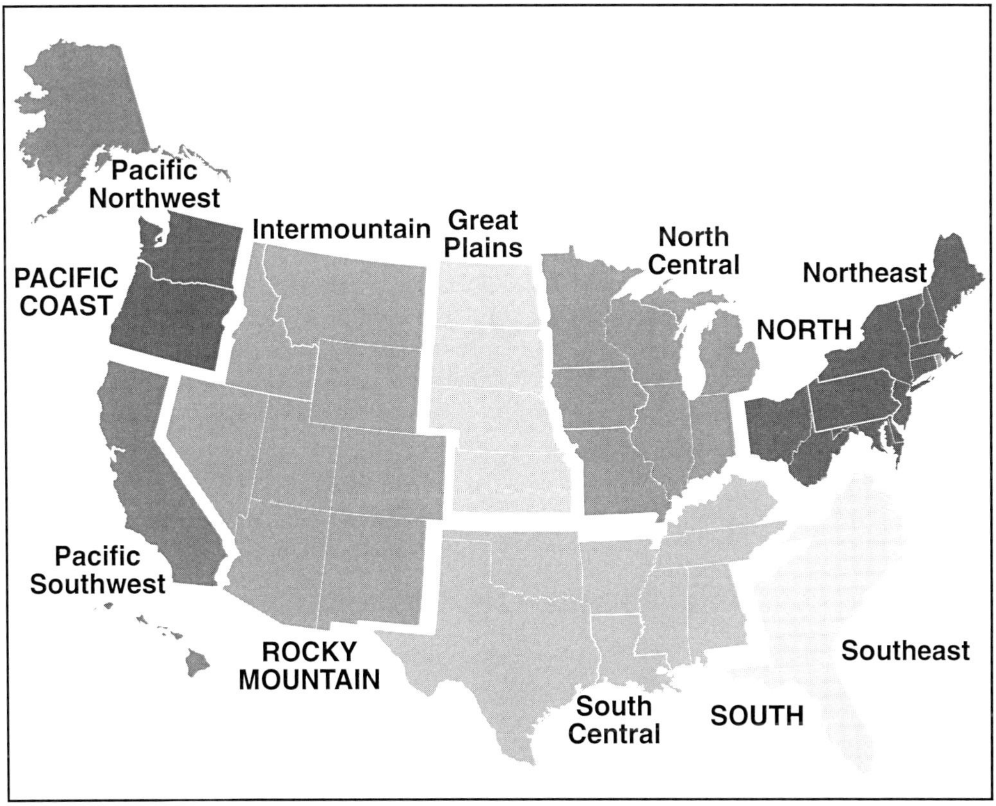

Figure 3-2 Forest Service planning regions and subdivisions.

Table 3-1 States in regions and subdivisions for Resources Planning Act analysis.

NORTH	SOUTH	ROCKY MOUNTAINS	PACIFIC COAST
Northeast	**Southeast**	**Great Plains**	**Pacific Northwest**
Connecticut	Florida	Kansas	Alaska
Delaware	Georgia	Nebraska	Oregon
Maine	North Carolina	North Dakota	Washington
Maryland	South Carolina	South Dakota	**Pacific Southwest**
Massachusetts	Virginia	**Intermountain**	California
New Hampshire	**South Central**	Arizona	Hawaii
New Jersey	Alabama	Colorado	
New York	Arkansas	Idaho	
Pennsylvania	Kentucky	Montana	
Rhode Island	Louisiana	Nevada	
Vermont	Mississippi	New Mexico	
West Virginia	Oklahoma	Utah	
North Central	Tennessee	Wyoming	
Illinois	Texas		
Indiana			
Iowa			
Michigan			
Minnesota			
Missouri			
Ohio			
Wisconsin			
20 States	**13 States**	**12 States**	**5 States**

of woodland (area with trees that at maturity will not grow taller than 16.4 feet in height) exist in the United States, primarily in public holdings in the western states of Texas, New Mexico, Arizona, Nevada, and Utah. Total forest land area has remained stable in recent years (Oswalt et al. 2014; Sleeter et al. 2013).

Public Forest Land

Public forest land, concentrated in the West, falls into three major categories: federal, state, and local. Federal ownership dominates public forest land holdings. The United States Department of Agriculture's Forest Service owns and manages the majority of federal forest land. However, other agencies, including the Department of Interior's Bureau of Land Management, Fish and Wildlife Service, National Park Service, and the Department of Defense also manage large amounts of forest land. Of the federal forest land base, National Forests tend to have the most interspersed holdings, meaning that Forest Service lands are interspersed with other non-federal holdings. Intermingling of ownerships greatly complicates and increases the costs of management because different owners have different objectives and are subject to different laws and regulations (discussed in more depth in Chapter 7).

East of the Mississippi River, state and local forests dominate the public lands. In the North, five states—New York, Pennsylvania, Michigan, Minnesota, and Wisconsin—account for over 59% of the almost 23 million acres of state forest land (Table 3-3). In the South, Florida alone accounts for over 36% of all state forest land. Although not nearly as dominant as in the West, National Forests total almost 12 million acres in the North and another 13 million acres in the South. Michigan leads the North with 2.7 million acres of forest land (much of which is interspersed with state forests) and in the South, Arkansas has almost 2.5 million

Table 3-2 Forest land in the United States by region and subdivision.

REGION Subdivision	LAND AREA Thousands of Acres	FOREST LAND Thousands of Acres	FOREST LAND Percent
NORTH			
Northeast	126,507	84,846	66
North Central	286,162	90,730	32
Total acres	**412,669**	**175,576**	**43**
SOUTH			
Southeast	146,884	89,644	61
South Central	386,148	154,872	40
Total acres	**533,032**	**244,516**	**46**
ROCKY MOUNTAIN			
Great Plains	194,173	6,724	3
Intermountain	547,691	124,614	23
Total acres	**741,864**	**131,338**	**18**
PACIFIC COAST			
Alaska	365,616	128,557	35
Pacific Northwest	103,964	52,222	50
Pacific Southwest	103,809	33,805	33
Total acres	**573,389**	**214,584**	**37**
U.S. TOTAL	**2,260,954**	**766,014**	**34**

Source: Oswalt et al. 2014.

Table 3-3 Acres (thousands) of public forest land by region.

Region	Federal Land	State	Local	Total Public	All Ownership	Percent of Forest Land
North	14,570	22,995	8,895	46,461	175,575	27
South	21,857	7,937	2,951	32,745	244,716	13
Rocky Mountain	91,590	4,480	195	96,265	131,338	73
Pacific Coast	110,452	32,838	1,179	144,469	214,604	67
TOTAL U.S.	**238,469**	**68,250**	**13,220**	**319,940**	**766,233**	**42**

Source: Oswalt et al. 2014.

acres of National Forests (Oswalt et al. 2014). Only six states east of the Mississippi River—Connecticut, Delaware, Maryland, Massachusetts, New Jersey, and Rhode Island—have no National Forests.

Private Forest Land—Corporate and Non-Corporate

Private **corporate** forest land holding is just over 19% of the total forest land in the country. The largest number of acres of these lands (48 million) is in the Pacific Coast Region; the highest proportion of private corporate holdings (26.7%) is in the Southern Region. Maine has both the most acres (10 million) and highest proportion (57.5%) of private corporate forest land (Table 3-4). Weyerhauser Corporation, headquartered in Seattle, Washington, is the largest publicly traded private forest land holder in the United States, owning some 12.8 million acres. Other major forest land owners in this category include Rayonier (2.3 million acres), family-owned Sierra Pacific Industries (1.9 million acres), and Potlach (1.6 million acres) (Landthink 2017).

Also referred to as "non-industrial private" or "small private" landowners, the private **non-corporate** sector totals over twice as many acres of forest land as that held by private corporations. The South has the most acres in this category—almost half of the total. Texas leads all the states with over 30 million acres in private non-corporate forest land—just over 10% of the national total of all lands in this category (Table 3-4). Of the estimated 61 million private owners, 61% own less than 10 acres of forest land. Much of this land is interspersed with agricultural land and private corporate forest land, especially in the South.

Table 3-4 Acres (thousands) of private and tribal forest land by region.

Region	Private Corporate	Private Non-corporate	Total* Private	All Ownership	Percent Private Forest Land
North	29,161	99,954	129,115	175,575	74
South	65,246	146,726	211,972	244,716	87
Rocky Mountain	5,001	28,845	33,846	131,338	26
Pacific Coast	48,061	22,074	70,136	214,604	33
TOTAL U.S.	**147,469**	**297,599**	**445,069**	**766,233**	58

*Includes tribal land.

Sources: Oswalt et al. 2014.

Tribal Forest Land

At about the time of Columbus the population of Native Americans, in what is now the United States, was approximately 5 million people and there was no need to practice conservation because the Indian was not a destructionist (Kinney 1937). By 1890, the population of Native Americans declined to around 250,000 while at the same time the white population increased to 75 million (Taylor 1991). As the U.S. population continued to increase, the ancestral lands of Native Americans decreased. Native American tribal forest lands (commercial and non-commercial)—both those owned outright by the tribes and those held in trust—totaled an estimated 7.8 million acres in 2013 (Table 3-5). The largest number of acres is in the Northwest (Gordon et al. 2013). The largest tribal holdings include the Menominee in Wisconsin and the White Mountain Apache in Arizona.

One of the characteristics of Indian reservations is the unique **trust relationship** between the U.S. Government and the tribes that determines how reservation activities from health care to natural resources are administered and managed. This trust relationship also affects the cooperation that tribes extend to the surrounding private, state, and federal lands and vice versa. Considerable legal documentation by the federal courts exists from the date of the Constitution to the present that incrementally clarifies the relationship between Indian tribes and the U.S. Government. The development of a trust relationship can be traced back to the discovery of America and its colonization by the British (Veeder 1968).

The **trust responsibility** to administer Indian lands resides almost exclusively with the Secretary of the Interior who executes the will of Congress (5 U.S.C. 22). In the past, Congress has also provided that the Commissioner of Indian Affairs shall, under the direction of the Secretary of the Interior, and agreeably to such regulations as the President may prescribe, have the management of all Indian affairs and of all matters arising out of Indian relations (25 U.S.C. 2; and 25 U.S.C. 1, 1A). Under the Secretary of Interior, the Bureau of Indian Affairs is primarily responsible for managing Indian lands.

Native Americans living on reservations approach the problem of land management direction for the future a little differently than most large organizations (either private, state, or federal) where there is generally a set of **goals** with specific objectives that serve to guide the actions and commitments of the organization to a common cause. Organizational goals and objectives are generally written by the leadership of that organization and are included in a document to set policy and direction for the future. While Native American reservation administrative documents generally contain well-written goals and objectives for each resource, there may yet be another document or probably an unwritten but well-understood philosophy that reflects a deeper feeling and commitment by tribal

Table 3-5 Acres (thousands) of Native American forest trust land classifications.

Location	Commercial Forest Land	Non-commercial Forest Land	Total Forest Land
Northwest	2,667	796	3,463
Lake States	1,091	193	1,284
Southwest	1,718	725	2,443
Eastern	311	30	341
Alaska	175	51	226
Total Trust	5,962	1,795	7,757
Total Trust and Fee Lands	6,051	1,812	7,863

Sources: Gordon et al. 2013.

members to their life style. This philosophy is sometimes referred to as a **vision** (IFMAT-I 1993).

Visions have strong meaning because they are the product of all tribal members—not just the thoughts of elected or appointed leaders. These expressed feelings result from knowledge about plants, animals, and the environment that have been passed along by generations of medicine men and family elders and are often the core and set the tone of a major policy statement. For example, the Intertribal Timber Council (ITC), which has a membership of 65 tribes, published Vision 2000 with the following guiding values:

> Forest resources are vital to the economic and social welfare of many Indian Nations and Native Alaskan Corporations. The management of these valuable and renewable resources not only provides income and employment opportunities for our people, but also affects our lives in many other ways. The harmony of man, trees and other vegetation, soil, water and wildlife, which collectively comprise the forest community, influence our very emotional and spiritual well being. (ITC 2007, pg. 2)

While economic and social values are mentioned, the key words in this statement are "**emotional and spiritual well being**" because they emphasize a priority that permeates both the oral and written word of Indian goals, objectives, and management plans. These same words do not occur, at least not very often, in non-Indian state or federal writings and legislation on natural resource management. Wildlife habitat relationships, ecosystem management, integrated resource management, forest health, sustainable resource management, and multiple use are terms most likely to be used in official tribal documents; they can, however, be accounted for by a deeper spiritual understanding in traditional Indian knowledge through the philosophy of caring for the heart and soul of the earth—the plants and animals.

The **Self-Determination and Education Assistance Act of 1975** (PL 93-638) provided a new opportunity for tribes to manage their own resources, including wildlife, on Indian reservations. Under the Act, BIA activities can be contracted to tribes. This allows tribes to take responsibility for managing their own affairs. Since the Act was passed many reservations have created Natural Resource, Forestry, and Fish and Wildlife Departments that are staffed and administered by Native Americans. The important factor, however, is that these tribes have control over management of lands within their boundary and can set policy and management direction.

A 638 contract (refers to the Act) to a tribe for resource management does not eliminate the requirement to conform to federal regulations such as the Endangered Species Act. This is because federal funds from the Interior Department are passed to BIA and then to tribes under contracts. Although many tribes have endangered and threatened species programs and try to conform to federal regulations, they generally want to do so in a manner consistent with their own culture (IFMAT-1 1993). In addition, tribes are not well funded to inventory and monitor populations and habitats because most of the funds for this purpose are given to the states.

A system of sustained-yield forest management using selection harvesting was initiated by the **Menominee** Tribe after the reservation was established in 1854. The vision was one of a forest to be harvested at a rate that would achieve a balance between growth, mortality, and the annual production of timber. The Menominee tribal leaders believe that resource development must be part of a larger social and economic plan. They believe that good forest management practices based on sustained yield determined by a continuous forest inventory will produce diverse conditions, good wildlife habitat, water quality, and aesthetics values (Pecore 1992). The success of the Menominee Tribe in maintaining a sustained yield is based on a structure that has not changed over time: a regulated forest using area and volume control with even- and uneven-aged management, a large number (884) of CFI (Continuous Forest Inventory) plots contained in 109 compartments, and limits on the allowable cut not to exceed 7% annually, 3% for a five-year period, and 2% over a 15-year cutting cycle.

Forestry has been the major force setting the direction for Menominee resource management. The current members have inherited a forest that has not been abused; for this reason, no major wildlife problems have been noticed by the tribe. As the human population increases there could be more demand for certain featured species, such as deer, that probably will remain at a constant density level.

Challenges

Just considering the diversity of ownership, its geographic and size heterogeneity, varying objectives, and history, forest lands in the United States pose complex management challenges. Adjacent parcels of the same forest type may have experienced different treatment histories and with different ownerships may be subject to different laws and regulations. Even adjacent forest lands, of the same forest type, stand structure, and ownership class may have different applicable laws and regulations depending on their size. Coupled with the immensely variable habitat requirements of different wildlife species the challenges multiply. However, other social, political, economic, and ecological forces add yet more complexity to managing forest lands for wildlife.

3-2 Forests and the Wildland Urban Interface

From the end of World War II (1945) to 2018, the population of the United States grew from 139.93 million to 327.45 million—an increase of 234% (USDC 2018). This dramatic rise in population and policies that encouraged the growth of the suburbs pushed the boundaries of cities and towns farther and farther into agricultural and forest lands. Millions of acres of forests, farms, orchards, and ranches became tracks and tracks of single-family homes. This growth inevitably led to the point where housing backed directly to forest land. Also, given the intermingling of ownerships, housing developments often occurred within a larger forested landscape.

This situation led to what is now known at the Wildland Urban Interface/ Intermix (WUI) where housing is near (interface) or intermingled with (intermix) dense wildland vegetation (Martinuzzi et al. 2015). Approximately 190 million acres fall into the WUI in the United States—just under 10% of the country's total land area. Higher population densities and landownership patterns (high number of small private forest land tracks) create the situation where the South (50% of the total WUI acres) and East (36%) dominate the area of WUI in the country (Cohen 2008). North Carolina has the largest number of WUI acres (13.4 million), while Connecticut has the largest percentage of its land in WUI (65.6%).

However, WUI land area and the number of houses located in the WUI do not correspond. For example, in 2010, approximately 44 million homes were in the WUI—one-third of the entire housing stock in the country. In parallel, about 30% of the population lived in the WUI (Radeloff et al. 2018). The low density of housing puts a high demand on infrastructure, such as roads, water, power, sewers, and waste collection. More prominently, the growth of the WUI, along with changes in forest structures arising from past fire suppression activities, has profoundly changed fire management strategies. More people and human activity increases the risk of ignition and, once started, an emphasis on saving structures changes fire-fighting tactics.

The increase in the WUI also affects wildlife populations and habitat. The type and severity of these effects vary depending on location, vegetation, types and density of development, and the wildlife species of interest. For example, WUI development can disrupt migration routes for white-tailed or mule deer. Clearing of land for construction and access roads may reduce hiding and thermal cover and increased road traffic may lead to more vehicle collisions and deaths of deer. Domestic animals, especially dogs, may harass deer forcing them to move into less desirable or otherwise occupied areas. On the other hand, landscaping and gardens associated with houses may provide excellent sources of food and water, and humans who like seeing deer often provide supplemental food to attract animals. In addition, human populations tend to have large negative effects on the large predators required to regulate deer herd size, reducing predation. In the Northeast, these effects have led to an incredible increase in white-tailed deer numbers to the point that some municipalities have authorized deer hunting within their city limits to reduce herd sizes (Kilpatrick et al. 2007).

With other species, notably the Northern Spotted Owl, WUI development may have only negative effects with few—if any—benefits. For this species, development that takes out roosting or nesting trees will seriously degrade the habitat. WUI development that results in high levels of human activity during roosting season may also degrade habitat quality. Activities in the WUI have decidedly mixed effects on wildlife on a species-by-species basis and to generalize will miss the subtleties and nuances needed to effectively manage forests and wildlife habitat.

3-3 Wildland Fire

Fire in the natural world is regulated by fuels, weather, and topography—the basics of **fire behavior**. All other things equal, fire spreads to available fuel depending on its amount, arrangement, and moisture content. Weather both prepares the fuel for burning/extinction and drives spread (e.g., wind). Topography provides the surface on which the fuel resides and modifies the effects of weather (aspect and slope). Forest and rangeland fuels are typically available to burn multiple times each year in the temperate regions of the world, depending on season and weather patterns. Wildland fire is the interface between the readiest of an area to burn and the probability of an ignition. That interface also defines the opportunities for prescribed fire (Tappeiner et al. 2015).

Most every forest type in the United States evolved with fire and are therefore created and maintained by fire as well as destroyed by it. Some types, such as longleaf and ponderosa pine or the Sierra Nevada mixed-conifer types, evolved with frequent, low-intensity fires. Trees have thick bark, high branches, and well protected buds to resist damage from surface fire. Jack and lodgepole pine and aspen types evolved with infrequent high-intensity fires. The cones of both jack and lodgepole pine often require very high temperatures to open so that the seeds can disperse—temperatures only achievable in nature with fire and in the post-fire environment. Aspen re-sprouts vigorously from dense subterranean root systems.

The combination of fire intensity and fire frequency, with associated effects on system ecology and including common seasonality and spatial characteristics, is called a "fire regime" (Agee 1993). Frequent fires tend to be low intensity because a short interval between burns does not allow for the accumulation of much fuel. Fire is typically limited to surface fuels, which creates open park-like stands with multiple age classes and canopy layers. High-intensity fires can only happen with relatively low frequency because it takes time to re-accumulate the fuels necessary to sustain such intensity and match up with a climatic cycle to prepare those fuels to burn. These fires create new even-aged stands.

Native Americans actively managed many forest types to assist in agriculture, hunting, travel, and protection from human and non-human dangers (Bonnicksen 2000). Frequent burning in the pine/oak types in the Southeast kept forests open to create better habitat for game animals, increase the ease of hunting, help clear and keep clear lands for crops, such as corn, beans, and squash, and to favor orchard-trees, such as oaks for acorns. In California and the Southwest, frequent burning also helped maintain certain grasses, forbs, and shrubs for basket weaving and for medicinal uses. Native Americans also burned the Great Plains to help renew grasses, the primary food for the bison. Recently burned areas with young, nutrient-rich plants attracted bison such that other areas could grow fuel, attracting future fire—thus maintaining a cycle.

The arrival of European settlers began a series of events that made epic changes in our forested landscape, not only reducing its size and extent but also, in many cases, greatly altering structure and reducing resilience. These activities included cutting trees for cash crops, which created massive volumes of logging slash—branches, twigs, leaves, needles, tops, broken trees, and unmerchantable logs and trees. Subsequent fires burned at an intensity and size that native seed sources sometimes could no longer re-colonize the areas. As a result, many areas moved into ecological pathways far different than had been the norm for millennia. In other words, the successional pathway occurred. Combined with efforts to permanently convert forest lands to agriculture and widespread over-grazing of forested lands resulted in changes noted above. This pattern of aggressive and mostly unregulated harvesting and agricultural conversions of forest land continued until the late 1800s. A suite of incredibly devastating forest fires, a set of severe floods, concerns about a "timber famine," and the announcement of the "Closing of the American Frontier" by the U.S. Census Bureau began to focus attention to the role and negative effects of fire in forest land.

Two of the most devastating forest fires in the United States started on the same day: October 8, 1871. The Peshtigo Fire in northeastern Wisconsin and Michigan's Upper Peninsula resulted in over 1.5 million acres of forest land totally and another 2.3 million acres partially burned; the fire killed 1,200 to 1,500 people. A combination of fires across Michigan's Lower Peninsula, from Holland on the western side near Lake Michigan to Pt. Huron on the east to the "Thumb" region, burned over 1.2 million acres and killed approximately 200 individuals. A third fire also started this same day (Nelson 2013). Although the Michigan and Peshtigo fires dwarfed this third fire in terms of extent (2,000 acres) and deaths (200 to 300) its location caught the attention of the popular press of the time, relegating coverage of the other two fires to relative obscurity. The third fire: The Great Chicago Fire.

Linked with the larger conservation movement of the time and the sociopolitical views of politicians such as Theodore Roosevelt, these fires focused attention on the management—or more specifically, the lack of management—of forest land across the country. A key result of these forces was the establishment of the USDA Forest Service, an agency charged with the protection and management of hundreds of millions of acres of forest land. Another fire set a policy direction that has had profound effects on the management of forest resources, current forest conditions, and wildlife habitat on both private and public forest lands.

The Big Burn of 1910 impacted over 3 million acres of forest land in the states of Washington, northern Idaho, and western Montana and killed 87 people (Nelson 2013). The magnitude of this fire, and the inability of the Forest Service to effectively fight it, led to calls to disband the agency. In response, the Forest Service instituted the "10:00 am policy" that called for any fire to be suppressed by 10:00 am on the day following its reporting (van Wagtendonk 2007). Guided by this policy, the Forest Service became extremely effective in suppressing forest fires.

Unfortunately, an unintended consequence of such effective suppression excluded fire from essentially all forest ecosystems. As a result, the frequent/low-intensity regimes shifted to infrequent/high-intensity regimes and infrequent/high-intensity regimes shifted to much more infrequent/and much higher intensity regimes. Such changes led to radical altering of stand structures. For example, in the Southwestern ponderosa pine type, the historic fire frequency was 7 to 10 years and forest-supported stocking of approximately 23 trees per acre compared to current conditions of over 800 trees per acre with basal area doubling (Covington and Moore 1994). Thus, forests throughout the West went from an open park-like setting with relatively few large trees to a dense structure with a multitude of small trees crowded together (Stephens et al. 2015).

A second unintended consequence of the 10:00 am policy was how it shaped the debate on fire ecology and the use of fire in forest management. Despite the views of many supporting the judicious use of fire as a management tool, including conservation icon Aldo Leopold, research on the role of fire was not well received. Advocates of fire in the West, such as Harold Biswell (van Wagtendonk 1995), and in the Southeast, Herbert Stoddard (Johnson et al. 2002), faced strong opposition to their theories. Unfortunately for their critics—and more importantly for forest management—their work, based in part on the practices of Native Americans and early Anglo settlers, was sound, their conclusions correct, and their predictions of the effects of fire suppression accurate.

By the late 1960s efforts by foresters, fire managers, and researchers had begun to change management direction with respect to fires in attempting to mimic natural fire regimes. In 1968, the National Park Service began to allow limited "let burn" policies where naturally caused (and in some instances human-caused) fires were allowed to burn on a limited basis. The idea here was that such fires would mimic the ecological roles fires had played for thousands of years. In 1978 the Forest Service abandoned the 10:00 am policy (van Wagtendonk 1995).

However, such policies tended to ignore the fact that effective fire suppression caused massive changes in stand structures with increased biomass and fuel ladders. When fire was moved into such areas, stands followed the infrequent/high intensity path resulting in massive conflagrations, rather than low intensity ground fires (Walker et al. 2018). The structural changes did not result in major changes in fire sizes until the late 1980s. The Yellowstone Fires of 1988 burned almost 800,000 acres and resulted in two fatalities (USDI 2016). Although small in comparison to the Peshtigo Fire or the Big Burn, this fire dramatically pointed out the dangers associated with let burn policies.

Changing stand structures with much higher tree densities, along with long-term droughts in the West, also predisposed trees to water stress and a reduced ability of trees to resist diseases and insects—especially bark beetles. Bark beetle infestations increased in size and intensity resulting in millions of acres of tree mortality. This mortality added to fuel loads. Such conditions over millions of acres of forest land, coupled with increased human use of forest lands, including growth in the WUI, all contributed to increases in the number of all forest fires and the acres burned (Table 3-6). In addition, fire sizes have increased (Dennison et al. 2014). Nationally, since 1997, nine of the 10 largest forest fires have occurred since 2004 (Table 3-7). In California, fourteen of the state's largest wildfires have burned since 2000 (CalFire 2018). Although based on extremely small sample sizes, fire seasons appear to have extended. For example, in 2017 California experienced three massive forest fires, including one in December, historically a wet winter month.

Firefighting costs have also increased dramatically over time. Federal fire suppression costs alone topped $2 billion in 2015 and reached almost $3 billion in 2017 (Table 3-6). These costs do not include costs incurred by state or local agencies nor do they reflect damages resulting from these fires (property, timber, or habitat loss). In fiscal year 2018, the Forest Service budget included $1.75 billion for the management of National Forest System lands and $2.5 billion for Wildland Fire Management (USDA 2017). In other words, almost 60% of the agency's budget went into fire suppression activities. Changes in the budgeting process scheduled for fiscal year 2020 will fund suppression activities separately from

Table 3-6 Total forest fires, acres burned, and suppression costs, 2008–2017.

Year	Number of Fires	Acres Burned*	Forest Service $	Dept. of Interior**$	Total $
2017	71,499	10,026,086	2,410,165,000	508,000,000	2,918,165,000
2016	67,595	5,503,538	1,603,806,000	371,739,000	1,975,545,000
2015	68,151	10,125,149	1,713,000,000	417,543,000	2,130,543,000
2014	63,312	3,595,613	1,195,955,000	326,194,000	1,522,149,000
2013	47,579	4,319,546	1,341,735,000	39,919,000	1,740,934,000
2012	67,774	9,326,238	1,436,614,000	465,832,000	1,902,446,000
2011	74,126	8,711,367	1,055,736,000	318,789,000	1,374,525,000
2010	71,971	3,422,724	578,285,000	231,214,000	809,499,000
2009	78,792	5,921,786	702,111,000	218,418,000	920,529,000
2008	78,979	5,292,468	1,193,073,000	392,783,000	1,585,856,000

*Includes all private, state, and federal lands.
**Bureau of Indian Affairs, Bureau of Land Management, National Park Service, and U.S. Fish and Wildlife Service.
From National Interagency Fire Center (n.d.).

Table 3-7 Ten largest forest fires, 1997–2018.

Year	Fire Name	State	Total Acres
2004	Taylor Complex	AK	1,305,592
2006	East Amarillo Complex	TX	907,245
2017	NW Oklahoma Complex	OK	779,292
2007	Murphey Complex	ID	652,016
2009	Railbelt Complex	AK	636,224
2004	Eagle Complex	AK	614,974
1997	Inowak	AK	610,000
2012	Long Draw	OR	557,628
2004	Solstice complex	AK	547,505
2011	Wallow	AZ	538,049

Source: National Interagency Fire Center 2008.

the rest of the agency's budget. The history of fire in the forest lands of the United States has left a legacy of conditions that has increased both fire hazard and suppression costs, and left forests in unhealthy conditions. This situation threatens the forest, adjacent human lives and infrastructure, and of course wildlife habitat. But the situation we now find ourselves in, due to our complex forest fire history, has yet more complicating conditions.

3-4 Climate Change and Drought

Earth surface temperatures and precipitation have varied over time, from annual and decadal patterns associated with ocean currents to geologic time. For example, anthropologists theorize that severe and extended drought conditions in the American Southwest in the mid-1200s to the mid-1300s, as documented in tree ring records, strongly contributed to the Anasazi people abandoning their sophisticated settlements in the Four Corners region (Benson and Berry, 2009). Despite expressed opinions to the contrary, data indicate that Earth surface temperatures have also increased since the late 1800s (Karl et al. 2015; Figure 3-3). Recent data

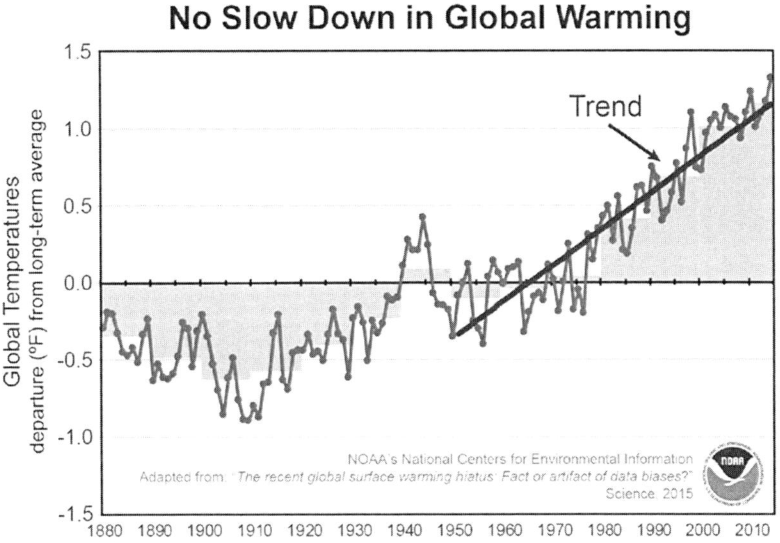

Figure 3-3 Annual global temperature departure from long-term average.

Source: Karl et al. 2015. Figure from NOAA (n.d.) National Centers for Environmental Information.

also point to major changes in precipitation amounts and patterns. These changes are uneven, however. Parts of the Northeast and Upper Midwest have experienced increased precipitation while droughts have occurred in the Southwest, Southeast, and California (NOAA 2016; Figure 3-4).

Such changes have mixed effects on forests. Warmer and drier conditions would seem to force forests to move up in elevation or in latitude to maintain desired conditions. Such movements will be slow, at least in human terms, for most tree species but punctuated by major disturbances like wildfire. Species at or near mountain tops may run out of places to move up to, and species more suitable to the new temperature and precipitation regimes may replace them if there is a source of seeds. Animal species with very specific habitat requirements could face great difficulties as their habitat disappears. More generalist species may have an easier time adapting, especially if they have mobility to follow the forest or can adapt to changing conditions.

Climate changes can have deleterious secondary effects on forests, as well. Warmer and drier conditions stress trees, leaving them more susceptible to disease, especially to bark beetles (Bentz et al. 2010). Stressed, dead, and dying trees make excellent fuel sources, thus increasing fire hazard, and increasing the potential for large wildfires. Warmer drier conditions also increase the difficulty in suppressing fire, leading to larger fires. Massive wildfires open areas for invasive plants to establish that will slow down reforestation and change wildlife habitat (usually for the worse).

Decreased precipitation has led to droughts, which have tertiary costs to forest and wildlife. As droughts continue, human populations will look for additional water supplies. Forest lands serve as the source of a massive portion of surface run-off and groundwater infiltration. As humans capture more and more water for their own use, available water to support riparian areas decreases.

Riparian areas only occupy a small portion of the overall landscape. Riparian areas make up approximately 100 million acres in the contiguous United States (5%) (Dahl and Allord 2018), but in the more arid West riparian areas total only 1–2% of the land area. However, their importance to water quality and wildlife habitat (including connectivity corridors) and species diversity far exceeds their small area extent (Conservation Reserve Enhancement Program n.d.). These long distances to available water may result in the extirpation of some species that do not have the ability to travel to the more scattered water locations. Concentrated water sources will tend to result in animals congregating in larger numbers at fewer locations. Good for predators, but not so much for prey.

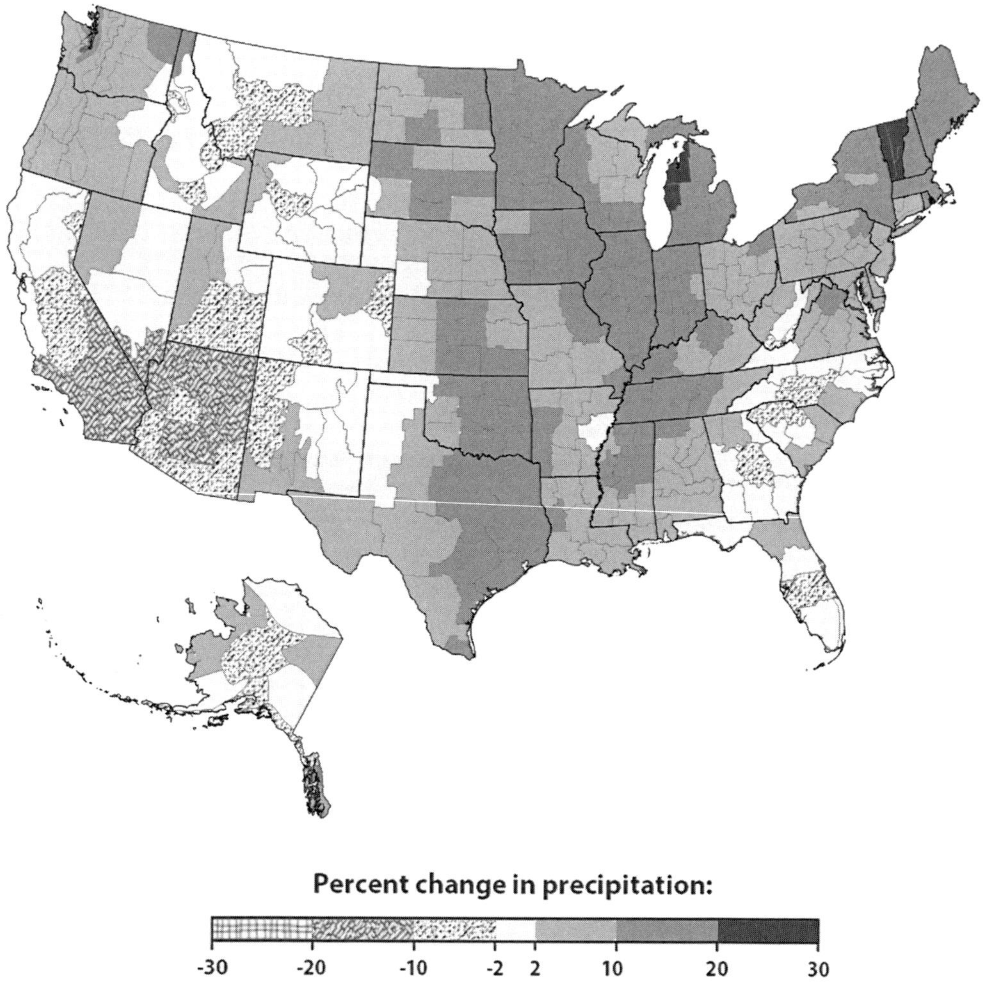

Figure 3-4 Percent change in precipitation since the early twentieth century.

From NOAA 2016 National Centers for Environmental Information.

Complex climate change models exist to project potential climatic patterns. The National Oceanic and Atmospheric Administration (NOAA) has an online interactive that provides users with the opportunity to look at historical trends in precipitation and temperature available at: https://www.ncdc.noaa.gov/temp-and-precip/us-trends. A simple climate change model from the Center for Science of the University Corporation for Atmospheric Research (UCAR) education is available at: https://scied.ucar.edu/simple-climate-model.

3-5 Exotic and Invasive Species

Humans have deliberately brought hundreds of different plants and animals to the New World. These include wheat, rye, hops, barley, oranges, avocados, almonds, cattle, horses, pigs, sheep, cats, goats, kudzu, and tamarisk. Many other species have arrived by accident; e.g. tumbleweed, cheat grass, Dutch elm disease, chestnut blight, and rats. Regardless of intent or source, the introduction of invasive species has had profound effects on forest land, forest land management, wildlife habitat, and wildlife populations. For example, cats—both domestic and feral—have a unique role in their effects on wildlife. Instead of damage to forest lands and wildlife habitat, their effects derive from predation. The domestic and feral cats, collectively known as "free-ranging," kill an estimated 1.3–4.0 billion birds and 6.3–22.3 billion mammals annually (Loss et al. 2013).

The terms exotic species, alien species, non-native species, introduced species, and invasive species are essentially synonymous. These are species deliberately or accidentally introduced into an area beyond their endemic habitat. Thus, these species may be from as far away as a different continent or as close as a different part of the same country. For our discussion here, we use the term invasive species. A subset of invasive species is a feral species. These domesticated plants or animals escaped from and then survived and thrived outside of captivity. Frequently, feral species pick up the name of "wild." For example, horses and burros that escaped (or sometimes were intentionally released by their human owners) from captivity and inhabit large areas of the West are correctly called "feral" and not "wild" horses and burros.

A staggering number of invasive species currently live in the United States. Experts have estimated the number of invasive species at approximately 50,000 (Pimentel et al. 2000). In some way every invasive species has effects on the environment where it now thrives. Only a relatively few, however, cause the most damage to forest and other wildlands. Some species have almost totally replaced native species in a given environment, e.g., the replacement of native bunch grasses with Mediterranean annual grasses in the Coast Ranges and Sierra Nevada foothills of California. Others have almost driven native species to extinction, e.g., the chestnut blight has almost eliminated American chestnut from North America.

Originally brought to the United States in the late 1800s as an ornamental, kudzu's introduction into the South during the Great Depression to control erosion and as food for domestic cattle initiated its rapid expansion. Kudzu grows rapidly (hence one of its nicknames—yard-a-night—as in, "it grows a yard a night") taking over open fields and draping itself over buildings, power lines, and trees. Currently, kudzu has taken over hundreds of thousands of acres in the South and has spread as far north as Michigan and Massachusetts and as far west as Oregon and Hawaii.

Humans have sometimes tried to undo the damage caused by invasive species by introducing another invasive species. For example, tamarisk (aka saltcedar) was introduced to the Southwest to stabilize severely eroded streambeds and gullies. In the absence of any natural control, it spread rapidly, replacing native species, such as willow, including most of those along the Colorado River in the Grand Canyon. To control this invasive species, a biological control agent, the tamarisk leaf beetle, was introduced in the late 1990s (Lewis et al. 2003). The beetle has successfully reduced the extent of tamarisk but this has led to yet another wildlife management issue: some native birds, such as the southwestern willow flycatcher, have adapted to the tamarisk-dominated areas and its elimination threatens the habitat of these birds (Shafroth et al. 2005). A similar situation occurred in the late 1800s.

In 1883, the mongoose was introduced into Hawaii to control rat populations, another invasive species. Unfortunately, rats are active at night and mongoose hunt during the day. Instead of controlling rat population, the mongoose turned to native species and along with the night hunting rats, have severely decimated populations of ground nesting birds, such as the Nene (Hawaiian) goose, the state bird of Hawaii (Maui Invasive Species Committee 2012).

One invasive species particularly damaging to forest lands in the United States deserves special attention: feral hogs. Since the time of the first Spanish explorers who brought them to what is now the United States, domestic pigs have escaped captivity and gone feral. Wild boar was later imported as a game species. The species interbreed and collectively—feral, wild, and crosses—have received the name "feral hogs." These hogs inflict incredible damage to forests. Their trails cut into the forest floor and when rains occur these gullies serve as a major source of erosion and sedimentation. In rooting for food, the hogs destroy understory plant and tree regeneration. As omnivores, they eat just about anything, reducing food sources for wildlife. Ferocious, aggressive, and armed with large tusks and sharp hooves, they pose dangers to other species—including humans.

Although concentrated in the South, discontinuous feral hog populations range into New Hampshire, as well as west to California and into Oregon (USDA 2018). Control efforts, including trapping, hunting, toxicants, fertility control, and fencing, have had limited effectiveness in controlling feral hog populations (Campbell and Long 2009). Invasive species cause billions of dollars in damage each year, in direct damage plus control costs. Cost estimates in 2000 totaled almost $137 billion with the greatest invasive animal costs associated with feral cats and rats (Pimentel et al. 2000). Although "top" lists vary, some of the worst invasive species include the Norway rat, European starling, water hyacinth, and kudzu (De Santi 2016; Grebenstein 2013) (Table 3-8).

Table 3-8 Top invasive species in the United States.

Plants	Animals
Purple loosestrife (*Lythrum salicria*)	Asian carp
Japanese honeysuckle (*Lonicera japonica*)	Asian long-horned beetle
Norway maple (*Acer plantanoides*)	Zebra mussel
English ivy (*Hedera*)	Brown tree snake
Kudzu (*Puerarian montana var. lobate*)	Burmese python
	Killer bees
	European starling
	Northern snakehead

Sources: De Santi 2016; Grebenstein 2013.

3-6 Outdoor Recreation

"With the goal of having a clear understanding of the term outdoor recreation it would be edifying to be able to present a common definition of outdoor recreation to be used by all researchers, but none exists" (USDI 1968, pg. 21). More recently, Phillips (1991) defined the term broadly in that it "includes activities that occur outdoors in an urban and man-made environment as well as those activities traditionally associated with the natural environment" (pg. 1). The U.S. Federal Government defines outdoor recreation as:

> all recreational activities undertaken for pleasure that generally involve some level of intentional physical exertion and occur in nature-based environments outdoors. The broad definition includes all conventional outdoor recreation activities and a range of additional activities. More formally, the broad definition includes all recreational activities undertaken for pleasure that occur outdoors. (USDC 2017, pg. 1)

Alternatively, "**outdoor recreation** or **outdoor activity** refers to leisure pursuits engaged in the outdoors, often in natural or semi-natural settings out of town" (Northern Michigan University 2018, pg. 1). All these definitions focus on *doing* to rebuild some internal value that no longer exists in an individual. In contrast, Hammitt (2005, pg. 1) presents the theory that outdoor recreation is a process for "restoring people's mental well-being that has been fatigued."

Although a universally agreed upon definition of the term does not seem to exist, very pragmatically, lists of outdoor recreation activities exist. For analytical purposes, the U.S. government (USDC 2018) created an extremely detailed list of such activities that range from those that most individuals could most likely readily agree upon (e.g., hiking, camping, hunting) to some that require more careful thought, such as attending amusement parks and recreational flying (Table 3-9). The Wilderness Society (n.d.) has generated perhaps a more satisfying list of major outdoor recreation activities that in addition to hiking, camping, and hunting includes fishing, rock climbing, and canoeing. Among the most popular are camping, hunting, and hiking (Table 3-10).

Regardless of the definitions and examples used, outdoor recreation has increased immensely in the past 70 years due to a combination of increased population, greater disposable income, more leisure time, better equipment, and greater access. For example, between 1944 and 1994, recreation use on the National Forests alone increased by almost 2.5 times (Table 3-11).

Table 3-9 Major outdoor recreation activities.

Identified Activities	
Conventional outdoor recreation	Bicycling
Boating/fishing	Canoeing/kayaking
Sailing	Other boating/fishing
Camping/climbing/hiking	Equestrian
Hunting/shooting/trapping	Hunting
Other shooting/trapping	Motorized vehicles
Motorcycles	RVs
Other motorized vehicles	Recreational flying
Skiing	Snowboarding
Other conventional outdoor recreation activities*	Multi-use apparel and accessories (conventional)**
Other outdoor recreation	Amusement parks/water parks
Festivals/sporting events/concerts	Field sports
Game areas (including golf and tennis)	Guided tours/outfitted travel
Productive activities	Other outdoor recreation activities**

*Consists of air sports, boardsailing/windsurfing, dog mushing, geoarching/orienteering/rock hounding, ice skating, land/sand sailing, races, running/walking/jogging, Scuba diving, skateboarding, snorkeling, snowshoeing, sand-up paddling, surfing, tubing, wakeboarding, water skiing, whitewater rafting, and wildlife watching, birding.
**Consists of agritourism, augmented reality games, beachgoing, disc golf, hot springs soaking, kite flying, model airplane/rocket/UAV, paintball, photography, star-gazing astronomy, swimming, therapeutic programs, water polo, yard sports.

Source: USDC 2018.

Table 3-10 Most common and popular outdoor recreation activities.

Most Common Outdoor Recreation Activities	Most Popular Outdoor Recreation Activities
Hiking and camping	Running
Hunting and fishing	Biking
Canoeing, kayaking and rafting	Camping
Sailing and motorboating	Fishing
Biking	Hunting
Rock climbing	
Horseback riding	
Skiing	
Restoration and conservation volunteering	

Source: The Wilderness Society (n.d.).

Use of wilderness constitutes a special part of overall outdoor recreation. Wilderness is defined in *The Dictionary of Forestry* (Helms 1998) as "any area where the earth and its community of life are untrammeled by man where man himself is a visitor who does not remain" (pg. 199). The wilderness system in the United States has increased from about 9.4 million acres in 1964 when the Wilderness Act passed, to over 110 million acres in 2017. Although the Forest Service manages the largest number of separate wilderness areas (113), the National Park Service manages the largest number of wilderness acres (44 million) (Wilderness Connect 2017). The Wilderness Act prohibits grazing of domestic livestock (almost entirely) and the use of motorized vehicles within wilderness areas.

Table 3-11 Outdoor recreation use, selected years.

Year	Total Visitor Days' Use (in thousands)
1924	4,660
1934	8,580
1944	7,152
1954	40,304
1964	133,742
1974	192,916
1984	227,554
1994	330,348

Source: USDA 2018.

In a systematic review of 274 articles that reported on the interaction between outdoor recreation and wildlife worldwide, Larson et al. (2016) found that recreation tended to decrease species survival and diversity and increased stress on animals. They also found that non-motorized vehicle use created more disturbance than motorized activities and that snow-related activities had the largest effects. Although the articles focused on birds and mammals, they also included results from studies of amphibians, reptiles, and fish. Increased human use has put multiple pressures on wildlife. Increased traffic increases the threat of animal death or injury by motor vehicle. More roads fragment habitat and block migration and dispersal routes. More humans in the forest results in more disturbance of key habitat components, such as nesting and roosting sites and sources of water. More human use also increases the potential for wildfire. Of course, increased hunting puts direct pressure on wildlife populations.

Although wilderness areas can provide refuges and habitat for forest wildlife, overall outdoor recreation use reduces useful habitat and increases stress on some species. Hunting can limit populations to maintain a balance with habitat. However, a reduction of large predators, e.g., mountain lions, wolves, and bears, can result in a population increase where competition for food can cause a rapid decline in a wildlife population, e.g., deer (see Decimating Factors in Section 6-3). In general, outdoor recreation results in negative effects on some wildlife species.

3-7 Forest Economics

No single activity has had greater effects on forests in the United States than has timber harvesting. Regardless of objective, whether for timber products, to convert land for agricultural land, or create infrastructure, economics drove these harvesting activities. Although now tempered by increasing concerns about sustainability and ecosystem services, including the protection of wildlife species and their habitats, water quality, and the conservation of soil, economics continue to provide a major driving force in the management of forest lands.

Although having only 5% of the world's population and 10% of the world's timber inventory, the United States consumes 28% of the world's industrial timber products (USDA 2014). In absolute terms, the total harvest of industrial timber products (all species) in the United States increased from about 12 billion cubic feet in 1962 and peaked at 15.9 billion cubic feet in 1989. Harvest dropped to just under 13 billion cubic feet in 2011 during the Great Recession (USDA 2014). Harvest from public lands peaked in the mid-1970s at about 6 billion cubic feet, declined to about 2 billion cubic feet by the mid-1990s, and has remained at that level since. Harvest from private lands peaked in the mid-2000s at about 14 billion cubic feet and declined to about 10.5 billion cubic feet by 2009 (USDA 2014). Per capita consumption of wood products has remained at about 75 cubic feet per year since the mid-1960s (Alvarez 2007). In total, approximately 11 million acres of forest land are harvested annually with partial harvests becoming an increasing proportion of the total acres cut (USDA 2014).

Lumber (53%) and paper products (32%) consume the largest share of timber products. For context, a standard "stick built" house of about 2,000 square feet in size requires about 12,600 board feet of lumber (House Designers 2018). Although nominally a cubic foot contains 12 board feet, empirically, depending on log size and utilization assumptions, the actual conversion is 5:1 to 6:1. Thus a stick-built house will require about 2,000–2,500 cubic feet of wood. A ton of paper (about one pallet of copy paper) requires approximately 100–150 cubic feet of wood.

Economically, the forest products industry carries major economic value. It employs over 900,000 individuals and generates a payroll of over $54 billion—about 4% of the country's Gross Domestic Product (AF&PA 2017). In addition, much of the employment in the forest products industry occurs in more rural areas and thus become major economic drivers in rural communities. Thus, when mills close, entire communities face major economic setbacks. Of course, these figures just reflect the forest products industry and do not include the economic activity generated by other uses of the forests, such as outdoor recreation, grazing, water yields, etc.

3-8 Summary

Humans have used forests as a source of food, fuel, and building materials for hundreds of thousands of years worldwide and for at least 10,000–15,000 years in the Western Hemisphere. Only in the past 200–300 years, and especially in the last 100 years, have humans used forests in ways that have put major pressures on forest wildlife. As noted earlier in this chapter, the unfettered cutting of timber, overgrazing, and concomitant conversion of forest land to agriculture along with wildfires of epic proportions have all massively and profoundly changed the forests as habitat for wildlife, interrupting the long-term species balance. Market hunting and predator control further interrupted this balance.

More recent events, e.g., fire suppression and exclusion, increases in outdoor forest recreation, and the expansion of the Wildland Urban Interface (WUI), have further changed forest structures and distribution. Road access also negatively affect wildlife. In 2008, the National Forests alone contained almost 375,000 miles of road (USDA 2008)—over nine times the mileage in the U.S. Interstate Highway system. On the other hand, significant areas of forest lands have been set aside in wilderness areas and a suite of other environmental laws have moved to improve forest lands for wildlife. At the same time, forests in the United States serve as an important economic driver in the country and the largest source of freshwater.

The growing population, wealth (overall), changing decisions about leisure spending of time and money, and the growth in awareness of non-monetary environmental values have resulted in increased demands on our forests and differences in how they are valued. Unfortunately, many of these demands and values conflict and, with a limited (and finite) forest land base, require decisions about trade-offs. In the perfect world, such decisions would require data acquisition and analysis, and the recognition of different desires and values. Eventually, decisions reflect political and social values and power. We address these issues in subsequent chapters.

References

Agee, J.K. 1993. *Fire ecology of Pacific Northwest Forests.* Island Press, Washington, DC.

Alvarez, M. 2007. *The state of America's forests.* Society of American Foresters, Bethesda, MD. 68 p.

American Forest & Paper Association (AF&PA). 2017. *State industry economic impact, United States.* Available at http://www.afandpa.org

Benson, L., and M.S. Berry. 2009. Climate change and cultural response: In the prehistoric American southwest. USGS Staff, Published Research. 725. 32 p.

Bentz, B. J., J. Régnière, C.J. Fettig, E.M. Hansen, J. Hayes. 2010. Climate change and bark beetles of the western United States and Canada: Direct and indirect effects. *BioScience* 60(2):602–613.

Bonnicksen, T.M. 2000. *America's ancient forests from the ice age to the age of discovery.* John Wiley & Sons, New York. 594 p.

CalFire. 2018. *20 Largest California wildfires.* Available at www.fire.ca.gov/communications/downloads/fact_sheets/Top20_Acres.pdf; accessed May 30, 2018.

Campbell, T.A., and D.B. Long. 2009. Feral swine damage and damage management in forested ecosystems. *Forest Ecology and Management* 257:2319–2326.

Cohen, J. 2008. The wildland-urban interface fire problem: Consequences of the fire exclusion paradigm. *Forest History Today* Fall:20–26.

Conservation Reserve Enhancement Program. (n.d.). Functions of Riparian Areas for Wildlife Habitat. Fact Sheet #4. New York State. 7 p.

Covington, W.W., and M.M. Moore. 1994. Southwestern ponderosa pine forest structure: Changes since Euro-American settlement. *J. of For.* 92(1):39–47.

Dahl, T.E., and G.J. Allord. 2018. History of wetlands in the conterminous United States in national water summary on wetland resources. U.S. Forest Service Geological Survey Water Supply Paper 2425. Available at https://water.usgs.gov/nwsum/WSP2425/index.html

Dennison, P.E., S.C. Brewer, J.D. Arnold, et al. 2014. Large wildfire trends in the western United States, 1984–2011. *Geophysical Research Letters* 41(8):2928–2933. Available at https://doi.org/10.1002/2014GL059576

De Santi, A. 2016. 10 most invasive species in North America. *Monsters & Critics.* Available at https://www.monstersandcritics.com/list/amazing-world/10-most-invasive-species-in-north-america/

Gordon, J., J. Sessions, J.D. Bailey, et al. 2013. *Assessment of Indian forests and forest management in the United States final report.* The Third Indian Forest Management Assessment Team, Intertribal Timber Council, Portland, OR.

Grebenstein, E. 2013. Escape of the invasives: Top six invasive plant species in the United States. *Smithsonian Insider.* Available at https://insider.si.edu/2013/04/top-six-invasive-plant-species-in-the-united-states/

Hamilton, A. 2016. Early American expansion: A Primer from Dred Scott. *National Vanguard.* Available at https://nationalvanguard.org/wp-content/uploads/2016/12/Map-1.-General-overview-of-U.S.-expansion-pre-Dred-Scott.png

Hammitt, W.E. 2005. A Restorative Definition for Outdoor Recreation in *Proceedings of the 2004 Northeastern Recreation Research Symposium,* Bricker, K., (comp., ed.). USDA Forest Service, Gen. Tech. Rep. NE-326. Northeastern Research Station: 1-5, Newtown Square, PA.

Helms, J.A. (ed.). 1998. *The dictionary of forestry.* Society of American Foresters, Bethesda, MD.

House Designers. 2018. *How many trees does it take to build a house?* Available at https://www.thehousedesigners.com/articles/how-many-trees-does-it-take-to-build-a-house.asp

IFMAT-I. 1993. *An assessment of Indian forests and forest management in the United States.* Indian Forest Management Assessment Team, Intertribal Timber Council, Portland, OR.

ITC. 2007. *Strategic planning and interim action plan.* Intertribal Timber Council, Portland, OR.

Johnson, A. Sydney, and P.E. Hale. 2002. The historical foundations of prescribed burning for wildlife: A southeastern perspective. P. 11–23 in *The role of fire in nongame wildlife management and community restoration: Traditional uses and new directions.* USDA Forest Service, Nashville, TN.

Karl, T.R., A. Arguez, and B. Huang. 2015. Possible artifacts of data biases in the recent global surface warming hiatus. *Science* 348(6242): 1469–1472. DOI: 10.1126/science.aaa5632

Kilpatrick, H.J., A.M. Labonte, and J.S. Barclay. 2007. Acceptance of Deer Management Strategies by suburban homeowners and bowhunters. *J. Wildl. Manage.* 71(6):2095–2101. Available at https://doi.org/10.2193/2007-058

Kinney, J.P. 1937. *A continent lost—a civilization won: Indian land tenures in America.* John Hopkins Press, Baltimore, MD.

Landthink. 2017. *Timberland ownership and management in North America: 2017 update.* Available at https://www.landthink.com/timberland-ownership-and-management-in-north-america-2017-update/; accessed May 31, 2018.

Larson C.L., S.E. Reed, and A.M. Merenlender. 2016. Effects of Recreation on animals revealed as widespread through a global systematic review. *PLOS ONE* 11(12):e0167259. DOI:10.1371/journal.pone.0167259

Lewis, P.A., C.J. Delach, and A.E. Knutson. 2003. Biology of *Diorhabda elongata deserticola* (*Coleoptera: Chrysomelidae*), an Asian leaf beetle for biological control of saltcedars (*Tamarix spp.*) in the United States. *Biological Control.* 27(2):101–116.

Loss, S.R., T. Will, and P.P. Marra. 2013. The impact of free-ranging domestic cats on wildlife of the United States. *Nature Communications,* Article 4:1396. DOI:10.1038/ncomms2380

Martinuzzi, S., S.I. Stewart, D.P. Helmers, et al. 2015. The 2010 wildland-urban interface of the conterminous United States. Research Map NRS-8. USDA Forest Service, Northern

Research Station, Newtown Square, PA. 124 p. [includes pull-out map]. Available at https://doi.org/10.2737/NRS-RMAP-8

Maui Invasive Species Committee. 2012. *Moving on from the mongoose: The success of biological control in Hawaii.* Available at http://mauiinvasive.org/2012/04/18/moving-on-from-the-mongoose-the-success-of-biological-control-in-hawaii; accessed May 30, 2018.

National Interagency Fire Center. 2008. *Total wildland fires and acres (1926–2017).* Available at https://www.nifc.gov/fireInfo/fireInfo_statistics.html; accessed May 28, 2018.

National Interagency Fire Center. (n.d.). *National fire and aviation web database.* Available at https://www.nifc.gov/fireInfo/fireInfo_stats_lgFires.html; accessed May 28, 2018.

Nelson, B. 2013. 10 of the worst wildfires in U.S. history. *Mother Nature Network.* Available at https://www.mnn.com/earth-matters/wilderness-resources/stories/10-of-the-worst-wildfires-in-us-history; accessed May 30, 2018.

Northern Michigan University. (2018). *Outdoor recreation leadership and management program.* Available at https://www.nmu.edu/hhp/OutdoorRecreation; accessed June 2, 2018.

NOAA. 2016. *Percent change in precipitation.* Available at www.ncei.noaa.gov; accessed February 2016.

NOAA. (n.d.). *The recent global surface warming hiatus.* Available at ncdc.noaa.gov/news/recent-global-surface-warming-hiatus; accessed June 2019.

Oswalt, S.N., W.B. Smith, and P.D. Miles. 2014. *Forest resources of the United States, 2012: A technical document supporting the Forest Service 2015 update of the RPA assessment.* USDA Forest Service Gen. Tech. Rep. WO-91, Washington Office, Washington, DC. 218 p.

Pecore, M. 1992. Menominee sustained-yield management: A successful land ethic in practice. *J. For.* 90(1):12–16.

Phillips, M.L. 1991. Definitions of outdoor recreation and other associated terminology in Proc. of conf. on National Conference for Outdoor Leaders, Public, Commercial, and Non-Profit.

Pimentel, D., L. Lach, and R. Zuniga. 2000. Environmental and economic costs associated with non-indigenous species in the United States. *BioScience.* 50(1): 53–65.

Radeloff, V.C., D. P. Helmers, H.A. Kramer, et al. 2018. Rapid growth of the US wildland-urban interface raises wildfire risk. *Proceedings of the National Academy of Sciences* 115(13):3314–3319. DOI: 10.1073/pnas.1718850115

Sacks, B.N., H.B. Ernest, and E.E. Boydston. 2006. San Francisco's Golden Gate: A bridge between historically distinct coyote (*Canis latrans*) populations? *Western North American Naturalist.* 66(2):16. Available at https://scholarsarchive.byu.edu/wnan/vol66/iss2/16

Shafroth, P.B., J.R. Cleverly, and T.L. Dudley. 2005. Control of Tamarix in the Western United States: Implications for Water Salvage, Wildlife Use, and Riparian Restoration, *Environmental Management* 35(3):231–246. DOI: 10.1007/s00267-004-0099-5

Sleeter, B.M., T.L. Sohl, T.R. Loveland, and R.F. Auch. 2013. Land-cover change in the conterminous United States from 1973 to 2000. *Global Environmental Change* 23:733–748.

Stephens, S.L., M.P. North, and B.M. Collins. 2015. Large wildfires in forests: What can be done? *Action Bioscience.* Available at http://www.actionbioscience.org/environment/large_wildfires_in_forests_what_can_be_done.html

Tappeniner, J.C., J.D. Bailey, and T.B. Harrington. 2015. *Silviculture and Ecology of Western U.S. Forests,* Second edition. Oregon State University Press, Corvallis, OR.

Taylor, R. 1991. *The feral hog in Texas.* Texas Parks & Wildlife, Austin, TX. 22 p.

USDA. 2008. *The U.S. Forest Service: An overview.* USDA Forest Service, Washington, DC.

USDA. 2014. *U.S. forest resources facts and historical trends.* USDA Forest Service Publication FS-1035, Washington, DC. 64 p.

USDA. 2017. *Fiscal year 2018 budget overview.* USDA Forest Service, Washington, DC. 53 p.

USDA. 2018. National forest recreation use, 1924–1996. USDA Forest Service. Available at https://www.fs.fed.us/recreation/programs/facts/

USDC. 2017. Preparing the outdoor recreation satellite account. *Bureau of Economic Analysis.* Available at https://bea.gov/outdoor-recreation; accessed June 2, 2018.

USDC. 2018. *Monthly Population Estimates for the United States: April 1, 2010 to December 1, 2018 (NA-EST2017-01).* Available at https://www2.census.gov/programs-surveys/popest/tables/2010-2017/national/totals/na-est2017-01.xlsx

USDI. 1968. *A program for outdoor recreation research.* Bureau of Outdoor Recreation report on a study conference conducted by the National Academy of Sciences, Washington DC. 90 p.

USDI. 2016. Yellowstone 1988 fires. *National Park Service (NPS).* Available at https://www.nps.gov/yell/learn/nature/1988-fires.htm; accessed May 30, 2018.

van Wagtendonk, J.W. 1995. *Dr. Biswell's influence on the development of prescribed burning in California.* USDA Forest Service Gen. Tech. Rep. PSW-GTR-158. 11–15 p.

van Wagtendonk, J.W. 2007. The history and evolution of wildland fire use. *Fire Ecology Special Issue* 3(2):3–17.

Veeder, W.H. 1968. Federal trust responsibilities for managing Indian forest with points and authorities in support. Memorandum. USDI Bureau of Indian Affairs, Washington, DC.

Walker, R.B., J.D. Coop, and S.A. Parks. 2018. Fire regimes approaching historic norms reduce wildfire-facilitated conversion from forest to non-forest. *Ecosphere* 9(4):e02182. 10.1002/ecs2.2182.

Wilderness Connect. 2017. *Wilderness area database.* Available at https://www.wilderness.net/

The Wilderness Society (n.d.). *Outdoor recreation FAQs.* Available at https://wilderness.org/article/outdoor-recreation-faqs; accessed June 2, 2018.

Knowing the Land 4

The number of techniques available to inventory wildlife species and habitat has increased; a manager has a variety of tools from which to select. Although perhaps desirable, the availability of numerous techniques can create problems in selecting a cost effective and biologically sound alternative. As a result, biologists, foresters, and land managers must invest considerable time planning before starting fieldwork for an inventory.

Scientific management requires good data, which often requires the collection of field data—an inventory. Too often, however, inventory implementation proceeds before adequate planning and objective setting: the "fire, aim, ready" approach. The inventory process should start with the needs of management planning, which starts with setting management objectives. Experience has shown that without clear, well-defined objectives, one of three things happens: either too much or too little data or the wrong types of data are collected. All these outcomes result in the loss of time, money, and professional credibility.

The passage of resource management legislation (Chapters 6 and 7), has greatly increased the emphasis on inventory because of the multiple types of data needed to fulfill all the decision-making requirements. Historically, forest-based inventories focused almost exclusively on trees in terms of timber production. The emphasis now rests on inventorying all resources at the same time; in other words, multi-resource inventory (Reed and Mroz 1997; Schrueder et al. 1993).

4-1 Multi-Resource Inventory

Once a decision has been made to conduct an inventory then one task needing early attention is to determine if data can be collected as a part of another resource inventory, such as for timber or watershed management. If no such inventories are planned it will have to be initiated as a single-purpose inventory, such as surveys of wildlife populations. In short, do not collect data if it is already available. Make use of existing sources, such as timber surveys, range allotment analysis, recreation plans, and watershed and soil surveys, even if the data needs interpreting or revision. In any case, the actual data collection must be consistent. Use of the following guidelines will help to meet this goal:

- collect data using compatible methods over time,
- collect data at the same time of year, and
- collect data on the same plots if the sample was systematic as in previous inventories.

Use a format, either electronic or paper, that forces data entry into an allocated space for a given topic in the order it will be entered into an electronic database. This procedure reduces the possibility of mistakes by a data entry operator. Another alternative is to use an electronic field recorder programmed for data transfer into a database management system. Field data recorders (FDR) are portable, battery-operated microcomputers. Most of the new FDRs can be programmed with rules allowing only data with certain characteristics to be entered. The advantage of some types of recorders is that the data can be transmitted electronically to an office computer. On the negative side, FDRs, like all computers, may become damaged or run out of power in the field or be forgotten back at the office. Therefore, standard field gear should include paper forms, preferably with waterproof paper, as a backup.

Categories of Inventory

Category of inventory has to do with the purpose and use of the information collected. In general, inventories have multiple hierarchical categories that correspond to the organizations planning hierarchy (discussed further in Chapter 7). For the U.S. Forest Service, a three-tiered inventory—national, regional, and management unit—supports the hierarchical planning approach the agency uses.

National inventories usually provide limited information at a very large scale consisting of lists of resources used for formulating legislation, determining policy, setting management and research priorities, and supporting budget presentations, such as those described in Chapter 3. Information collected specifically for a national inventory generally cannot be disaggregated to a lower category. **Regional** inventories typically focus on a geographical area or major vegetation types that cut across political boundaries. This type of inventory supports the development of programs and the allocation and distribution of resources (such as budgets). The information becomes part of a planning document that can be revised and updated to reflect changing conditions.

The third category of inventory is a management unit that will vary in size. It can be as large as a national forest or as small as a watershed. **Management unit** inventories are the basis for management decisions and most often include detailed resource information. Information collected at the management unit level generally contains too much detail to allow for aggregation to higher categories. However, in almost all cases some information is suitable for regional use. The categories of inventory are additive from the local level to a national level such as local to regional to national. This scheme has many advantages particularly if all the data are collected at the same detail, format, and consistency at each level.

For a large forest products company, the levels might be corporate, regional, district, and project area. Furthermore, at each level, the inventory may receive additional subdivision based on management needs and constraints, such as state and local zoning, political subdivision, or other legal restrictions on use, such as tax status. For the small, non-industrial landowner the inventory may only have one level—the owner's individual tract of forest land. Regardless of the number of levels, inventory should provide the information needed for planning and decision-making and, when applicable, the information collected should have characteristics that will allow for its aggregation and disaggregation with the least amount of effort possible.

Intensity of Inventory

The intensity of an inventory must be controlled by the purpose and objective of the inventory, but all too often it is controlled by the funds available. Ideally, the level of detail increases with the level of intensity from low to high. **Low-intensity** inventories consist largely of lists of species, grouped by some characteristic of interest, such as threatened and endangered, game and nongame, taxonomic group, or economic value. Low-intensity inventory information, frequently used at the national level, may also serve as the first step in a high-intensity inventory for a management unit. Data for low-intensity inventories are largely derived from literature surveys or expert opinion.

A **moderate-intensity** inventory builds on the species list and classification by adding information on where a species feeds and breeds. It also can include an estimate of population abundance using general terms of low, moderate, or high, and a population trend of increasing, decreasing, or stable. Data for moderate-intensity inventories is derived from literature reviews, expert opinion, and limited field surveys.

High-intensity inventories are the most detailed, most expensive, and take the greatest amount of time to complete. Best practices call for data collection by field crews with knowledge of the animals and their habitat. Data include plant species composition and abundance, animal population characteristics (density, sex and age ratios, etc.), and quantity and quality of habitat. If data are collected properly, they can be aggregated with other management units to provide information for regional and national purposes. The intent and scope of high-intensity inventories require carefully planning with clearly defined priorities, precise cost estimates, and the desired data precisely identified.

An inventory of resources is the first step in the management process and without an inventory, there can be no management except by default. When considering the **category** and **intensity** of inventory, the manager can select from several combinations; however, some combinations are not as practical as others because the detail is not necessary or the cost is too high.

Data Acquisition

Once decisions have been made on data collection needs, the first place to start data collection is to find out what data already exist. Such data may come in various formats and locations from paper maps and field notes in file cabinets to detailed satellite imagery in sophisticated electronic databases. Although some of the data may be quite old, avoid summarily dismissing such data; historic field notes may include descriptions of important observations not captured in field data. In addition, historic field notes and data can be invaluable for defining past stand structures as a basis for determining reference conditions (Chapter 7). One of the first pieces of data necessary for the planning of an inventory is maps.

Maps and Scale. Maps serve as models of the world and they illustrate many different attributes of the world. The GPS maps available on smartphones include a variety of attributes—streets, highways, houses, businesses—and because of their electronic format have an interactive capacity. Through the phone, users can ask for certain information, such as a route and directions to a location. Other maps may only include geologic features or vegetation. Historically resource managers have used a variety of maps, but maps originally started with basic information such as topography.

The inventory process adds information to the map or establishes linkages between spatial representations on a map and detailed tabular data, such as records of the number and size of trees. Geographic Information Systems, discussed in the next section, perform this function but with a much larger capacity and ready access than humans can accomplish. Regardless of the type of format, a key element of using maps is a decision about scale that provides a level of resolution appropriate to the task at hand. **Scale** is the ratio of actual distance on the ground to the same distance on the map. Scale is expressed as a representative (unit-less) fraction where the map and ground distances are shown in the same units. For example, on a map with a scale of 1:63,360, 1 inch on the map represents 63,360 inches on the ground. Alternatively, 1 foot on the map represents 63,360 feet on the ground.

A scale of 1:24,000 is the standard map scale for 7.5 minute latitude (angular distance north or south from the earth's equator measured through 90 degrees) and 7.5 minute longitude (great circles running through the north and south poles) map. Each quadrangle (square sheet) covers approximately 40,215 acres. Large-scale maps such as 1:7,920 or 1:24,000 are useful when small areas need delineation with detailed information. Intermediate-scale maps range from 1:50,000 to 1:100,000, cover larger areas, and are suitable for land management planning (Steger 1986). Maps of 1:250,000, to 1:1,000,000 are small scale and cover very large areas on a single sheet.

Natural resource managers historically used maps at a scale of 1:7,920 and 1:15,840 because they provide acceptable levels of resolution for different levels of inventory, but they are not the only scales used by modern map makers (Table 4-1). These two scales are also commonly used in aerial photography in the United States because conversion from inches, miles, and acres is readily made, so detail from photographs is easily transferrable to the maps. The out-

Table 4-1 Map scale conversions.

Scale ratio	Feet per inch	Inches per mile	Acres per inch^2
1:500	41.667	126.720	0.0399
1:1,000	83.333	63.360	0.1594
1:2,000	166.667	31.680	0.6377
1:5,000	416.667	12.670	3.9856
1:7,920*	660.000	8.000	10.0000
1:15,840*	1,320.000	4.000	40.0000
1:20,000	1,666.667	3.168	63.7600
1:24,000	2,000.000	2.640	98.8270
1:31,680	2,640.000	2.000	160.0000
1:63,360	5,280.000	1.000	640.0000

*Commonly used.

line of the management unit drawn on the map provides the basic reference for sampling and determining area.

In the United States, the National Cartographic Information Center can provide information about maps, charts, aerial and space photographs, satellite images, geodetic control, and digital and other cartographic data. Maps can be accessed online or by contacting the national headquarters through mail.

Remote Sensing. In simple terms, remote sensing refers to the acquisition of "information about objects through the analysis of data collected by special instruments that are not in physical contact with the objects of investigation" (Avery and Berlin 1992, pg. 1). Human eyes, ears, and noses are organic remote sensing platforms that people have used for millions of years. In the early twentieth century, aircraft began to supplement human senses as remote sensing platforms with the production of **aerial photographs**. World War II spurred great advances in the production and use of aerial photos as a means of acquiring intelligence about troop movement, target identification, and bomb damage assessment.

Most commonly, specially modified aircraft with high resolution cameras mounted underneath them would fly in straight lines across the landscape taking a series of photographs. These photographs overlapped both along and on both sides of the flight lines. Photos that overlapped, called "stereo pairs," allowed foresters and wildlife managers to see the landscape in 3-D (Avery and Berlin 1992; Green et al. 2017; Burkhart et al. 2019). Aerial photos create a powerful tool to help identify tree species, tree density, and tree size. They also allow for the identification and determination of the spatial location of slope and topographic features, such a cliffs, landslides, and water features. Resource managers can delineate stands—areas of like vegetation in terms of size, species composition, and density—that increase the efficiency of inventories (Section 4-2).

Aerial photos come in a wide variety of scales (just like maps) and types (such as black and white, color, or color infrared) and are examples of analog, as compared to digital imagery. Photos are produced in "hard copy"—even if acquired online in a digital file—and the data captured in the photos cannot be manipulated. Aerial photography also represents a passive sensor in that the cameras rely on the reflectance of the object of interest to capture information.

Beginning in the early 1970s, more powerful and sophisticated remote sensing platforms have developed. NASA's Landsat 1, launched in 1972, was one of the first satellites that produced commercially available digital imagery. Landsat 1 carried a Multispectral Scanner (MSS) that captured and recorded the reflectance of objects across four spectral bands: green, red, and two infrared bands. The reflectance values from each band were represented by a set of numbers. Imagery, sent to ground stations, could be purchased for commercial use.

The imagery became a powerful tool in natural resources because of its capacity to be manipulated. For example, combining the reflectance values of the green band and red band were especially powerful in detecting vegetation. Different plants gave off different signatures (combinations of reflectance values) that could be used to identify species, sizes, and densities. In addition, imagery of the same location over time could detect changes in the landscape, such as changes in vegetation resulting from timber harvesting and fires.

Landsat 1 ceased operation in 1978, but its success has led to new and even more powerful successors. Landsat 1 carried a four band MSS with 79 meters by 57 meters (about 260 feet by 187 feet or approximately 48,620 square feet—about 1.1 acres) level of ground resolution, known as a pixel. In contrast, the latest platform—Landsat 8 launched in 2012—carries two different sensors containing 11 different bands of data, with resolutions ranging from 15 meters (for a panchromatic band) to 100 meters for the two different thermal infrared bands. In addition, users have free access, use, and redistribution to Landsat 8 imagery. The U.S. Geological Survey operates Landsat 8.

Additional satellite platforms exist, each with its own set of characteristics, including level of resolution, image availability, coverage, access costs, and spectral capture capabilities. These include Spot, Pleidas, and Sentinel 2 (OpenForests 2018). Other types of imagery and platforms exist beyond the MSS carried on satellites. Aircraft-based high altitude digital photography produces manipulatable imagery to a large extent but at a high level of resolution. Airborne Lidar (Light Detection and Ranging) sensors carried by aircraft (and increasingly by drones) emit an electromagnetic pulse that illuminates a specific point on the ground. Like traditional radar, the Lidar sensor records the return pulse—the "bounce." The aggregation of the return pulses (the point density) allows the determination of a point's latitude, longitude, and elevation. By calculating ground height at a given point, the height of an object at that point can also be determined, along with other object attributes such as crown diameter (Green et al. 2017).

All these remote sensing techniques create rich data sets that can help create maps of vegetation. However, the real value of these tools only comes into play with the field validation of the classification system created by manipulating the data (i.e., "unsupervised" classifications) to verify their accuracy, creating "supervised" classifications. Trust your classifications (to a point) but always verify. Remote sensing imagery alone can generate massive amounts of data and the data increases by orders of magnitude when other information, such as numerical vegetation descriptions, are associated with the spatial data. Fortunately, a powerful tool exists to organize, store, analyze, and display these massive amounts of data sets—Geographic Information Systems.

Geographic Information Systems

"A geographic information system (GIS) is a framework for gathering, managing, and analyzing data" (ESRI 2018, pg. 1). Figure 4-1 displays a system that performs the following four tasks:

- captures data though scanning, digitizing, or manual data entry, or by accessing a database,
- stores and manages data,
- manipulates, analyzes, and edits data, and
- displays data.

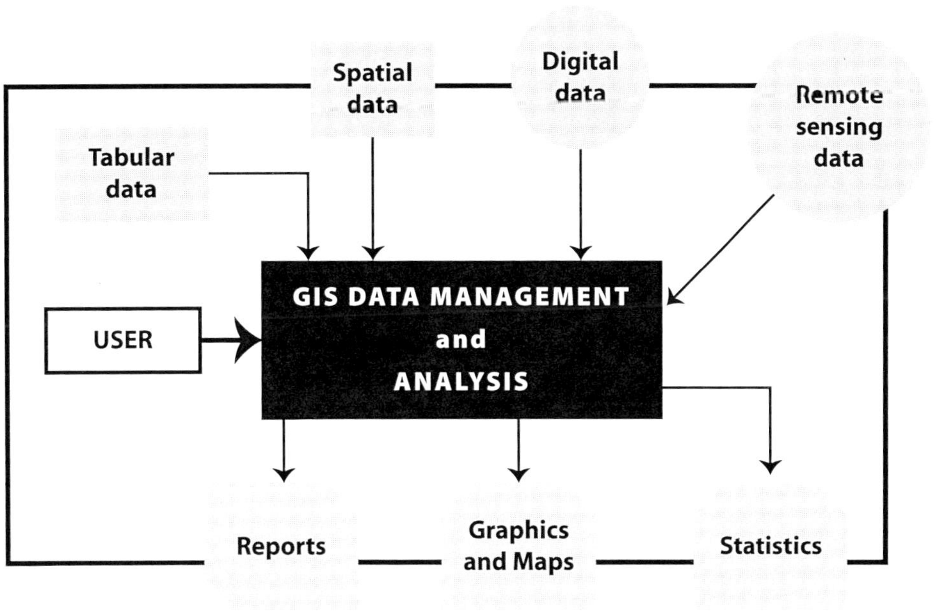

Figure 4-1 Geographic information systems can integrate resource data.

Adapted from Short and Burnham 1982, Avery and Berlin 1992.

At the most basic level, a GIS is a computerized means of making base maps with overlays. Foresters and forest biologists appreciate the value of overlays in decision-making because of the many hours spent using colored pencils to code habitat characteristics, delineate hunt unit boundaries, and locate water developments on acetate film. GIS analysis of spatial data increases the types and accuracies of analyses. Overlays present a simultaneous visual interpretation of the several factors involved that would otherwise be difficult to explain in written or oral form.

GIS maps are built from data transferred to a computer electronically by tracing on the map with a handheld digitizer that creates overlays; this requires but a fraction of the time expended using the manual method. A digitizer converts line data to points that can be connected and later converted back to lines on a map. One major advantage of a GIS is that revisions can be made without revising the entire map. Another advantage is that a combination of overlays can be used to show where characteristics coincide.

However, the production of maps uses only part of the capabilities of a GIS. When a GIS is coupled with a Database Management System (DBMS) that contains detailed information and relates it to items on the overlays, the total system becomes a powerful management tool. For example, a base map of boundaries of a group of timber stands is prepared by GIS. In an associated DBMS, the characteristics of the stands are recorded (species, acres, tree density, quadratic mean diameter, etc.). The DBMS also contains another file with quantified data on habitat conditions for a given wildlife species.

Knowing that the stands and habitat characteristics have been defined, overlays can then be plotted that identify stands with the defined habitat characteristics. Layers containing physical features, such as roads, springs, and lakes, can be used to identify key habitat corridors and obstacles to migration and distribution routes. If the DBMS also contains a file describing tree growth and yield, then stands can be manipulated in the computer to model habitat conditions for future dates and alternative management practices. Green et al. (2017) thoroughly describe the relationship between imagery and GIS and how they can be combined into a powerful management tool.

The value of a GIS is no more than the data residing in it. Remote sensing provides important data but these data are incomplete. Acquiring field data is the next step in the creation of a database suitable for managing forests. The high cost of collecting field data puts a high priority on designing an efficient collection process. In other words, a priority exists for good inventory design.

4-2 Inventory Design

Planning an efficient and effective inventory requires a strong understanding of advanced statistics. However, land managers usually do not design inventories; this task is appropriately left to biostatisticians who have advanced expertise (and often have earned a PhD degree) in the area. Instead, land managers need to understand basic statistical principles so that the inventory makes sense to them and so they can explain the basic approach to others inside and outside their organization. Because of this need, the next section covers basic statistics. Although the basics of much of the following discussion is described in more detail in both Harold Burkhart, Thomas Avery, and Bronson Bullock (2019) and in Frank Freese (1962), a short journey into statistics will help ground our discussion about sampling.

Basic Statistics

In forestry and wildlife management, managers often speak about completing an "inventory" of some attribute of the ecosystem, e.g., trees, herbage, or perhaps animals. However, these inventories differ considerably from the inventory completed in retailing or manufacturing. For example, in an appliance store, taking inventory would entail going into the warehouse and counting the number of washing machines or refrigerators stored there. In manufacturing, we might track through an inventory, the gallons of high fructose corn syrup or the pallets full of bags of chocolate. In these examples, we have identified and recorded the entire population of interest.

Accuracy, Precision, and Bias. In natural resources, we seldom have the luxury of time and resources to complete such inventories. Just think about how long it would take to count just the number of trees on a 100,000 acre forested tract in New Hampshire, let alone measure any attributes of those trees, such as height and dbh. Instead, natural resource managers rely on sampling the population and using those samples to estimate the population parameters of interest. Understanding these two very different uses of the word inventory can prevent much confusion and misunderstanding.

Sampling and estimating requires the use of statistics—a brief primer of which we present here along with other concepts of note. The first statistical concepts we will cover: **accuracy** and **precision**. Although often used interchangeably, these two closely related concepts differ importantly. Accuracy describes the closeness of an estimate to the true value. Precision, on the other hand, describes how close different samples are to each other. Sampling protocols are established to maximize accuracy *and* precision, given finite resources. A key to accomplishing this goal is to eliminate (or at least minimize) bias. **Bias** is a consistent and systematic distortion in measurement due to actions such as the use of wrongly calibrated instruments to flawed sampling procedures (Burkhart et al. 2019). Thus, a sample may produce an estimate that has both high accuracy and high precision, which of course has no bias.

Other combinations of accuracy, precision, and bias exist. (Figure 4-2). Of the different undesirable potential results, a sample that produces a precise but inaccurate estimate (in other words, has a high bias) is probably the easiest situation to remedy as sources of the bias are generally easy to find and correct. The most difficult situation arises with inaccurate and imprecise estimates. In this case, the entire approach may have flaws at a magnitude that greatly reduces the value of any of the data and where a remedy may necessitate a re-collection of data. Given the cost of acquiring the field data used in natural resource planning and management, such an outcome can greatly (negatively) affect budgets, professional reputation, planning, and management efforts.

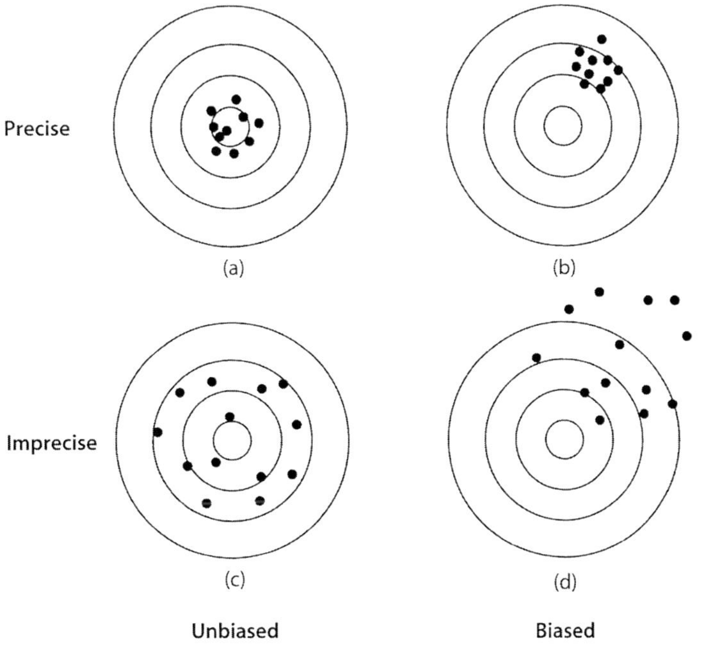

Figure 4-2 Accuracy and precision.

From Burkhart et al. 2019.

MEAN AND VARIANCE. Also known as the "average," the **mean** of a set of data is simply the total value of a set of items divided by the number of items. For example, consider the following tally of the dbh of 26 trees:

Tallied dbh (inches)

8	10	9	7	7
8	8	7	8	8
5	5	5	8	9
10	7	8	10	8
9	9	7	9	7
8				

Use the following equation to calculate the mean:

$$\bar{x} = (\Sigma x) / n$$

Where $\sum x$ = sum of all x values in the sample
n = number of samples

In our case, $\sum x = 8 + 8 + 5 + 10 \ldots 9 + 8 + 7 = 204$ and n = 26

Therefore, the mean dbh ($\bar{x}$) equals 204/26 = 7.85 inches
But knowing a single point estimate, the mean, can mislead as it provides no indication of the range of the data. For example, consider the following set of four tree heights (in feet) that has an average of 95 feet:

80, 40, 110, and 150

Now consider a second set of four trees, also with an average of 95 feet:

96, 87, 105, and 92

Clearly, although the two sets of trees have the same mean value, the variability of the tree heights within the sets differs considerably, with set one having a much wider range than set two. Using the mean, therefore, can mislead and thus a measure of the dispersion in the values from a sample is extremely important. Two measures of this dispersion are the **variance** (s^2) and the **standard deviation** (SD or s). The standard deviation is the square root of the variance and has the advantage of being in the same units as the mean. In addition, we use the standard deviation to calculate the variability of a sample relative to its mean (discussed below).

In words, the variance of a sample measures the square of the average difference between the mean of a sample and each individual sample value. Mathematically you can calculate the standard deviation of a sample using either one of the following two equations (although the first equation is easier):

$$s = \sqrt{\frac{\Sigma x^2 - (\Sigma x)^2 / n}{n-1}}$$

This is equivalent to the formula

$$s = \sqrt{\frac{\Sigma(x-\bar{x})^2}{n-1}}$$

Using the first equation, the variance of our example is:

$$s = \sqrt{\frac{1650-(204)^2/26}{25}} = \sqrt{1.98} = 1.41 \text{ in.}$$

In statistical terms, the larger the variance or standard deviation, the more variable the sample. However, the standard deviation does not allow us to compare the variability of two different samples when samples use different units of measure. For example, which sample has the greater variability: a stand with a mean of 3580 cubic feet per acre and a standard deviation of 644 cubic feet per acre or our 26 tree sample of trees having a mean dbh of 7.85 inches and a standard deviation of 1.41 inches? A statistical measure known as the **coefficient of variation** (CV) will help us answer this question.
The coefficient of variation measures the relative variability of a sample, expressed as a percent. For our dbh example, the CV equals:

$$CV = \frac{s}{\bar{x}}(100) = \frac{1.41}{7.85}(100) = 18\%$$

The CV for our volume comparator equals:

$$(644/3580)(100) = 18\%$$

Therefore, even though the magnitudes of the numbers differ greatly, these two samples have the same relative variability.

CORRELATION AND CAUSATION. Before we leave basic statistics, a short discussion of two other statistical concepts is in order: correlation and causation. Causation refers to a way of connecting. Correlation indicates that two (or more) variables have a relationship. Two types of correlations exist: positive (one variable increases as the other increases) and negative (one variable decreases as the other increases). We can measure this relationship using a statistic called the "correlation coefficient" that has a value that ranges from -1 (a perfect negative correlation) to +1 (a perfect positive correlation). However, a correlation does not indicate a causal link between the variables. In other words, you can have two variables that have correlation of +1 (perfectly positive) but have no causal relationship. Correlations prove valuable in modeling phenomena, but take care to not confuse the two concepts.

Inventory Plots

Plots are the most common system in use for all types of sampling of vegetation or animals. Selecting the number, size, and type of plot is related to the need to have information to make management decisions.

TYPES OF PLOTS. Four types of sample systems can be used in vegetation inventory:

- variable radius plot,
- point,
- fixed radius plot, and
- transect.

Variable radius plotting is used primarily for trees. This approach employs the sampling technique of Probability Proportional to Size (PPS). In other words, the larger the object of interest, the larger the chances are for that object to be selected. As a result, this system is not appropriate for developing stand structure, which requires inventory of all sizes of trees. The sample is obtained by using a calibrated prism to decide whether a tree is to be counted. This type of sampling is fast and accurate if the sample is sufficiently large to overcome variance. Points are commonly used for sampling herbaceous plants and grasses at given distances along a permanently established line. Only the species under the point is recorded. Fixed radius (i.e., circular) plots are probably the most common system in use for all types of sampling of vegetation or animals. Fixed radius plots have dimensions that facilitate the calculation of the density of organisms (e.g., trees per acre). The final type of sample system discussed here are transects. Five types of transects exist:

- a graduated line for use with points,
- a long line with a width of inches or feet,
- a line with plots of any size or shape at intervals along the line,
- a line intercept to select trees to measure, and
- a transect with nested plots.

Line intercept sampling works well in dense shrubby life-form trees (such as pinyon and juniper in the West). Using this method, trees are selected for measurement when a branch crosses the line (Fox and Raskob 1992). Given the many combinations of sample systems, sufficient time must be taken to determine the correct system for the task.

Vegetation sampling to measure the attributes of different life-forms of trees, shrubs, and herbs can be done using nested plots. Nesting means that a large plot may enclose a smaller plot inside (Figure 4-3). Nested plots are advantageous because they incorporate different plot sizes at one location for sampling different factors. The center of the large plot may or may not serve as the center for smaller plots. Smaller plots can be systematically located within the larger plot. An alternative to nesting is to cluster the subplots around the outside of the primary (largest) plot. The efficiency here stems from minimizing travel time to multiple plots.

SIZE AND SHAPE. Sampling vegetation in habitat surveys data is often collected from within circular plots on a transect. Circular plots have proved satisfactory in sampling because of their ease in use when deciding what is to be counted if an organism is partially in a plot. This is because the length of the circumference of a circle is smaller than the length of the sides of a square of the same size, thus requiring fewer border decisions. Circular plots are also easier to establish from a single point on the ground because it is one point in a circle and not four points for a square plot.

The plot size selected for sampling vegetation must be based on the spatial characteristics of the vegetation. In general, the radius for the sample plot is equal to the average distance between the items to be measured. A species area curve can be determined by completing a nested plot study where the number of species are counted on different plot sizes until no more new species are added and this plot size is the one selected. For habitat inventory work a 0.10 acre circular plot is adequate for trees, a 0.01 acre for shrubs (Table 4-2). A similar way to determine plot size is to once again complete a nested plot study, and then construct a density area curve (a modification of the species area curve) (Mosby 1963). For trees, four or five different plot sizes are nested with trees counted on each plot. The average of a sample of individual plot sizes is then expanded to a per unit basis and the densities compared. The smallest plot having the density closest to that of the largest plot is the most efficient.

However, other plot sizes have advantages for forbs and grasses because the weight in grams clipped from the plot can be directly converted to pounds per acre. For example, the weight of 1 gram of material from a 0.96 square foot plot (6.63 inch radius) is equivalent to 10 pounds per acre. The weight of material from the small plot (20.97 inch radius) in Figure 4-3 would convert to 100 pounds per acre. These two plot sizes are

Large =	0.1 ac	Trees, logs, stumps, etc.
Medium =	.01 ac	Shrubs, small trees, wildlife sign
Small =	9.6 ft^2	Herbaceous vegetation

Figure 4-3 Example of nested plots useful in habitat inventory.

Table 4-2 Circular plot radii and corresponding acres for sampling vegetation.

Size of plot (acre)	**Radius: Feet**	**Radius: Inches**
0.001	3	8.7
0.01	11	9.3
0.025	18	7.4
0.05	26	4.0
0.10	37	2.8
0.25	58	10.5
0.50	83	3.2
1.00	117	9.0

commonly used in estimating forage—species of browse, forbs, and grasses eaten by a certain animal—in allotment analysis in range management.

Sampling

RATIONALE. As discussed earlier in this section, vegetation-based inventories sample rather than enumerate the population of interest. Therefore, the decisions made about an approach to sampling have incredible importance in determining the efficiency—a combination of cost and producing unbiased population estimates. Multiple approaches exist for choosing a method for selecting samples.

APPROACHES. Overall, our resources and the information we need drive our approach, and the inventory design also depends to a large extent on the characteristics of the organisms we wish to measure. A good sample design can maximize the value of our inventory. Four variables help define the design:

- plot type,
- plot size,
- plot distribution, and
- plot number.

Deciding on the sampling design requires the careful consideration of all four of these variables simultaneously as they do have causal relationships. For sampling in forests, circular 1/10th acre plots are extremely common. This size and shape has many advantages, including that such plots are intuitively understandable and can be easily explained. After deciding on plot type and size, the decisions about number and distribution are left.

Once the sizes of plots have been selected the number of plots needed for a statistically valid inventory can be determined. The easiest approach is to make a pre-inventory sample and to estimate numbers using the variance formula (Snedecor 1956):

$$N = S^2 \times t^2 / d^2$$

Where N = number of samples required
S = standard deviation
t = confidence interval at given degrees of freedom
d = margin of error (arithmetic mean times designated accuracy)

For example, when counting browse plants on a pre-sample of 10 plots, $S = 6$, $t = 2.262$ (9 degrees of freedom at 0.95 confidence), and $d = 1.5$ (mean = 15, accuracy is designated at 0.10) then:

$$N = (36 \times 5.11) / 2.25 = 81.76 = 82$$

The number of sample plots needed as computed from the pre-sample data totals 82. There is a restriction in using the formula that the data must follow a normal distribution. The number of sample plots depends entirely on the variance of the item being sampled, but in many cases the amount of money available to do the job will be the deciding factor. Statistical techniques based on the cost of establishing plots as a factor in determining how many to establish are available. However, a general rule is that a sample size from 2 to 5% of the area is a minimum.

SCHEME. Random samples are the best of all possible choices because they eliminate bias. However, such sampling is often difficult to do in the field because of problems in transferring a sample drawn from locations on a map to finding this location on the ground. To select a random sample for inventory, a dot grid is placed over the standard map or on aerial photographs and each dot then becomes a sample point. Numbers are then drawn from a table of random numbers to select the points to sample.

A systematic sample is more convenient to execute in the field and most statisticians accept it with constraints, such as adding a random start component. The general procedure for a systematic sample is to randomly select starting points along the border of the area to be sampled and then locate the circular plots systematically at a given distance along parallel transects.

STRATIFICATION. One of the easiest ways to reduce variation in any sample scheme is to stratify the area into like units. For example, if two vegetation types are present in an area, statistics can be calculated for each type separately rather than including both as one set of sample data. Stratification of vegetation types can be done with the aid of aerial photographs and other imagery whereby the strata are identified based on one or more characteristics, such as vegetation types, species composition, physical size of trees in stands, administrative units, etc. The goal is to create strata where the variations in the characteristics of interest *within* the strata are less than the variations in these characteristics *between* strata. To this end, Lund et al. (1978, pg. 392) states that characteristics should:

- logically be related to the information that managers want sampled,
- exist in nature or have been artificially determined,
- represent a homogenous condition that can be defined and is easily recognizable, and
- have a definition that conforms to accepted standards.

For more detail, see *Forest Measurements* (Burkhart et al. 2019).

4-3 Information to Collect

There have been many attempts to collect wildlife data in multi-resource inventories in the past but the rate of success has been variable. Estimating the density of animals, including amphibians, birds, mammals, and reptiles, cannot in most cases be done on plots established for a multi-resource inventory because each animal group possesses unique behavioral traits requiring different sampling procedures.

An integrated, multipurpose approach works well when the objective is to collect habitat information applicable to many species, such as the number of dead trees, stumps, and rotten logs used for cover. The information to be collected in most inventories can generally be grouped into one of three categories. These categories do not contain the entire domain of items to be listed or sampled but are intended as a preliminary list from which to begin selection. Information for specific areas can be added to it.

Vegetation

- Dominant overstory species, basal area, percent crown cover, dbh, trees per acre, and degree of grouping. Crown cover is the area of crown of trees or shrubs measured as a percent of a plot.
- Understory measurements for shrubs, small trees, and herbaceous plants. Browse plant count, including form and age class, or a measure of basal area or crown density.
- Forage production estimate. Clip and weigh material from small plots. Separate by grass and forbs.
- The number of vegetation layers present and a measure of the amount of live foliage in each layer.

Physical Features

- Rock areas
- Cliffs
- Seeps and springs
- Caves
- Lakes, ponds, and streams
- Rainfall patterns
- Slope and aspect
- Topographic features (hills, valleys, etc.)
- Burned areas by date and acreage
- Hunt or management units
- Cultural features of roads, trails, buildings, campgrounds, and recreation areas

Signs of Wildlife Presence

- Fecal material or other signs (tracks, feathers, etc.)
- Tree nests (squirrels and birds)
- Ground nests
- Hollow trees
- Rodent mounds and tunnels
- Roost trees (wild turkey, pigeons)
- Squirrel middens (caches)
- Ground dens (rabbit, bear, badger, coyote)
- Snags (cavity-nesting birds and mammals)
- Elk wallows, deer rubs

4-4 Maps for Management

After habitat data has been collected, the next step is to consider how the information is to be analyzed and displayed. The availability of GIS increases the capacity to produce maps, graphs, histograms, and other data displays. The increased capacity comes with the danger of creating too many displays or displays in an awkward or uninformative format. Therefore, care should be taken to ensure that graphics well illustrate and support decision-making. However, probably the most useful way to transfer information is by maps. Habitat data collected on sample plots can be used to develop three types of maps: food, cover, and water. These maps should show at least the following items:

Food Map

- Shrub areas by species and acreage. Form and age class of woody plants.
- Isograms of forage production in pounds per acre based on data collected from sample plots. An isogram is a line connecting points of equal quantity, similar to an elevation contour on a map.
- Locations of big trees that are likely mast producers. Mast is all fruits, nuts, and berries.
- Concentrations of plant species that produce berries.
- Natural openings and wet meadow areas by acreage and forage production.

Cover Map

- Plant reproduction thickets by species and acreage.
- Location of caves, den trees, snags, and rock outcrops.
- Dominant overstory species by acreage. Information can be drawn from timber type maps.
- Roost trees and special features such as bear dens.

Water Map

- All streams (both perennial and intermittent).
- Location of current or potential pollution sources.
- All springs, wells, and permanent water bodies.
- Isograms of the zones of influence around all water sources. Water deficient areas will become apparent.
- Important amphibian habitats (moist areas, seeps, and springs).

These maps can be made as overlays on aerial photographs or existing topographic maps of suitable scale or, even better, the data can be digitized and incorporated into a GIS and DBMS.

4-5 Who Is Where?

Without doubt the most difficult job of any wildlife manager is to determine the presence or number of animals on a given unit of land. Until the details of an animal population are accurately known, the management of a unit lacks any clear direction. Numbers or an indication of abundance are needed to evaluate carrying capacity (discussed at length in Chapter 6), to determine the allowable harvest of game animals, to establish viable populations, and to determine the dynamics of the population.

A scale factor controls the estimation of animal populations just as it controls the making of habitat inventory maps. Recall that scale refers to the degree of resolution needed to meet management objectives. For example, a low-intensity habitat inventory generally requires the same intensity of sampling for animals. The same relationship is true of high-intensity surveys. There are two general ways of accounting for the presence of animals: directly by estimating their numbers or indirectly by identifying some physical evidence.

Documenting Presence

Direct Methods. A direct method depends upon making a true census, which is a count of all the animals in an area or an estimate of the number by sampling. A manager seldom has the time and personnel to do a complete census. Animals that depend on a territory and can be easily seen or heard, birds for example, can be censused by mapping a defended area. Aerial counts work well for large animals—such as deer, elk, and antelope—when in open habitats or for species that flock. Aerial photographs, from which animal numbers can later be checked against observations, should augment counts for flocks.

One direct way to estimate animal numbers is to use sample plots and mark-recapture methods. Formulas for estimating animal numbers are based on the equation known as the Peterson or Lincoln Index. The Lincoln Index is a good place to start when planning a survey that requires estimating small mammal populations because it is a simple ratio of marked to unmarked animals in a given time. An example is from trapping tree squirrels using the Lincoln Index formula:

$$M/N = m/n \text{ or } N = nM/m$$

Where M = number marked during the 1st trapping period
N = population estimate
m = number of marked animals recaptured in the 2nd trapping period
n = number of animals captured in the 2nd trapping period

35 squirrels were trapped in the first trapping period, and 20 marked and 5 unmarked (25 total) in the second trapping period. The population estimate for these data is:

$$N = 25 \times 35 \;/\; 20 = 44$$

All population estimation methods have advantages, disadvantages, and assumptions that must be considered when selecting the one method best suited to the objectives of the survey. Direct methods are only used when there is a need for a high level of resolution, such as for setting hunting seasons or for associating numbers with habitat factors for research purposes. Estimating numbers of wildlife using direct techniques is seldom done because arriving at a population estimate is time consuming, expensive, and the accuracy of the data is often not at a level that provides confidence in decision-making.

INDIRECT METHODS. Indirect methods rely on indexes developed from animal signs, such as tracks, fecal droppings, nests, or calls. One common technique is fecal (pellet) counts for deer, which commonly defecate a fixed number of times a day. The number of droppings in sample plots is rapidly converted into deer per unit area or deer days use per unit area for a given number of days, ranging from 365 for yearlong use to as few as several weeks if the time of residence (time between counts) can be determined. The pellet count method has been used mostly in the western forest regions to estimate days use per acre for deer, elk, antelope, and cattle. It is not now in common use except to record where an animal is documented to have been present in different vegetation types and stages.

The use of radio **telemetry** and **photography** is becoming more common as space-age technology has provided electronic components that are very small and powerful enough to make transmitters that can be used to track most animals. There are now many choices in selecting electronic devices for tracking animals and documenting their use of habitat. Placing cameras in various areas where animals are thought to live can record their presence there. In using a photographic technique, some animals that have been previously captured and marked can be identified. This technology does come with a high price that may not make it practical for common use. When using photography, the options include time lapse, where a photo is taken at given intervals, or motion detection camera when an animal is present.

Another way to identify the presence of animals is to use **road counts** where the road is considered a transect that passes through one or several forest types. Observations can be continuous or at stops made at a given time interval. To strengthen the results, observations should be made several times a year. Road transects are especially useful for documenting the presence of grazing animals, such as deer and elk, but black bears and turkeys may also be recorded.

In addition to road counts, personnel assigned to a specific administrative unit such as a district can have small note pads in their backpacks to create **field records** of wildlife observations in their daily routine. The field records can be turned in to their administrative office once a month. While road counts and field observations may not seem to be a worthwhile effort, these observations can accumulate over time, providing insight into opportunities for habitat management projects. The records often become an animal checklist for their administrative unit.

A **checklist** is an accounting of animals and plants that occur in a specific administrative unit and is the first resource component needed in preparing a management plan. The Renewable Resources Planning Act (RPA) of 1974 required the Forest Service to make an inventory of present and potential renewable resources every ten years. To meet the needs of an inventory, considerable time was required to identify plants and animals that were present in National Forests within regional administrative units. One by-product of the first RPA assessment was a checklist for animals and plants that became an updatable file for continuing use.

4-6 Evaluating Habitat

Habitat evaluation is an assessment of the "health" of food, cover, and space factors as they affect a single species or a group of species using the same resources. Before an assessment of the health of a habitat can be undertaken, a scale factor specifying the detail and levels of resolution must be determined. When the scale is decided, any number of the habitat evaluation techniques that have been developed over a long period of time can generally be used to make an assessment against set standards.

A basic problem in managing wildlife is a lack of ways to determine the effects of environmental change, both natural and human induced, on wildlife populations. Once effects are known and quantified, they can be used to evaluate land treatment alternatives. One way to approach the problem is through habitat capability models. These models are based on the premise that wildlife populations are a function of their habitat, and habitat quality in part determines population density. The need to develop techniques for determining the present or future wildlife habitat conditions is becoming more important as forest managers develop land-use plans. The need is there because vegetation is constantly being changed to benefit humans, but often at the expense of wildlife species, which need it for food and/or cover.

Habitat Capability

The ultimate tool for evaluating habitat is a quantified system relating animal populations to different combinations of food and cover for each ecosystem: oak-hickory, northern hardwoods, and so forth. Capability refers to the potential to maintain resources of food and cover. Habitat capability models are not without problems. One criticism is that by basing population density only on quality of habitat, other factors influencing survival and reproduction are not represented. These factors are predators, hazards, disease, competition within and between species for resources, and inherited characteristics, such as behavior (for example, territorial).

This shortcoming can in part be remedied by common sense. In habitat models, vegetation is assumed to be the most important environmental factor controlling populations. By deduction, populations at some level can be related to habitat quality. For example, the Abert's squirrel needs trees, but how many? One tree in the middle of a large clearcut is not adequate nor are two, three, or four. Obviously, such conditions are poor squirrel habitat. On the other hand, 10 acres with 200 trees per acre at an average diameter of 11 inches might be enough to maintain a viable population and so is something better than poor habitat. There is no doubt that other environmental factors, not just resources (food and cover), affect a wild animal population, but measuring and quantifying the effects has not been a very practical approach for managers.

A Habitat Evaluation Procedure (HEP) to evaluate the requirements of selected species has been developed for use by the Fish and Wildlife Service (USDI 1980). Once the species has been selected, then a habitat suitability index, based on research data and expert opinion, is calculated to compare differences between two or more areas at the same time or the relative value of the same area now and in the future. The assumption is that a habitat suitability index (HSI), a numeric value summarizing habitat suitability based on habitat quality with a decimal value of 0.0 to 1.0, can be developed for the selected species. One feature of the HSI is a series of equations that provide the quantitative value as:

$$HSI = [V_1 + V_2 + V_3]/3 \quad \text{Compensatory Model}$$

$$HSI = [2V_1 + V_2 + V_3]/4 \quad \text{Weighted Mean}$$

$$HSI = [V_1 \times V_2 \times V_3]^{1/3} \quad \text{Geometric Mean}$$

Each of the factors (V_1, etc.) pertains to some habitat component, such as forage, escape cover, or water, with a decimal rating. A number of habitat suitability indexes have been developed for both terrestrial and aquatic habitat and can be obtained from the U.S. Fish and Wildlife Service upon request. Quantified habitat models are very useful but also very expensive and time consuming to develop and validate sufficiently for management purposes. For this reason, ecosystem managers and biologists presently accept something less than this ultimate tool to get the habitat evaluation job done.

Condition and Trend

The terms condition and trend are well established in the range literature and have become part of wildlife management terminology as well, primarily for game animals and particularly for the food element in resources. Condition refers to present forage composition and production in relation to the kind and amount a site can produce under either accepted management practices or climax conditions (Holechek et al. 1989). The objective of a habitat condition survey is to determine whether the trend in plant development is up, down, or stable by predetermined criteria. Because of the complexity of plant communities, no single factor can be used for determining habitat condition. A host of interacting factors affect plant growth. Further, these factors never remain constant. These characteristics make it difficult to classify entire plant communities consisting of numerous species responding in different ways through time to combinations of factors.

Several methods of estimating wildlife habitat condition and trend based on key browse plants in selected areas have been developed (Dasmann 1948; Dasmann 1951; Cole 1958; Patton and Hall 1966). These methods evaluate use of only key browse species (preferred plants) and do not consider the contribution of herbaceous vegetation or other factors such as litter or soil stability. The assumption is that the condition of the key browse species is an indication of the state of all factors contributing to habitat condition and trend. However, the trend in habitat condition could be downward before the shift is reflected in key browse species alone. Factors determining habitat condition and trend are:

- plant species composition,
- soil properties,
- herbage production,
- degree of plant use, and
- available moisture.

Plant species composition is important in that there is considerable variation in forage value among and preference for different species. Also, nutritional requirements of big game animals are better met by a mixture of species. Therefore, the better habitats are those providing a variety of preferred forage species. The reasons for including an evaluation of soils, particularly erosion, is obvious—it is the medium from which terrestrial plants obtain nutrients. Herbage production depends on plant density, vigor, and soil moisture. Herbage is all species of browse, forbs, and grasses. All vegetation present must be considered, even though certain plant species are not important constituents of the diet of the targeted wildlife species. In contrast to herbage, forage constitutes only the species eaten by a given animal. Soil moisture is the amount of water between soil particles and is measured by a moisture probe that indicates water content or from samples using color and texture as indicators.

Methods of evaluating habitat condition based on ecological principles consider plant composition and use as the major criterion for judgment. Thus, a plant inventory list for the area being evaluated is a necessity. An essential feature of the inventory is the classification of species grouped by their successional position—increasers, decreasers, or invaders (Table 4-3).

These terms refer to the ways plants react to grazing pressure by ungulates (Figure 4-4) and are species specific. The terminology was developed in range management but has been extended to include wildlife species of deer, elk, and others.

Table 4-3 Some common terms used to indicate plant value and condition of vegetation.

Condition A	Condition B	Condition C
Decreasers	Increasers	Invaders
Preferred	Stable	Low Value
Desirable	Intermediate	Least Desirable
Choice	Fair	Unimportant

Increasers are plants that increase in numbers with increasing grazing pressure because other, more desirable plants are consumed first. The opposite is true for decreasers; with increasing grazing pressure, these plants decrease in numbers. Invaders are plants that result from increased grazing but were not part of the original climax vegetation (native). Each condition class is related to successional climax but it should be recognized that climax vegetation is not always the most desirable food or cover on a given site for a given wildlife species. For this reason, certain preferred or desirable plant species may indicate equal or better wildlife habitat condition than climax species.

One technique used for evaluating condition and trend for browsing animals is to estimate the degree of use of plant annual growth. Various types of transects or plots can be established and the number of twigs browsed compared to the number not browsed provides a utilization factor (percent) that is compared from year to year (Shafer 1963, Stickney 1966). Another approach is to enclose browse plants in small, protected plots to compare annual growth on similar plots not protected from grazing animals. This same procedure can be used for herbaceous species.

In addition to plant composition and abundance, there are other ways to determine habitat condition; the use of indicators is one. Indicators are only as

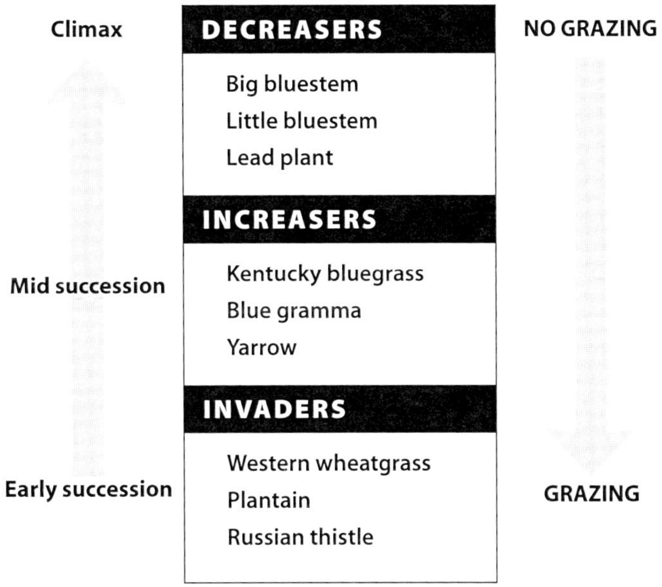

Figure 4-4 Successional changes related to grazing on the North American Prairie.

Adapted from Weaver 1954.

useful as the accuracy with which they reflect a cause-effect relationship. When applied to a specific plant, the terms increaser, decreaser, or invader are the indicator type. Another indicator often used is animal condition. Since animals are a product of their environment, animal condition should also be an indicator of habitat condition (cause-effect situation). Methods used to assess animal condition include indicators of productivity, such as offspring (demographic vigor), weight, and blood chemistry. One of the problems in using animal condition is that by the time the animal's condition begins to decline the decline in the habitat is already well advanced.

Habitat is often evaluated by using a score card, which is another form of habitat model. Score cards use present conditions and relate them to a described standard. For example, a score card using a rating from 1 to 5 indicates the physical conditions in ponderosa pine that is habitat for Abert's squirrel (Patton 1985). Because this squirrel is so closely associated with a tree species of major timber value, its welfare depends on managers being able to design timber harvests in a way that will maintain adequate habitat for a squirrel to survive and reproduce. Data to develop the score card resulted from an eight-year research project that included measurements of more than 7,000 trees—including 600 trees defined as feed trees, 500 nest trees—and trapping and marking 58 squirrels over six years. As a result of this research, a system was developed using tree density and size to estimate habitat capability based on squirrels per acre. The score card was field tested by wildlife biologists on three National Forests. Comments and suggestions from these biologists resulted in many revisions and field testing to arrive at the score card in Table 4-4. Consider that a timber inventory shows a tree average of 12 inches dbh and a tree density from 121–160 per acre. This combination has a rating of 3 and a value of Good with the capability of having 0.14 squirrels per acre.

Table 4-4 A system for rating Abert's squirrel habitat in Southwestern ponderosa pine in Arizona.

	Average tree diameter (dbh)						
Avg. trees per ac	**8**	**9**	**10**	**11**	**12**	**13–15**	**16–19**
<40	1	1	1	1	1	1	1
41–80	1	1	1	1	1	1[a]	2[a]
81–120	1	1	1	1	2	2[a]	3[a]
121–160	1	1	1	2	3	3	4
161–200	1	1	2	3	4	4	4
>200	1	2	3	4	5	5	5

Habitat capability: **1 (poor)** = 0.02 sq./ac; **2 (fair)** = 0.05 sq./ac; **3 (good)** = 0.14 sq./ac; **4 (very good)** = 0.37 sq./ac; **5 (optimum)** = 0.99 sq./ac.

[a]Add 1 if trees are in groups of 4 to 5.

Another evaluation technique frequently used in the West when there is a conflict in elk, deer, and cattle competition for browse in a grazing allotment is the classification of key shrubby plants (browse) into form and age classes (Table 4-5). Availability is the height of the shrub in relation to how high an animal can reach to obtain leaves and stems. The degree of hedging is a subjective judgement based on experience. Hedging is an estimate of the effect of browsing on the regrowth of stems after they have been severed. Several years of continued use cause the stems to clump.

The classification of browse plants into age classes varies according to the species. Ages are often assigned according to main stem diameter size; for example, young might be in the 1/8 to 1/4 inch class. Plants are considered decadent if more than 25% of their crown is dead. Plants classified by form and age class along a permanent transect are compared with previous measurements usually at three-year intervals to arrive at an evaluation for deer, elk, or cattle use of an area. The exam-

Table 4-5 Classification of browse plants by form and age class.

	Form Class		Age Class
1	All available, little or no hedging	1	S = Seedling
2	All available, moderately hedged	2	Y = Young
3	All available, severely hedged	3	M = Mature
4	Partly available, little or no hedging	4	D = Decadent
5	Partly available, moderately hedged		
6	Partly available, severely hedged		
7	Unavailable		
8	Dead		

Sources: Dasmann 1951, Cole 1958, Patton and Hall 1966.

ples given are just two of the many types of scorecards (habitat models) that have been and are still being used by wildlife biologists to evaluate habitat. While the criteria for each scorecard varies, they are almost always based on a system of total points to arrive at a rating using adjective statements (poor to good, etc.).

Monitoring Conditions

Monitoring is the process of checking, observing, or measuring the outcome of a process or the changes deviating from some specified quantity, such as in the scorecard example described previously. Detecting change has always played a significant role in wildlife management, whether in a low-resolution population analysis or some higher level of measurement aimed at determining habitat condition and/or trend. In recent years monitoring has taken on a new dimension since it has been incorporated into resource legislation and regulation. Managers needing to monitor an area or the outcome of a management plan are confronted with the problems of how much detail is needed to detect change and the cost of obtaining this level of detail. Monitoring is a feedback mechanism involving five steps (Salwasser et al. 1983):

1. Set objectives
2. Identify actions and impacts to be evaluated
3. Collect and analyze data
4. Evaluate results
5. Make corrections in the system

The management goal should be to set in motion a detection network to monitor the resources indirectly through habitat condition and trend. While the techniques to achieve this have already been developed, though not always refined, the means of constructing an integrated system to monitor all forms of wildlife has not. The effort to correct this shortcoming has led to some interesting and fruitful approaches in systems analysis.

The most direct way to detect changes affecting wildlife is to monitor vegetation quantity (acres) and quality (succession) over time. This can be done for each management unit on a map documenting acreage and location of vegetation by successional stage. Stand size and location are the factors monitored over time. Location is as important to monitor as is area size, for it depicts the juxtaposition of food and cover. Jared Verner (1983) has developed a flow chart depicting the relationships among elements of an integrated system for monitoring wildlife populations (Figure 4-5 on the following page). While the boxes cannot identify the predictive tools used in action items to be implemented and goals to be evaluated, getting the subsystems within each of the boxes to work will be a long-term process. The flow chart is, none the less, a realistic approach to solving the problem of an integrated monitoring system.

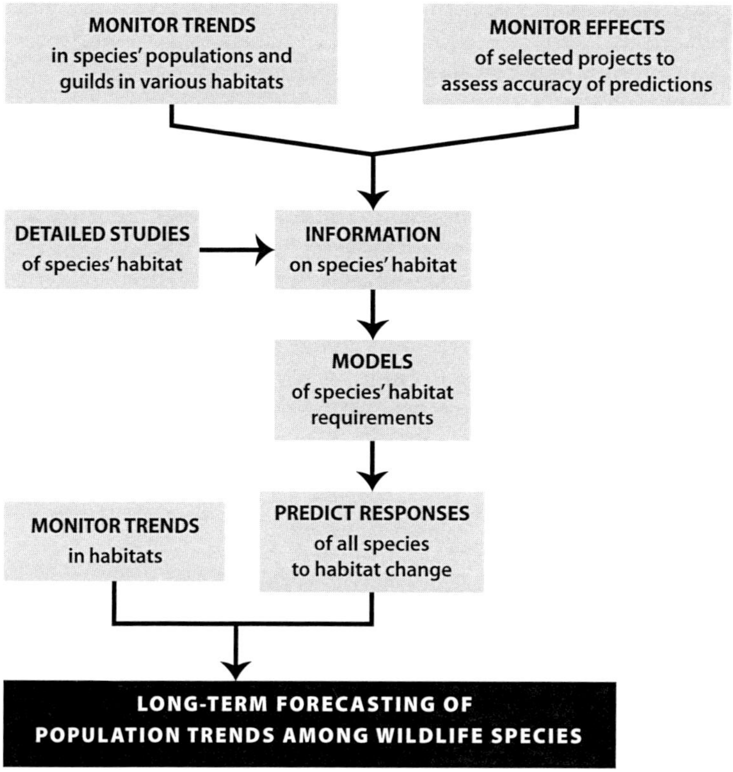

Figure 4-5
A systems approach to monitoring multiple wildlife populations.

Source: Verner 1983.

4-7 Summary

The number of techniques used to inventory wildlife species and habitat has increased so that a manager now has a variety of tools from which to select. This wealth of information necessitates spending more time in planning (discussed further in Chapter 7). Depending on the purpose of the inventory, in general, there are three categories of inventory: national, regional, and local. Any of these three inventories can be conducted at a low, moderate, or high rate of intensity, dependent on the level of resolution needed for decision-making or use. Resolution also pertains to map scale and sampling design—both important components of conducting inventory data collection to determine early in the planning process, usually at the same time as when selecting time, category, and intensity.

Information collected in the inventory will vary depending on management objectives but for forestry and wildlife management, generally consists of vegetation components and structure, physical features of the landscape, wildlife homes and signs, and wildlife observations. In addition to wildlife observations, the inventory may require an estimation of animal numbers. Such estimations may be made using either indirect or direct techniques. Indirect techniques use indexes developed from animal signs while direct techniques necessitate a census involving mark-recapture methods or aerial surveys for estimating a population.

Habitat data collected in an inventory can be used to develop food, cover, and water maps. All three of these types of maps are critical to the decision-making process. Aerial photography and geographic information systems are useful in compiling management maps. The inventory data together with compiled maps and overlays are used to evaluate habitat. Habitat evaluation is an assessment of the health of the habitat components of food, water, and cover based on ecosystem functioning, habitat quality, populations, or condition and trend. Habitat monitoring is the process of checking, observing, or measuring—a process to detect desired changes or unwanted deviation from stated standards. Many of the techniques necessary to monitor habitat and populations already exist and have been deployed, but they require refinement and integration into a coherent system for practical field application.

REFERENCES

Avery T.E., and G.L. Berlin. 1992. *Fundamentals of remote sensing and air photo interpretation*, Fifth edition. MacMillan Publishing Company, New York.

Burkhart, H.E., T.E. Avery, and B.P. Bullock. 2019. *Forest measurements*, Sixth edition. Waveland Press, Long Grove, IL.

Cole, G.F. 1958. Range survey guide. Federal Aid Project W-37-R. Montana Dept. of Game and Fish, Missoula.

Dasmann, W.P. 1948. A critical review of range survey methods and their application to deer range management. *Calif. Fish and Game* 34:189–207.

Dasmann, W.P. 1951. Some deer range survey methods. *Calif. Fish and Game* 37:43–52.

ESRI. 2018. *What is a GIS?* Available at https://www.esri.com/en-us/what-is-gis/overview; accessed July 14, 2018.

Fox, B.E., and P.E. Raskob. 1992. Comparing the efficiency of three inventory sampling methods to determine timber volumes in pinyon-juniper woodlands. *Western Journal of Applied Forestry* 7(4):110–113.

Freese, F. 1962. Elementary forest sampling. Agric. Hdbk. 232. USDA Forest Service, Washington, DC.

Green, K., R.G. Congalton, and M. Tuckman. 2017. *Imagery and GIS: Best practices for extracting information from imagery.* ESRI Press, Redlands, CA.

Holecheck, J.L., R.D. Piper, C.L. Herbel. 1989. *Range management principles and practices.* Prentice-Hall, Inc., Englewood, NJ.

Lund, H.G., V.J. LeBau, P.F. Ffolliott, and W.W. Robinson. 1978. *Integrated inventories of renewable natural resources.* USDA Forest Service Gen. Tech. Rep. RM-GTR-55. Rocky Mountain Research Station Publications, Fort Collins, CO.

Mosby, H.S. (ed.). 1963. *Wildlife investigational techniques.* The Wildlife Society, Washington, DC.

OpenForests. 2018. *Remote sensing data sources and specifications.* Available at https://openforests.com; accessed July 13, 2018.

Patton, D.R. 1985. A model to evaluate Abert squirrel habitat in uneven-aged ponderosa pine. *Wildl. Soc. Bull.* 12(4):408–414.

Patton, D.R., and J.M. Hall. 1966. Evaluating key areas by browse age and form class. *J. Wildl. Manage.* 30:476–480.

Reed, D.D., and G.D. Mroz. 1997. *Resource assessment in forested landscapes.* John Wiley & Sons, New York.

Salwasser, H., C.K. Hamilton, W.B. Krohn, et al. 1983. Monitoring wildlife and fish, mandates and their implications. *Trans. N. Amer. Wildl. and Nat. Resour. Conf.* 48:297–307.

Schrueder, H.T., T.G. Gregorie, and G.B. Wood. 1993. *Sampling methods for multiresource forest inventory.* John Wiley & Sons, New York.
Shafer, E.L. 1963. The twig-count method for measuring hardwood deer browse. *J. Wildl. Manage.* 27:428–437.
Short, H.L. and K. Burnham. 1982. *Technique for structuring wildlife guilds to evaluate impacts on wildlife communities.* Special Science Report Wildlife 244. U.S. Fish and Wildlife Service, Fort Collins, CO.
Snedecor, G.W. 1956. *Statistical methods.* Iowa State Press, Ames.
Steger, T.D. 1986. *Topographic maps.* USDI Geological Survey, Washington, DC.
Stickney, P.F. 1966. Browse utilization based on percentage of twig numbers browsed. *J. Wildl. Manage.* 30:204–206.
USDI. 1980. Habitat as a basis for environmental assessment. ESM 101, Release 4-80, Div. of Ecol. Serv., U.S. Fish and Wildl. Serv. Washington, DC.
Verner, J. 1983. An integrated system for monitoring wildlife on the Sierra National Forest. *Trans. N. Amer. Wildl. and Nat. Resour. Conf.* 48:355–366.
Weaver, J.E. 1954. *North American prairie.* Johnson Publishing Co., Lincoln, NJ.

Doing the Work of Conservation

5

Up to this point, we have addressed the basics of forests and forestry: definitions and classifications of the resources, the big issues, and how to understand what you have and what it is capable of providing over time and space. That is the *study* of the forest—its biology, physiology, and ecology among other things. In this chapter we begin to discuss how managers use that understanding to meet the wide range of landowner objectives and produce the goods and services for generations to come.

Recognizing your objectives is key to successful conservation and management. A single private landowner can likely state multiple and clear objectives for outputs or outcomes from their property; similarly, a company likely has a mission and policy statements, as well as administrative procedures and a company history, which specify management objectives for desired outputs and outcomes. Public lands have a broad set of objectives that fluctuate with time and sometimes conflict. Regardless, matching the objectives to the capability of the lands is crucial and even broad and conflicting objectives can be satisfied over time and space with large land bases.

But the work of conservation is action. Aldo Leopold (1949) aptly states that we ultimately leave our mark on the land with the axe and the shovel. Conservation is about the intent and activities (and sometimes lack of action) that we take to leave that mark during our short tenure here on this earth.

5-1 Forest Conservation Management

Conservation management, forestry in general, and silviculture specifically have evolved with time (Tappeiner et al. 2015; D'Amato et al. 2017). The roots of silviculture can be found in *Principles of Handling Woodlands* when Henry Graves (1911) defined it as "the art of establishing, developing, and reproducing forests." That early focus was strongly on a secure timber base for an expanding nation, but consideration was given for basic sustenance (e.g., fuelwood), forage for domestic stock and wildlife (game species), and water supply. The philosophy of the time was on utilization of a massive and presumed limitless resource; it persisted and dominated through the first half of the twentieth century.

The writings of Leopold, among others, promoted the notion of a land ethic and conservation—the wise use of resources while sustaining the productive capability of the land for generations. Leopold (1949) contrasted the agricultural/utilitarian model with an emerging organic/biotic/ecological model: "We see repeated the same basic paradoxes: man the conqueror *versus* man the biotic citizen; science the sharpener of his sword *versus* science the searchlight of his universe; land the slave and servant *versus* land the collective organism" (pg. 260).

Leopold rejected those paradoxes; he saw that an ethical relationship with the land and all its associated parts, recognizing all their inherent value was the only way to move forward with land management—something well beyond utilization of the resource for economic gains. His perspective set the groundwork for much of our subsequent thoughts about sustainable land management: actions that are economically feasible, ecologically resilient, and socially acceptable over time. It also set the stage for resurgence of the preservation philosophy. The second half of the century saw the Wilderness Act and Endangered Species Act (Chapter 6). These laws, as well as others, and the preservation/environmental proponents behind the movement encoded a flawed premise that land is best left reserved and untouched. Though preservation is an excellent stopgap tool at small temporal and spatial scales to protect a thing, it fails as a long-term strategy over large landscapes because of normal ecological dynamics—much less the huge new challenges of human population expansion, other exotic species and processes, changes in land ownership patterns, and global climate change (D'Amato et al. 2017).

The late twentieth century saw the emergence of "ecosystem management" and the twenty-first century has seen "ecological forestry"—both are the evolution of old forestry and silviculture into a new era (Tappeiner et al. 2015; Franklin et al. 2018). This time period is reversing the federal land management pendulum from preservation back toward conservation. The mind-set recognizes Leopold's vision of doing things that are "right" for the land; actions that sustain the integrity and natural stability (**resistance** and **resilience**) of the land while meeting the needs of the landowners; and it harkens back to Pinchot's founding mission of the Forest Service to provide the greatest good for the greatest number over the longest time. Finally, it speaks to any landowner's dream to someday leave their estate to their children and grandchildren. These priorities make for good silviculture and good forestry.

Ecosystem Management

Long-term sustainability of forest habitats can be accomplished by adopting a position of ecosystem management: examining the whole instead of individual parts. This **holistic approach** proposes that some entities, such as a unit of forest wildlife habitat, have a value independent of and greater than the sum of its parts. The thought is similar to the idea of holistic medicine, which is to treat the whole person instead of separate body organs. It considers a unit of land as a system (e.g., a watershed). The parts of the system are the plants, animals, soil, climate, water, air, topography, and human influences comprising the whole (Figure 1-2). Any given factor in the system can be out of balance, which in turn affects other factors in a cycle of events. The cycle of events over time is the core of ecosystem management that must be understood and in most cases the cycle of events transcends our own ecological longevity. Ecosystem management involves time and space attributes; it is more relevant and acceptable to solving problems now than in the past because of all new information from research and experience that make it possible to apply.

THE TIME ATTRIBUTE. The time attribute encompasses changes in plant and animal species composition, interactions, and densities (Chapter 2). Four measures of time are useful in understanding ecosystems and forest wildlife community changes:

- **Instantaneous time** is used in studying population dynamics, particularly for those species characterized by a high reproduction rate. This measure of time is concerned with rates of change of populations.
- **Present time** refers to time affecting animals in the near term (1–5 years). It is the "here-and-now" time used by managers in most management plans.
- **Ecological time** is associated with the long-term (decades to centuries) successional changes in plant and animal communities and is the time scale most commonly used in describing plant community replacement and stand dynamics.
- **Geological time** provides a framework for understanding the past and ways of forecasting what might happen in the future. Geological time is the tens and hundreds of thousands of years required to influence regional and global changes in topography and climate.

THE SPACE ATTRIBUTE. The space attribute of natural ecosystems is not predetermined because ecosystems are open systems. Boundaries exist because humans have decided they exist and are necessary for characterization and management, but this boundary condition should not become a barrier to understanding and solving management problems. Space within ecosystems exists for plants and animals as they have individual boundaries that can be defined. Ecosystems can be as small as a single leaf on a tree or a fish pond or as large as a watershed or vegetation type, but the whole earth is considered the tertiary ecosystem. Locations and boundaries of ecosystems are important because they become the basis for further disaggregation for classification, inventory, monitoring, and management purposes. The concept, however, focuses on the biological, chemical, and physical processes and not specifically on space or geographical area.

THE PROCESS. There are three aspects of the process that must be considered:

- management and planning are modified to consider management units across larger areas than are currently being reviewed;
- the time perspective is directed to longer management periods; and
- there must be a process of **integrated resource management planning** to develop alternatives for decision makers.

Most natural resource managers think of forest stands as being small, which was acceptable and has worked well in the past to meet management objectives. But small stands, while creating tremendous diversity in *edge*, may also fragment the forest and reduce habitat for some animal species (e.g., those requiring interior forest conditions). Stands of various sizes need to be delineated within the boundaries of watersheds and landscapes in the tens of thousands of acres. While ecosystems can be any size and shape, there are practical limitations to how small and how large an area can be to be managed as a unit.

Natural ecosystems can be classified into five spatial categories (Billings 1970): local ecosystems, formational ecosystems, continental or ocean-basin ecosystems, the earth ecosystem, and the solar ecosystem. Local and formational ecosystems are of the most interest to land managers. Local ecosystems are stands, vegetation and forest types, or portions of those types delineated by topographic features or other definable boundaries. An example would be an Oak-Hickory forest type within a watershed or a forest district. Formational ecosystems are large landscape areas needed to get the mix of goods and services that cannot be provided in smaller units. The Society of American Foresters (SAF) (1993) has suggested the scale of 100,000 to 1 million acres. Management at the landscape scale may mean the redefining of boundaries of forests and management units to conform to natural features, such as watersheds instead of roads, streams, and fences (property lines).

Forest ecosystems and associated wildlife habitat often develop over long time scales, following natural succession as modified by silvicultural practices. Most current planning horizons, however, are based on 5–10 years of here-and-now time. This timescale is far too short when planning for a sustainable forest that requires the ecological time of succession from 60–250 years or longer.

Integrated resource management planning (IRMP) is the process that puts ecosystem management to work at the project level or with groups of spatially or temporally related projects. The importance of the IRMP process is in the collective thoughts of all the interdisciplinary team (ID Team) members to develop management alternatives that are based on the best information available at the time. However, for the IRMP process to work, as one part of maintaining sustainable forest ecosystems, the ID Team should include members outside of the agency that are responsible for the process. In his Policy Letter (1992) on ecosystem management, then U.S. Forest Service chief Dale Robertson asked regional foresters and experiment station directors to:

- focus on desired present and future conditions of the land and its human communities;
- integrate thinking and actions at multiple spatial and temporal scales;
- be especially careful in sensitive areas;
- employ the ecological capabilities and processes of the land;
- get people involved in planning and carrying out project work;
- involve scientists through adaptive management; and
- integrate resource management for operational efficiency.

If a healthy, sustainable forest with diverse wildlife habitat can be attained over a landscape, then probably all the amenity and commodity desires of people will be met. The natural components in such a landscape are linked one with the other in areas sufficiently large enough to serve the public. The human component is linked to the ability of the system to produce and the willingness of people to accept that limits exist and some components, such as wildlife habitat and populations, old forest conditions with large trees, and available timber volume, will change in time and space. This need sets the stage for good silviculture.

5-2 Silviculture and Prescriptions

Modern silviculture is therefore more broadly defined as the art and science of manipulating forest vegetation, and the growing space/conditions for that vegetation over time in order to meet the objectives of the landowner(s). John Tappeiner et al. (2015) lists a range of definitions and Ralph Nyland (2016) most recently and simply defined silviculture as "establishing and maintaining healthy communities of trees and other vegetation that people deem important" (pg. 1). In this book and for our purposes, we will focus on silvicultural practices that mainly promote the development of wildlife habitat—the structure, composition, and processes that sustain populations of animals (see Chapter 6). Those practices can include mechanical, chemical, and burning tools and techniques.

Silvicultural Systems

Managers typically manipulate communities of trees and associated vegetation as a unit or stand rather than individually (see Chapter 1). This is analogous to the efficiencies and advantages of teaching classrooms of students rather than individuals. Like classrooms, we broadly characterize them as either even-aged, where most individuals are within a narrow age range, or uneven-aged. As such, age defines the silvicultural system. **Even-aged** stands of trees have and are managed as a bell-shaped distribution of age and size (diameter, height, crowns, or volume) with a mean or median tree condition and small-to-modest range of variability about that mean (Figure 5-1). A small number of individuals (<10%) can fall outside of that range without major influence on the behavior of the main cohort; the average tree has meaning and the conditions of the stand are tightly linked to that average tree description.

Uneven-aged stands of trees consist of and are managed as a wide range of tree ages coexisting and developing over time, much like a one-room schoolhouse that mixes old and young students together in one environment. The trees therefore have a wide range of diameters, heights, tree crown characteristics, and standing volumes such that averages have no meaning; choosing a tree characteristic to describe the stand is wrong. Diameter distributions are typically described as "inverse-J" since there are more younger/smaller trees than larger trees (Figure 5-1). This declining number with age and size is based on presumed mortality and slowing growth with age/size. Stands are best described by the ranges and arrangements of tree characteristics and conditions.

Two-aged and multi-aged stands are intermediate to these two classic conditions. Two-aged stands are characterized by a small number of relatively uniform overstory trees of one condition, with a relatively uniform understory of another

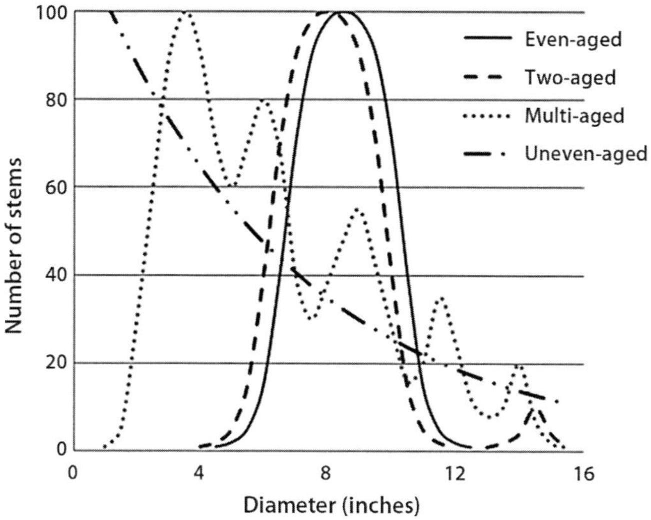

Figure 5-1 Archetypical diameter distributions for four classic age/size structures for stands: even-aged (bell curve), two-aged, multi-aged, and smooth uneven-aged.

Each distribution represents an approximately equal number of stems per acre and amount of basal area per acre.

condition (sizes, species, and vigor). Multi-aged stands are composed of three or more identifiable layers or conditions; these stands grade into irregular inverse-J distributions (Tappeiner et al. 2015).

This continuum of age/size structures is created and maintained by stand dynamics and ecological disturbances over time (Chapter 1). Stand initiation events create even-aged stands, which then persist through stem exclusion as single-aged stands. Silvicultural practices can shorten or lengthen these stages of stand development depending on landowner objectives. An industrial landowner typically wants to shorten the stand initiation phase, driving a fully stocked young plantation quickly to crown closure and stem exclusion to establish a new stand and save years of time on the rotation (and their return on investment). Thereafter, these landowners manipulate tree spacing and stand density to produce the species composition, growth rates, sizes, heights, crown characteristics, and branch sizes desired for a future product at rotation's end. Conversely, a landowner interested in early-seral stand conditions that are rich is surface forage for wildlife would want to slow the pace to crown closure and stem exclusion.

Understory reinitiation creates a second and/or third and/or fourth cohort through regeneration episodes in order to develop two- and (subsequently) multi-aged stands leading to the old-growth structural stage. Similarly, silvicultural tools and techniques can be used to promote or avoid these stages (Tappeiner et al. 2015). Heavy thinning, with some spatial variability in the harvesting (i.e., small openings for wildlife habitat), promotes understory development that includes tree regeneration. Those established individuals can be released with subsequent harvest entries to promote multiple cohorts, multiple canopy layers, and both vertical and horizontal structural diversity. Sporadic heavy entries into a stand foster multi-aged diameter distributions; regular and repeated entries that create only minor disturbances foster smoother, uneven-aged distributions.

Harvests and Regeneration Methods

Regeneration methods correspond well with designated silvicultural systems (Figure 5-2). **Clearcutting** with scattered residual legacies (living and dead) create stand initiation conditions, emulating large-scale and high-intensity disturbances. Larger harvest areas are often referred to as "clearcuts" with smaller harvest units called "patch cuts." Both are even-aged systems if the opening is tracked as one unit. Openings can be any shape but, when the opening is narrow to aid in natural regeneration for surrounding seed trees, it is referred to as "strip cutting." Forest types that naturally regenerate from sprouting (e.g., aspen) following clearcutting is referred to as "coppicing." The conditions created by such heavy disturbance support the rapid regeneration of a single cohort of trees that dominate subsequent stand dynamics.

Plantation forestry is the epitome of these stand dynamics, with management practices that strongly regulate regeneration (rate of establishment, species composition, and spatial pattern through planting) and early competition among individuals and with associated vegetation. Landowner objectives tend to focus on accelerating the stand through stand initiation to crown closure and the beginning of stem exclusion. Hence, a focus is placed on site preparation, planting, and early vegetation management to ensure full stocking and rapid early development.

Seed tree and **shelterwood** regeneration methods mimic clearcutting when/if the overstory residual trees (left for seed production and shelter) are ultimately

SILVICULTURAL SYSTEM	REGENERATION METHOD
Even-aged	Clearcutting/clear felling with minimum reserve trees; patch cuts, strip cuts and coppicing
	Seed tree and shelterwood harvests with traditional overwood removal
	Clearcutting with abundant reserves (individuals and clumps) for habitat structure
Two-aged	"Irregular" seed tree and shelterwood harvests (residual overwood retained)
	Heavy thinning entry with openings and residual dense patches for structural diversity
Multi-aged	Multi-aged group selection harvests
Uneven-aged	Single (or individual) tree selection harvests

Figure 5-2
Silvicultural systems and their corresponding regeneration methods, arranged along a disturbance gradient from heaviest to lightest.

removed after the site has regenerated naturally (Tappeiner et al. 2015). However, a stand becomes two-aged if those residual trees are retained in the long term as legacies for compositional and structural (vertical and horizontal) diversity and overall habitat quality. Though the number of trees retained may be relatively low, their influence on future stand dynamics and the development of regeneration under them can be substantial: reduced growth rates and suppression mortality from shading and competition, physical damage from litter/branch fall, and inoculation with pathogens.

A third harvest entry into a stand that already has two age classes, with the intent to regenerate a third cohort, is typically not referred to as another seed tree or shelterwood harvest. Indeed, with such an objective to develop vertical and horizontal diversity through multiple age classes (and perhaps species composition), it is most appropriate to use uneven- or multi-aged terminology: **single-tree selection** or **group selection** (Tappeiner et al. 2015). These two approaches form a continuum rather than a contrast; selecting individual, single trees for removal from a stand does not preclude marking a small cluster or group of trees to create the residual structure and regeneration environment needed, nor does focusing on removal or retention of groups of trees preclude the marking of single trees. Both approaches leave a partially disturbed forest with sufficient residual structure to meet objectives and the potential to regenerate and grow for future desired conditions. There is no thinning *per se* of a main cohort, and no ultimate rotation end planned. Rather, there are regular entries into the stand to promote long-term continued development of desired composition and structure.

Tending the Stand

Uneven- or multi-aged management approaches inherently provide for most or all manipulation of stand composition and structure over time given regular, lighter harvest entries. The details of these entries can be varied each time to account for unexpected stand dynamics (e.g., regeneration failure) and/or new management issues (e.g., climate change and exotic species). Regardless, the beauty of multi-aged management is the continuous forest cover for all interests and objectives: wildlife habitat, watershed protection, environment, aesthetics, and future wood supply. It does not produce or retain the maximum amount of biomass or carbon for the site, but it is therefore less susceptible to disturbance agents that are attracted to dense, layered, carbon-rich forests (e.g., wildfire). In addition, it creates wonderful flexibility for landowners and managers over time and space.

For even-aged stands, managers often must manipulate species composition and structure (spacing) between the major regeneration events by altering site productivity and stand density. The inherent productive capability of a site can be altered with treatments that make light, nutrients, and moisture more readily available to desired vegetation (e.g., prescribed burning, fertilization, and chemical or mechanical vegetation control). Stand density is typically manipulated through planting, seedling protection, chemical vegetation management, and precommercial/commercial thinning of the overstory as it develops. All these treatments/options have associated impacts (costs and benefits) that must be weighed against landowner objectives (Table 5-1).

Burning is a natural ecological tool with a long history of use (millennia) to manage ecosystems. However, prescribed burning is a relatively blunt and unpredictable instrument, and the general public is averse to smoke impacts. Fertilization and chemical vegetation control are often more precise in their application and therefore have compelling specific applications to land management (Tappeiner et al. 2015); but they are less natural in addressing productivity and density issues, and the general public often questions chemical use in the forest. Mechani-

Table 5-1 Intermediate stand treatment options and their ecosystem impacts, organized by the three broad areas of silvicultural toolbox: prescribed burning and chemical and mechanical treatments.

Intermediate Treatments	Implementation Options	Ecosystem Impacts
Prescribed burning	• Pile burning of accumulated biomass and activity fuel from thinning • Broadcast burning • Pile burning followed by broadcast burning	• Fine surface fuel reduction; increased canopy base height; reduced wildfire hazard • Altered species composition through understory mortality • Nutrient cycling and stimulation of plant and animal diversity • Creation of snags and openings through limited overstory mortality • Increased water availability and watershed production • Stimulation of understory plant communities and forage
Chemical	• Fertilization, typically following thinning	• Stimulation of tree leaf area and stem growth • Understory development in the near term before being shaded out
	• Vegetation Management (understory spraying for weed control)	• Targeted alteration to species composition • Stimulation of tree growth through nutrient and water availability
Mechanical	• Thinning from below, or "low" thinning	• Captured small-tree mortality • Increased canopy base height; wildfire hazard reduction
	• Thinning from above, selection thinning, crown thinning, and heavy free thinning	• Captured self-thinning mortality • Stimulation of residual tree growth and understory • Structural development vertically and horizontally • Promotion of old-growth characteristics over time

cal site/understory treatments can be most targeted in their application but are typically very labor intensive (and expensive) and create collateral and undesirable site disturbances, such as erosion and the introduction of weeds.

The application of regeneration methods and intermediate stand treatments can be strategically packaged in a **silvicultural prescription** to guide management of a site over time and through changes in personnel. Prescriptions can be structured many ways, but the basic components include:

- a thorough site description including physical site characteristics, soils, vegetation communities, management history, and landscape context;
- a concise statement of landowner objectives and internal/external limitations to management;
- anticipated stand development in the absence of treatment, demonstrated through a quantitative model when possible;
- one or more alternative developmental pathways based on a set of proposed silvicultural actions, also as quantitative projections; and
- a summary comparison of respective outcomes and outputs from those alternatives, including the no-action alternative, with costs and benefits detailed.

Silvicultural prescriptions structure the thoughts of a silviculturist, communicate information to colleagues and the decision maker (for a preferred alternative), and provide a basis for monitoring and adaptive management over time.

5-3 Forest Regulation and Scheduling

Just as trees make up a stand, stands make up a forest or ownership. With sound conservation management, the regular and sustainable flow of outcomes and outputs over time can be guaranteed by viewing the whole ownership and scheduling activities strategically. One stand cannot be all things at once—young and old, open and dense, conifer and hardwood, layered and simple. A collection of stands, however, can provide a range of stand conditions that balance landowner objectives (often for a range of conditions) and regulate the production of outcomes and outputs over time and space (Davis et al. 2001).

As a demonstration, assume a large ownership with 60 stands of approximate equal size and productivity. A simple forest management schedule with annual harvests of one stand at the oldest age of 60 provides for regulation of outcomes and outputs in perpetuity. Every age class with all their associated stand dynamics and habitat values (up to age 60) would be represented. Larger ownerships with 180 stands could annually harvest three stands; smaller ownerships with 20 stands could harvest only one stand every three years, depending on finances and work schedules. Few situations are this simple, of course, but the regulated flow philosophy provides a framework for more complex situations and the planning process detailed in Chapter 7.

In reality, all ownerships have a range of stand sizes, histories (ages and density), and productive capabilities depending on current species composition, microclimate, soils, and prevalent disturbance agents (among other things). Therefore, managers must schedule stands proportional to their ability to provide regular outcomes and outputs. Those proportions can change with time depending on stand improvements with ongoing management or unplanned disruptions to the schedule. For example, much of an ownership might lack large snags for habitat value requiring more and larger stands during certain years in the near term until those features can be developed with time and carried from one rotation to the next. Once those stands are more productive, the pattern for regulation can be changed and the total productive capability/expectation of the ownership increased.

Forest regulation and scheduling must be monitored over time, however, for accuracy. Under the best of conditions, it is still only a plan based on assumptions; therefore, the plan needs to be adaptable to an uncertain future because disturbance agents (e.g., fire) can alter the best laid plans. If a wildfire severely burns a third of an ownership, then it creates a disproportionate amount of stand initiation conditions and the schedule of operations (harvests and intermediate treatments) must be adjusted to accommodate a new long-term sustainable flow of outcomes and outputs. Once the new schedule is established, a batch of new silvicultural prescriptions must be developed to efficiently move the new current condition of the stands to a desired future condition and set of stand dynamics.

5-4 The Ever-Present Role of Time and Space

Landscapes and centuries form the largest temporal and spatial scales for conservation management. It is exceedingly difficult to work at those scales given our job tenures and life spans, and the enormity and complexity of natural landscapes with a multitude of existing conditions, landowners, management objectives, and values. However, good conservation management always views "our" actions on "these" lands within a greater context.

Stands vs. Landscapes

Landscape ecology and management recognizes the relationship and adjacency of stands to one another. Large-scale physical patterns (e.g., topography and watersheds, habitat and movement corridors) set the context for smaller-scale activities and, conversely, those smaller-scale changes can have major consequences for larger-scale processes. Beyond the physical and ecological reality of these processes, the public (and perhaps our own) **perception** of land management activities is highly influenced by the landscape context (Franklin et al. 2018).

Diversity and complexity are fundamental to this concept of scale and its impacts on wildlife habitat, as well as a pressing management objective for many landowners (Fahey et al. 2018; Webster et al. 2018). Small-scale diversity and complexity, or **alpha diversity**, is the variability in species composition and vertical/horizontal structure within a stand; **beta diversity** is that variability among adjacent stands that form a condition on a hillside, for example; and **gamma diversity** is all the variability within a landscape. Even-aged stands are relatively simple in terms of age classes, tree sizes, structure, and often species composition—we consider them to have low alpha diversity. Conversely, multi-aged stands have high alpha diversity given the range of conditions within such complex stands. A collection of adjacent stands can be very similar to one another (e.g., all young even-aged stands or all older multi-aged stands) or markedly different from one another—low and high beta diversity, respectively. A collection of stands managed and maintained for different objectives and perhaps including a riparian reserve area and/or an open meadow would have high beta diversity. At the landscape scale, managers can similarly produce and maintain high or low gamma diversity, which is determined by the collection of dozens to hundreds of stands and how different they are among one another. Industrial forest landscapes can have low or high gamma diversity depending on harvest patterns and scheduling; federal forest landscapes can have low or high gamma diversity depending on past human and natural disturbances.

Adaptive Management

Diversity and complexity arise over time just as they do over space, and managers can substitute time and space to meet the accumulation of objectives for a given landscape or ownership (Tappeiner et al. 2015). Just as one stand cannot meet all objectives at one point in time, it cannot meet the same objectives all the time as it progresses through natural stand development, is managed, or is disturbed by forces beyond human control. This reality also speaks to the need for long-term planning and monitoring, which are at the heart of ecosystem management. Indeed, uncertainty fundamentally drives the need for planning (Chapter 7)—and specifically to plan for the unexpected. Land managers do that through the creation and maintenance of diversity across their ownership (Webster et al. 2018). It speaks to the wisdom of not having all of one's eggs in a single basket, and to Leopold's encouragement to "keep every cog and wheel" when we are tinkering with our natural resources to best conserve them over time.

Because of uncertainty about how a land management decision might affect future outcomes and outputs, managers need some process that will allow for adjustments over time. C. S. Holling (1978) and Carl Walters (1986) initially suggested a process now known as **adaptive resource management** and a thesis that additional scientific understanding will come from the experience of management as an ongoing, adaptive, and experimental process. Walters explained that there are three essential ingredients:

- modeling to determine uncertainties (unknowns) and generate alternatives,
- statistical analysis to determine how uncertainties are likely to propagate over time in relation to policy choices, and
- formal optimization combined with game playing to seek better problem solutions.

Ecosystem management is adaptive in character over space and time because land managers understand that knowledge of a complex world is limited and vague (Chapter 4), and uncertainty is the normal result when trying to project future outcomes and outputs. However, those same managers must make informed decisions about resource objectives, initiate the proposed actions, observe the outcomes of the actions, and then make changes that will meet the objectives (Tappeiner et al. 2015). Ecosystem management in many respects is similar to the commitment of some organizations to total quality management (TQM). If ecosystem management is adopted, can it be practiced on the ground the same as in concept?

Current **adaptive management** formalizes the process of ecosystem management—planning, monitoring, and learning as an organization over time. It has four basic tenets:

- specify the management issue and question to be addressed by some new treatment approach or idea;
- implement multiple (three or more) examples, or **replicates**, of that new idea rather than just one;
- keep an untreated control for comparison; and
- record details about the initial treatment and results, as well as monitoring data and information over time.

Borrowed from experimental methodology (and the timeless advance of civilization through science), these four tenets improve the likelihood that current and future managers can learn from their actions, adapt their treatments, and ultimately defend their professional judgements. This may be essential when there are high levels of social uncertainty and political pressure (Franklin et al. 2018).

Planners who claim to use adaptive management need to clearly specify the issues and questions to ensure that proposed actions will address those issues and questions. Too often, a general question about the role of some factor (e.g., stand density) on some stand attribute (e.g., habitat value) fails to provide sufficient context to resolve a management issue. A more specific question might be, for example, about threshold levels of basal area that support successful red tree vole reproduction. Such detailed framing provides guidance to specific treatment options that generate, for example, discrete ranges of basal area across a unit that then allow an evaluation of impacts (Irwin et al. 2018).

Replication and controls then provide some level of assurance that any observed results are not unique to one site or one example at one time but, rather, might be transferable to other similar locations and situations. Managers can casually observe the differences/similarities among examples and contrast them with controls to advance their understanding of stand responses to planned and implemented management actions, without sophisticated experimental designs and statistical analyses. They can partner with researchers for that more detailed work and subsequent publication wherever needed.

Finally, adaptive management does not really happen without good record keeping about the question, proposed actions, actual implementation, early results, and stand development over time. This requires dedication to **monitoring** and record keeping that promotes organization and allows communication with current colleagues and future colleagues after a manager's tenure. There are abundant tools currently to collect, organize, and store data and information for one's own organization, as well as for posterity. Unfortunately, it is all too often neglected when time and resources (budgets) become tight.

5-5 Summary

The core premises of conservation management and silviculture are: knowing what you have, the trajectory on which it is currently traveling, and what you want in the long term (management objectives and landowner preferences), then applying the tools and techniques that promote stand development in the direction that the landowner desires, if not the actual endpoint. This is the basis of silvicultural prescriptions; silviculturist know that prescriptions succeed or fail based on their ease of application, which means entering the stands at the right time with the right tools and in ways that steer the natural processes rather than attempt to override them.

Both the current forest condition and the trajectory on which it is moving are constantly changing. This is a management challenge, but it is reality and why preservation is fundamentally flawed. Human population expansion, exotic species invasions, and climate change only exacerbate these challenges. A sound management approach utilizes forest planning and scheduling that builds opportunities for the future and is resistant and resilient to emerging disturbances while responding to landowner objectives and preferences.

Public ownership creates a wide range of competing objectives for the land and its natural resources; and these interests constantly change over time, as well. Both speak to the need for collaborative learning and planning involving stakeholders, multi-party monitoring of the various management approaches applied, and long-term adaptive management.

When it is all said and done, however, there is no alternative to good silviculture that moves a stand and a landscape in the right direction as best you can and that creates opportunities in the long run for your successors. In a sense, we are developing a role for what might be called "muffin forestry." Borrowing from the cooking world, much of our forestry heritage is akin to precise cookbook forestry that can optimize one or more outcomes or outputs, when a recipe is followed closely. The resultant product may require precision in ingredient measurements, mixing order, and/or cooking time. Complex, shifting objectives and large landscapes with uncertain futures speak more to flexible, imprecise recipes like those for muffins—variable ingredients, some mixing here and there, cooked for some length of time until they look about right.

References

Billings, W.D. 1970. *Plants, man, and the ecosystem.* Wadsworth Publishing Co., Inc., Belmont, CA.

D'Amato, A.W., E.J. Jokela, K.L. O'Hara, and J.N. Long. 2017. Silviculture in the United States: an amazing period of change over the past 30 years. *J. For.* 116(1):55–67.

Davis, L.S., N.K. Johnson, P. Bettinger, and T. Howard. 2001. *Forest management,* Fourth edition. McGraw-Hill Higher Education, New York. 804 p.

Fahey, R.T., B.C. Alveshere, J.I. Burton, A.W. D'Amato, and Y.L. Dickinson. 2018. Shifting conceptions of complexity in forest management and silviculture. *Forest Ecology and Management* 421:59–71.

Franklin, J.F., K.N. Johnson, and D.L. Johnson. 2018. *Ecological forest management*. Waveland Press, Inc., Long Grove, IL.
Graves, H. S. 1911. *Principles of handling woodlands.* John Wiley and Sons, New York.
Holling, C.S. (ed.). 1978. *Adaptive environmental assessment and management.* John Wiley and Sons, Chichester, UK.
Irwin, L.L., R.A. Riggs, and J.P. Verschuyl. 2018. Reconciling wildlife conservation to forest restoration in moist mixed-conifer forests of the inland northwest: A synthesis. *Forest Ecology and Management* 424:288–311.
Leopold, A. 1949. *A Sand County almanac.* Oxford University Press, New York.
Nyland, R.D. 2016. *Silviculture concepts and applications,* Third edition. Waveland Press, Long Grove, IL.
Robertson, F.D. 1992. Ecosystem management of the National Forests and Grasslands. Letter to regional foresters and station directors. 1330–1331. USDA Forest Service, Washington, DC.
Society of American Foresters. 1993. Sustaining long-term forest health and productivity: Executive summary of the task force report. *J. For.* 91:32–35.
Tappeiner, J.T, D.A. Maguire, T.B. Harrington, and J.D. Bailey. 2015. *Silviculture and ecology of western U.S. forests,* Second edition. Oregon State University Press, Corvallis.
Walters, C. 1986. *Adaptive management of renewable resources.* Macmillan, New York.
Webster, C.R., Y.L. Dickinson, J.I. Burton, L.E. Frelich, and M.A. Jenkins. 2018. Promoting and maintaining diversity in contemporary hardwood forests: Confronting contemporary drivers of change and the loss of ecological memory. *Forest Ecology and Management* 421:98–108.

All Creatures Great and Small

6

What is man without the beasts? If all the beasts were gone, men would die from great loneliness of spirit, for whatever happens to the beast, also happens to man. All things are connected.

—Chief Sealth, Duwamish Tribe, 1855
(Letter written by Chief Sealth of the Duwamish Tribe of the State of Washington; sent to President Franklin Pierce concerning the proposed purchase of the tribe's land)

A forest contains many different species of beasts (animals), from small ants and spiders to large deer and bears. Animals do not exist alone but are always part of a community organization that contains environmental factors that influence an individual's ability to survive and reproduce. Therefore, all species exist in a population within a habitat in a functioning ecosystem.

As functioning ecosystems, 22 forest types in seven forest regions across the continental U.S. and Alaska (Figure 2-7) contain habitats for over 200 species of wildlife (birds and mammals). Energy flowing through these forest ecosystems is facilitated by factors in the up and down side of the energy circuit (Figure 1-1). The many food chains and webs involving a large number of species create a challenge for natural resource managers with fiscal constraints and federal regulations to conserve and manage populations while providing for human use compatible with sustainability of the resources.

6-1 Conserve and Manage

The conservation of forest resources and associated habitat for game species in the U.S. started in 1905 when the Department of Interior, at the request of President Roosevelt and Congress, transferred the nation's forest reserves to the Department of Agriculture with the objective to not "restrict" but combine forest use with preservation. The most important change in wildlife conservation came not because of this or other federal (or state) legislation, but from Leopold's publication: *Game Management* (1933). Leopold outlined management theory, techniques, policy, and administration that in totality set the philosophy and direction of how game (and later wildlife) management would be taught and practiced in the U.S. His discussion about a biotic pyramid was an example of how ecosystems work and indicated how animals were involved in ecosystem functioning. Individual animals create populations that facilitate energy flow through the system and an understanding of population dynamics that are central to ecosystem management and planning (Figure 2-1).

Authority and Direction

Since the time following Leopold's publication of *Game Management*, the U.S. Congress has passed legislation that deals with the conservation of forests and wildlife. Seven congressional Acts give the public a way to challenge and be part of a planning system. These Acts, listed below, have had and continue to have a major impact on how federal forest lands are to be managed.

MIGRATORY BIRD CONSERVATION ACT OF 1929. This Act provides for the protection of migratory birds. It implements regulations to control the selling, taking, importing, and transporting of migratory birds. Amendments to the Act authorized the acquisition, development, and maintenance of refuges.

FISH AND WILDLIFE COORDINATION ACT OF 1934. The Act authorized the Secretaries of Agriculture and Commerce to provide assistance to and cooperate with federal and state agencies to protect, rear, stock, and increase the supply of game and fur-bearing animals, as well as to study the effects of domestic sewage, trade wastes, and other polluting substances on wildlife. In addition, this Act authorized the preparation of plans to protect wildlife resources, the completion of wildlife surveys on public lands, and the acceptance by the federal agencies of funds or lands for related purposes provided that land donations received the consent of the state in which they are located.

THE MULTIPLE-USE SUSTAINED-YIELD ACT OF 1960. This Act was signed by President Eisenhower and directed that National Forests are to be comprehensively administered and developed for outdoor recreation, range, timber, watershed, and wildlife and fish purposes. It defines multiple-use as "the management of all the various renewable surface resources of the National Forests so that they are utilized in the combination that will best meet the needs of the American people (Sec. 4a)."

THE NATIONAL ENVIRONMENTAL POLICY ACT OF 1969. Better known as NEPA, this Act was enacted to preserve national heritage and to support diversity and the quality of renewable resources. A main focus of the Act was to have an environmental impact statement prepared for projects on federal lands to include a description of the actions and their effects on the environment. NEPA includes some very specific actions that need to be addressed in the impact statement before a project can be implemented.

THE ENDANGERED SPECIES ACT OF 1973. Provides for the conservation of species that meet certain requirements to be listed as endangered and to designate habitats to protect and recover imperiled species and the ecosystems upon which they depend. Species may be listed as either endangered or threatened. "Endangered" means a species is in danger of extinction throughout all or a significant portion of its range. "Threatened" means a species is likely to become endangered within the foreseeable future. All species of plants and animals, except pest insects, are eligible for listing as endangered or threatened. For the purposes of the ESA, Congress defined species to include subspecies, varieties, and, for vertebrates, distinct population segments.

THE RENEWABLE RESOURCES PLANNING ACT OF 1974. The legal document with amendments provides direction and guidelines for managing and protecting resources on National Forest lands. The Act authorizes long-range planning by the Forest Service to ensure a future supply of resources while maintaining a quality environment.

To plan and prepare for the future, a forest management plan with options and public review is prepared and approved every ten years. The Act has four major assessment components:

- an analysis of present and anticipated uses,
- an inventory of present and potential renewable resources,
- a description of programs and responsibilities, and
- a discussion of important policy considerations.

Amendments to the Act now include two new requirements:

- an analysis of the potential effects of global climate change and
- an analysis of the rural and urban forest opportunities.

The report focuses on four regions (North, South, Pacific Coast, Rocky Mountain) and eight sub-regions (Pacific Northwest, Pacific Southwest, Intermountain, Great Plains, North Central, Northeast, South Central, and Southeast). These regions and sub-regions do not exactly conform to the forest and range regions documented on the National Atlas map (USDA 2000).

THE NATIONAL FOREST MANAGEMENT ACT OF 1976. This Act provides for balanced consideration of all resources in U.S. National Forest land management planning. It directs the Secretary of Agriculture to protect and improve the future productivity of all renewable resources on forest land, including improvements of maintenance and construction, reforestation, and wildlife habitat management on timber sales. The Act specifies that habitat diversity includes variety, distribution, and relative abundance of plant and animal communities in a multiple-use context.

Survive and Reproduce

Once the direction is in place to conserve and manage the nation's renewable natural resources, the factors involved in bringing about a process to support the concept of ecosystem management are reviewed in Chapter 5. This concept recognizes the phenomenon of life (diversity) in developing plans for the sustainable use of our forests and wildlife along with the associated amenities (i.e., recreation and aesthetics). To understand how wildlife is part of the Up-circuit and Down-circuit (Figure 1-1) of energy flow and nutrient cycling in an ecosystem, we have to have knowledge about the biological and ecological factors involved in the ability of animals to survive and reproduce (Figure 1-2).

6-2 Biological Factors

The phenomenon of life (Principle 10) includes the many internal and external factors that are necessary for plants and animals to exist as individuals or in a community setting (Figure 1-1). A working knowledge of these factors increases the ability of a manager to make logical and reasonable decisions.

Innate Characteristics

Breeding season, age of sexual maturity, gestation (incubation), longevity, and fecundity are all elements of a species' reproductive characteristics. These characteristics are innate and have a wide range of values. **Breeding season** refers to the time of year mating can take place and is variable among species. Breeding season within species occurs at the same time so that there is maximum opportunity for the species to perpetuate itself. Time of breeding is controlled by day length (photoperiod), but breeding is also influenced by temperature and the general physical condition of an individual.

Breeding season within the geographical range of a species may differ; for example, in West Virginia the white-tailed deer breeds in November but in Arizona breeding occurs in January. Animals producing many young (mice, rats, etc.) generally breed in the spring and the young are born in the spring or summer when the environmental conditions are most favorable for their survival. Animals producing only one or two young at a time, such as deer, breed in the fall with the young being born the following spring or summer.

Two general types of breeding behavior are evident within a breeding season. A **monogamous** species is one in which a single male and female pair for the breeding season and maintain strong social bonds. This contrasts to **polygamy** where one male or female breeds with more than one female or male. In polygamous species a social bond forms but is not as strong as in monogamous species. One type of polygamy is **promiscuity** in which one male breeds with many females or *vice versa,* but no social bond is formed. For example, the bald eagle is a monogamous species, elk are polygamous, and cottontails are promiscuous.

Age at sexual maturity determines whether an animal can breed within a breeding season, although poor nutrition may delay its onset. Small mammals generally breed as juveniles. Deer, in contrast, are born in the spring and generally breed the fall of the next year (1.5 years of age) if they are in good physical condition. The length of time a female is pregnant defines the **gestation** period. Species with short gestation periods are capable of producing the greatest number of young and some species can breed more than once in a breeding season. Gestation period in mammals is analogous to the incubation period in birds. The longevity (age) of an animal species determines how many times and how many young a female produces in a life time. Longevity is of two types: physiological and ecological. Physiological longevity is a species' maximum lifespan under optimum conditions. Some examples of animal ages acquired from zoo records are presented in Table 6-1. The ages included have reached physiological longevity as the animals within the zoo have been provided optimum conditions.

Ecological longevity is the actual or realized longevity of a species in the wild. Long-lived animals generally produce the fewest number of young per breeding season. Take for example the number of young per litter, number of litters per breeding season, minimum breeding age, and ecological longevity of the cottontail rabbit contrasted with that of white-tailed deer (Table 6-2). The number of offspring produced per female of a species is **fecundity** (fertility). Some relative comparisons can be made about fecundity of animals (Table 6-3). The two categories of low and high fecundity are known as "**r-selection**" and "**K-selection**" (MacArthur and Wilson 1967). The former refers to a high intrinsic rate of increase (r) and the latter refers to the ability to maintain a population that fluctuates around a carrying capacity (K).

Table 6-1 Physiological longevity from zoo records.

Species	Years
Deer mouse	4
Cottontail	10
Mountain lion	20
Cottonmouth	21
Elk	22
Bullfrog	30
Grizzly bear	32
Great horned owl	68

Source: McWhirter 1981.

The **behavior** of animals affects their interactions with members of the same species, particularly during the breeding season. Studies on birds led to the idea of a **territory** as a defended area which is part of an animal's home range (Howard 1920; Nice 1941). More recent definitions of territory explain that it is an exclusive area (Schoener 1968) or any spacing of animals that is more even than random (Davies 1978). Different definitions have arisen because territory is better defined for some life-forms, such as birds, than it is for others, such as mammals. Mammals do have territories but they are more difficult to detect and are not always associated with breeding as has been demonstrated for mountain lions in Idaho (Hornocker 1969). In this study, mountain lions had winter territories where all lions shied away from each other but they often crossed territories.

Table 6-2 Contrasting characteristics between the cottontail rabbit and white-tailed deer.

Factor	Cottontail	White-tailed deer
Young/litter	4–6	1–2
Litters/year	2–4	1
Breeding age	3 months	1.5 years
Ecological longevity	3 years	8–10 years

Because territory is an expression of space, there has to be some way for the territorial animal to mark and identify the space. In mammals, this is done with fecal material or urine; in birds, territories are typically defended by displays (an act of aggression) or calls to mark a boundary. Physical violence among birds is rare, however, because one individual almost always breaks off the encounter. Territories can be

Table 6-3 General factors relating to low and high fecundity.

Low fecundity	High fecundity
High mobility	Low mobility
Long-lived	Short-lived
Large body size	Small body size
Population stable	Population fluctuates
Late succession	Early succession

horizontal as is the case of some mammals or vertical for birds. Some large carnivores, such as the wolf, maintain a group territory and defend it against intruders (Mech et al. 1971). There is evidence that the nature of territories changes as animal density changes (Brown 1969).

Territories are in preferred habitat at low animal densities, while at high densities some individuals are forced into less suitable or marginal habitat. Animals at low density defend larger territories but when density is high the territory is smaller (Kendeigh 1941). Closely related to territory is hierarchical dominance, which develops in social animals that form groups. These types of hierarchies are found in amphibians, birds, fishes, invertebrates, and mammals (Wilson 1975). Hierarchical dominance establishes a "pecking order" within the social group. This order assures the dominant males with mating opportunities (as is the case with wolves) (Fox 1980), and with food (as is the case with white-tailed deer) (Severinghaus 1972). In the case of white-tailed deer, the adult bucks dominate does while does dominate fawns in the order of feeding in winter deer yards.

Territories and dominance hierarchies are elements in the social behavior of some animal species and both appear to provide survival value, though this is hard to document. A question concerning population biologists at present is whether these two factors limit animal populations as has been suggested by Vero Wynne-Edwards (1962). In spite of the difficulties of sorting out behavioral problems associated with territories and hierarchies, there are some advantages for animal species possessing these traits:

- Limited resources are preserved for the small number of individuals in the defended area.
- Males are assured of reproducing with one or more female(s).
- The defending males are the dominant males of the species; therefore, genes of the strongest are passed on to the young.

Defining the role of animals in the functioning of ecosystems sometimes leads to complex theories that are difficult to explain and understand. These difficulties often result in confusing usages among practitioners. One such theory is **ecological niche** (Grinnell 1943), derived from the way an animal uses its habitat and the special adaptations either in physical structure or behavioral characteristics that facilitate this use. For example, squirrels climb trees, deer browse woody plants, and some rodents create burrows. The niche of each species is developed through evolution. Each niche therefore is unique and distinct from all others. This leads to an ecological principle that no two species can occupy precisely the same niche; this is commonly stated as the principle of **competitive exclusion**. Criteria defining niches are hierarchical and often overlap at higher levels.

One consequence of the principle of competitive exclusion is that if two species compete at the niche level then one species ultimately displaces the other or evolves to develop a new niche. Probably the easiest way to understand a niche is to examine how a particular species uses a resource that is also used by another species in the same geographical area. For example, the Abert's squirrel eats the inner bark of ponderosa pine clipped from twigs. The porcupine also eats inner bark but from the upper trunk and limbs. Although the two species use the same food resource they differ in where they eat and in the size of stems used. The squirrel and porcupine both feed in trees. At the next level they feed on bark, so these two species have at least two levels of overlap in their requirements, but as the levels are more narrowly defined the two species part company.

Within the geographical range of a certain species, local and regional movements occur coinciding with the onset of fall and spring. This two-way movement is **migration.** Migrations are in some cases vertical as in elk migrating up mountains in summer to avoid heat or to obtain high-quality food and thereafter descending to lower elevations in winter to snow-free areas where food is again available. Mule deer and moose also follow vertical movement patterns in many parts of North America. Probably the most known migrations are the yearly north/south flights of waterfowl (Linduska 1964). These yearly movements of waterfowl between breeding grounds in northern Canada and their winter grounds—largely in the United States and Mexico—have resulted in the delineation of regional management areas based on seven migratory flyways: Atlantic, Mississippi, Central, Pacific, Alaska, Canada, and Mexico.

Population Dynamics

A **population** (**P**) is a social, interbreeding group of animals occupying space and characterized by such qualities as natality (birth rate), mortality (death rate), and density, which are not characteristic of individuals. The ability to produce new offspring (fecundity) underlies the natality of a population. **Natality** is the number of new individuals added per year in a population and is expressed either as a crude birth rate or a specific birth rate. A **crude birth rate** describes the number of individuals born per unit of population; for example, 10 births per 100 individuals. **Specific birth rate** adds a factor such as age; for example, the number of young produced per year by 3-year-old females. The number of individuals in a population dying during a given period of time, usually a year, is the **mortality** of that population. Mortality is expressed as a rate per unit, such as number per 100, 1000, etc., and can be given for the total population or specific age classes. **Immigration** and **emigration** obviously refer to animals that join a group from an outside breeding population or leave a breeding population to join another. If natality and immigration in combination are higher than **mortality** and **emigration** then a population will increase; if the opposite is true, the population will decrease (Figure 6-1).

Two types of density can be defined and it is important to distinguish between them. **Crude density** is the number of animals per unit area regardless of what resources and other factors the area contains. An example is the number of deer per square mile even though part of the area may include a town and roads that are not deer habitat. **Ecological density**, on the other hand, includes only those areas considered suitable habitat for a species, such as the number of deer per acre in an oak-hickory forest type. The density of a population is given in number of animals per unit of area but there is no simple measure for distribution. Three different areas may have the same population density but one may be homogeneously distributed, another randomly distributed, while the third may

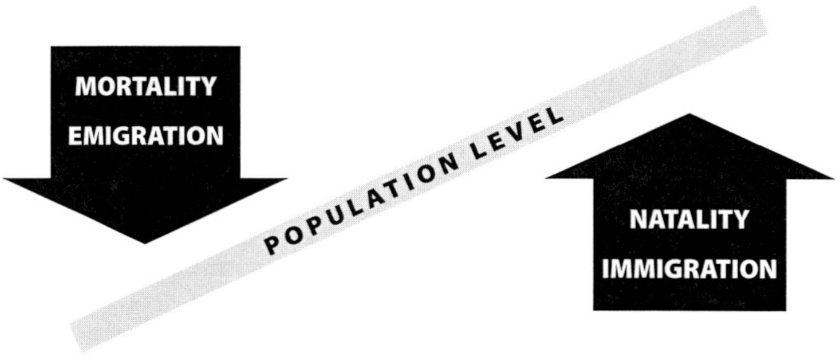

Figure 6-1 Populations can increase or decrease as a result of four factors.

be distributed in a randomly clumped manner (Figure 6-2). The pattern of animal distribution is important because pattern influences the manager's choice of inventory technique.

Populations are of two types: natural and local (Andrewartha and Birch 1984). A local population is a group of animals restricted to the use of a particular area. Local populations may exist on the side of a mountain or on several mountains, but use the same place consistently over time. Local populations probably arise out of the imprinting of young to the area thereby enhancing survival and reproduction. A natural population is the set of all local populations that are not separated by geographical or ecological barriers that prevent breeding. Natural populations may encompass large areas, such as the species' entire geographical range, but as long as there are no barriers to breeding between local populations the definition holds. The natural population typically displays a roughly steady-state density without large yearly fluctuations despite the fact that local populations may be high in one area and low in another. Thus, it is possible for a local population to become extinct while the natural population continues to survive and reproduce.

Federal legislation has introduced the term "viable population" but no clear definition is given. Biologically, a viable population is the number of individuals sharing a common gene pool who are able to reproduce and maintain the population from one generation to the next (as in a local population). The question that must be answered is: What is the number of local populations necessary to maintain a natural viable population? The question is important for managing wildlife because of the trade-offs that may be necessary to maintain viable populations of other species in the same area. The study of populations is fascinating but can be

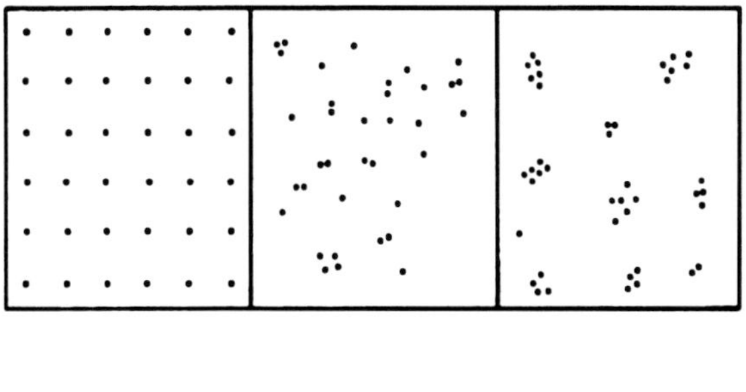

Figure 6-2 Population with the same density but three distributions.

complicated. A detailed population analysis requires mathematical tools that deal with rates of change and limits. In this section we approach population dynamics based on algebraic equations. The question is: How does the mating of several individuals produce a population?

To provide a theoretical answer to this question and to demonstrate how a population might grow, a hypothetical animal population can be developed by imagining the outcome of placing 1 male and 1 female (initial population of 2) with a physiological longevity of 10 years in a large enclosure. The enclosure is predator proof, free of all hazards and diseases, and possessing the optimum resources for the species. Two young are produced in June each year (female was bred the previous fall) with a 50:50 sex ratio at birth, and one breeding season per year. At the end of the first year the enclosure will contain 4 animals consisting of 2 males and 2 females (Table 6-4). In the second year both females breed and produce 2 young each for a total of 4 new animals.

Table 6-4 Uninhibited growth (exponential) for a hypothetical population with a 50:50 sex ratio, one breeding season per year, two young per year, and a physiological longevity of 10 years.

End of year	Adult M	Adult F	Young M	Young F	Sub-total M	Sub-total F	Total Population
1	1	1	1	1	2	2	4
2	2	2	2	2	4	4	8
3	4	4	4	4	8	8	16
4	8	8	8	8	16	16	32
5	16	16	16	16	32	32	64
6	32	32	32	32	64	64	128
7	64	64	64	64	128	128	256
8	128	128	128	128	256	256	512
9	256	256	256	256	512	512	1024
10	512	512	512	512	1024	1024	2048

Uninhibited Growth

At the end of 10 years there are 2,048 animals in the enclosure. A population increasing in this manner develops a characteristic **J shape** when plotted over time (Figure 6-3) and has been termed the **uninhibited growth model**. Uninhibited or **exponential growth** (also called "geometric" because the exponent is in the series 1, 2, 4, etc.) results when animals are introduced into unoccupied habitat called **pioneering populations** or under laboratory conditions. Individuals born into a population at the same time are referred to as a **cohort**. In our hypothetical animal population, a cohort turnover occurs at 11 years when the first pair introduced dies. It is important to notice in this animal population model that although there is a constant growth rate, 2 offspring for each mature female, the number of

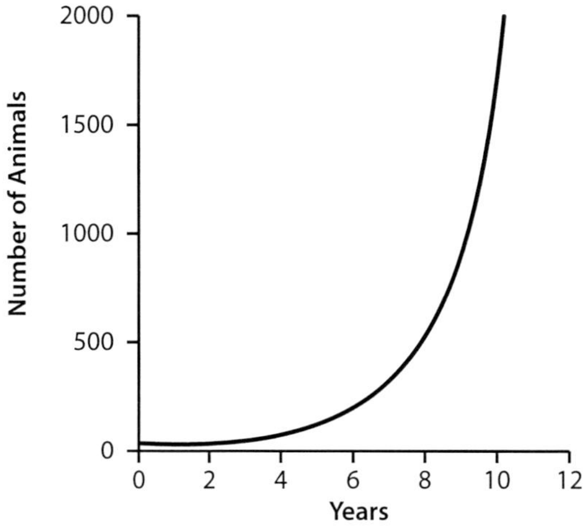

Figure 6-3 Uninhibited growth model.

individuals added each year changes dramatically. Adding 10% to a population when it is low ($0.10 \times 100 = 10$) is very different than adding 10% when it is high ($0.10 \times 1000 = 100$) and may in the same environments be catastrophic. This factor leads directly to a consideration of rates of population change.

Population (P) increases are a function of the number of young born to females of specific ages. The **net reproductive rate (R)** is the average number of females born per female in each age group. The value of R is included in fecundity tables but it is difficult to determine and is almost never available to wildlife managers. However, an approximation of R is provided by the finite rate of change. The **finite rate** (Greek letter **lambda = λ**), or the rate of change of population size, is the total yearly change in a population and can be expressed as a growth rate in the ratio of population numbers from one year to the next. Finite rates are the observed or actual rates obtained from population estimates. For example, using information from Table 6-4, the finite rate of change in population from Year 2 to Year 3 is:

$$\lambda = (P_3 / P_2) = (16 / 8) = 2 \qquad \textbf{(6.1)}$$

The base population of 8 changed by a factor of 2 (100%) in one year. Subsequent years can be calculated by the general equation: $P_{t+1} = \lambda \times P_t$

Where P_t = population for a given year
$t+1$ = population for the next year

For Year 5 the calculation is: $P_5 = (\lambda \times P_4) = (2 \times 32) = 64$.

Notice that the rate of change in our hypothetical population is constant between years. A population increasing at a constant growth rate follows a sequence of increasing power of lambda (λ) such that:

$$P_t = P_o (\lambda)^t \qquad \textbf{(6.2)}$$

Where P_o = initial population
P_t = population in Year t

This equation assumes that the species will breed only once during the breeding season and does not have overlapping generations. For example, the animal population in Year 5 with a λ of 2 is: $P_5 = 2 \times 2^5 = 64$.

A field example of exponential growth occurred when 2 male and 6 female pheasants were introduced on Protection Island in 1937 (Einarsen 1945). The population increased from 8 to 2,000 pheasants in six years. Other examples of uninhibited growth include deer introduced into the George Reserve (McCullough 1979) and reindeer on St. Matthew Island (Klein 1968). To simplify rates, it is convenient to use delta (Δ) notation to express a change in quantity to distinguish **finite changes** from **instantaneous changes (d)** used in derivatives in calculus. A change in quantity can occur in either the time or quantity. Another way to express rate of change of a population is to determine, via equation 6.1, an average rate per individual (r_o) per unit of time as:

$$r_o = \Delta P \ / \ P_t (\Delta t) \qquad \textbf{(6.3)}$$

Where r_o = the rate of change per individual per unit of time or specific growth rate
ΔP = the change in population
P_t = the population in Year t
Δt = the change in time

The rate of change for our animal population changing at a constant yearly rate is, (select any two years, such as Year 2 to Year 3) (Table 6-4): $r_o = (16 - 8)/8\ (3 - 2) = 1$.

The interpretation is that every animal (male and female) in existence in Year 2 will account for one animal to be added to the population in Year 3. The rate of population change per unit of time can be rewritten as: $\mathbf{\Delta P / \Delta t = r_o P_t}$.

Inhibited Growth

Populations cannot continue to expand indefinitely. Sometime during the uninhibited growth of a new population, the rate of increase must start to decline. This change results from the interactions and consequences of the environmental factors that affects an animal's chance to survive and reproduce (Figure 6-4). After these factors become fully effective, a point is reached where the population levels off and reaches an equilibrium with its environment (Figure 6-5). Exponential growth moving to a steady state over time is reflected by an S-shaped curve. The upper part of the curve, however, can take several shapes in addition to the one shown. The inhibition of the growth of a population can be modeled by inserting a variable in the uninhibited growth formula to reduce the value of **r** as the density increases. This expression is:

$$(K - P_t) \ / \ K \qquad \textbf{(6.4)}$$

Where K = carrying capacity
P_t = population at time t shows the same "S-shaped" sigmoid (logistic) curve

If the enclosure containing the hypothetical population has sufficient resources to support only 40 animals, the closer the population approaches 40 the closer the value of the expression approaches and eventually equals 0 as a limit. At carrying capacity (K), population growth is zero. To calculate the annual population for column 2 of the logistic growth curve in Table 6-5 the following equation is used:

$$\Delta P / \Delta t = (r_0 P)(K - P/K) \tag{6.5}$$

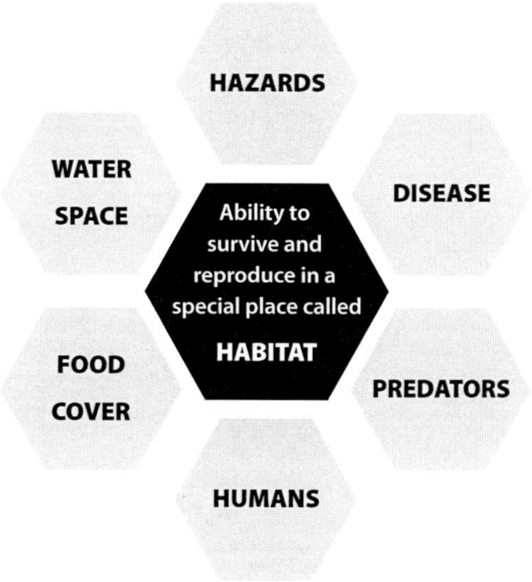

Figure 6-4 Environmental factors affecting and animal's ability to survive and reproduce in a special place (forest type).

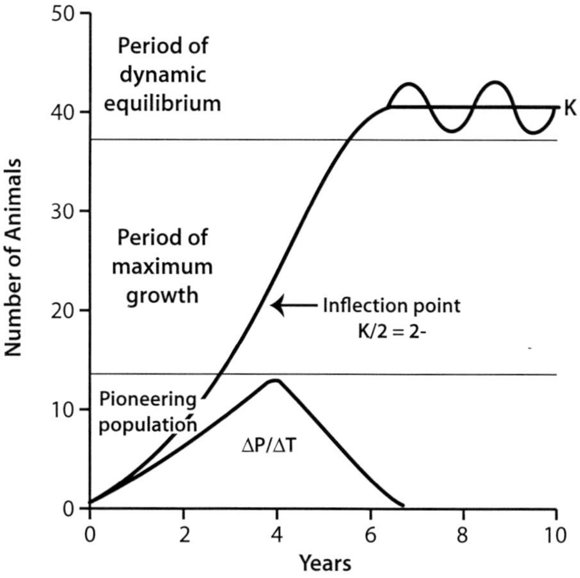

Figure 6-5 Inhibited growth model.

Table 6-5 Inhibited growth of the population in Table 6-1 with a K of 40.

End of Year	Exponential Growth*	Logistic Growth*	Yearly Addition
1	4	4	4
2	8	8	6
3	16	14	9
4	32	23	10
5	64	33	6
6	128	39	1
7	256	40	0
8	512	40	0
9	1024	40	0
10	2048	40	0

*Column 3 is the yearly addition for logistic growth. Columns 2 and 3 are rounded to the nearest whole number.

As an example, the logistic growth (yearly addition) from Year 6 to Year 7 is:

$$r_o = \Delta P \ / \ P_t \Delta t = (256 - 128) \ / \ 128 \times 1 = 1.$$

Substituting figures for K = (1 × 39) (40 – 39/40) = .97 = 1 rounded.

One of the interesting features of the logistic curve is that maximum production occurs at about K/2, which approximates the **inflection point** where the population is still increasing but at a decreasing rate. For our hypothetical population an inflection point occurs somewhere between Year 4 and 5. The sigmoid growth curve is a visual representation of a population model of what could happen—but contains many questionable assumptions, such as:

1. Mortality is entirely dependent on population density.
2. Natality and mortality are continuous, not discontinuous functions.
3. The environment never changes.
4. All individuals in the population react in the same way to changes in density.
5. The expression K – P/K has a biological basis.

The causes underlying the change from a J-shaped to an S-shaped growth curve have led to several differing schools of thought on the effects of animal den sity on animal populations. One way to analyze the effects is to categorize the influences into **density-dependent** and **density-independent** factors.

Density-independent factors are those affecting the population regardless of size—hazards, diseases, and predators. Density-dependent factors on the other hand are those factors affecting the ability of individuals to reproduce—stress, behavior, and the shortage of resources as a result of competition. In the wild the rate a population increases is reduced not by any single density-dependent or -independent factor but by a combination of both. Population regulation describes a combination of feedback mechanisms that bring about resilience or equilibrium of a population. Density-independent factors do not regulate populations but can reduce numbers on a yearly basis. The important point is that populations of any species cannot increase indefinitely without some regulating factor or factors intervening to control the density. Royal Chapman (1931) defined the intervening factors as environmental resistance:

$$P = f\,(Bp - Er) \tag{6.6}$$

Where Bp is the theoretical number an individual female can produce, such as young per year, when resources (food, cover, water) are unlimited. **Environmental resistance (Er)** is a factor or factors that limit biotic potential thus reducing wild animal populations. With so many assumptions and difficulties in measuring factors it hardly seems worthwhile to view the logistic equation and sigmoid growth curve as a useful tool for wildlife management. Nor will most wildlife managers have occasion to use the formulas or to determine an r value because virtually all of the populations they work with are already somewhere between K/2 and the carrying capacity.

Despite the theoretical difficulties, growth model concepts provide the wildlife manager an intellectual framework suitable to roughly determine the limits within which management activities can be conducted and to understand new approaches as they are developed. Intuitively, the initial growth in both the uninhibited and inhibited models make sense because animals reproduce and add young to a population but eventually something happens to restrict that growth. The growth models provide the foundation for concluding that if a given population was once high and is now low, then when presented the right conditions that population can return to high densities rather rapidly. Wildlife management seeks to discover what is that mix of right conditions? One of the great practical benefits of understanding growth curves and their characteristics is that they give the wildlife manager a place to start thinking about and discussing population problems.

Carrying Capacity

Critical to the inhibited growth model is the concept of **carrying capacity (K)**. The idea that an area can support only so many animals is well established in wildlife literature but an exact meaning of the term has defied attempts at clarification (Edwards and Fowle 1955). Carrying capacity is a balance between vegetation and animals, characterized by an expression of animal density (Caughley 1979). While the idea of carrying capacity is simple, the actual definition is more complicated because some type of constraint is needed for quantification. Three descriptions have been primarily used in the past (Dasmann 1981):

- The number of animals of a given species actually supported by a habitat, measured over a period of years.
- The upper limit of population growth in a habitat, above which no further increase can be sustained.
- The number of animals a habitat can maintain in a healthy, vigorous condition.

Most of the ideas about wildlife carrying capacity were first developed by range managers where the emphasis was on density of animals. In the case of livestock the term grazing capacity seems more appropriate (Stoddart et al. 1975). Carrying capacity is largely a concept rather than a real, quantifiable entity. As a population produces young each year, animal numbers (as for deer and elk) may exceed the year-long carrying capacity until some environmental factor checks the population increase, such as hazards, disease, predators, or lack of resources. Carrying capacity is not a fixed factor of the land and will change as food, cover, and water changes with the seasons or over time; the process is dynamic, not static as is often suggested by the many definitions. For herbivorous animals, such as deer, carrying capacity is difficult to determine and if capacity is exceeded for several years, food and cover can be affected adversely. The problem is that of detecting when animal density is starting to exceed carrying capacity. Most ungulate populations cannot be maintained at a level above carrying capacity for long periods without damage to the resource base.

In addition to reducing the impact of the factors that reduce the value of r in the logistic equation, the ecosystem manager's job is to keep animal numbers at a level that will not damage the basic soil and plant resources. In reality, animal populations fluctuate above and below carrying capacity because the environmental factors of hazards, predators, disease, resources, and humans are never constant. A continued fluctuation suggests a process leading to a balance (or **equilibrium**) and feedback between resources (primarily vegetation) and animal density. This fluctuation of population numbers might be termed the **process of dynamic equilibrium** (the same term is used in plant and animal succession and the river continuum concept) in place of the term carrying capacity (Caughley 1979). There are three parts to the process of reaching dynamic equilibrium for herbivores:

- Growth response of plants: The rate of increase of plant biomass as a function of plant density and growth rates.
- Functional response of herbivores: The rate of intake per animal as a function of plant and animal density.
- Numerical response of herbivores: The rate of increase in animal numbers as a function of plant biomass and reproduction.

When an equilibrium is reached between plant growth and animal density in the absence of hunting, then the **ecological carrying capacity** has been reached (Caughley 1979). By interjecting hunting, a new state of equilibrium is achieved that is below the ecological capacity, this is termed **economic carrying capacity**. The economic carrying capacity is determined by quantifying the density of animals desired to meet certain objectives.

Life Table

So far, the discussion of population dynamics has centered on natality, but what about mortality? What tools are available to help understand the changes in a cohort of animals over their lifetime? The study of human populations (demography) has provided techniques to deal with mortality by specific age classes. These techniques were developed by life insurance companies that were forced to estimate how long a human in a certain age class will live. Wildlife managers can use these same techniques to deal with questions of mortality among wildlife species.

A life table follows a cohort of animals born at the same time through its various outcomes until the last member has died. A life table developed in this manner is an **age-specific** or dynamic life table. However, biologists seldom have the luxury of studying a group of animals all born at the same time. The next best option is to reconstruct from either census data or death records what happens to groups of animals. These data lead to a **time-specific** life table in which animals in one period of time are captured and their distribution of ages determined. The mortality factor is calculated from the distribution of the age classes, based on the assumption that a stable age distribution exists.

A third kind of life table is a **composite** life table developed from mark-recapture data for two or more periods of time. Techniques for constructing this type of life table can be found in Silvy (2012). By way of example, Table 6-6 shows findings for a cohort of 58 Kaibab squirrels trapped for a period of eight years, two of which lived to be 6 years old (fx column). The Kaibab squirrel (a subspecies of Abert's) has a litter averaging about 3.5. While fecundity is high, so is the mortality in the 0–1 age class, which is 55%; the mortality decreases during mid-age (2–3 years old), and then increases again in the older age classes (4+ years). This mortality rate indicates an unhunted population in which some adults are able to approach the limits of physiological longevity.

Graham Caughley (1966) suggested that mortality in mammals follows a U-shaped trend with age; mortality in the Kaibab squirrel life table follows this pattern. Life tables, while a useful tool in analyzing a population, are difficult to construct and the data are expensive to collect; therefore, they are almost never used in real management situations. However, wildlife biologists in the United States and elsewhere have been counting cow:calf ratios, doe:fawn ratios, number of eggs hatched, hen:poult ratios as measures of productivity. When done consistently in the same environmental conditions, these types of ratios and counts can be a good indicator of **demographic vigor** and can form the basis for practical management decisions.

Table 6-6 Composite life table for the Kaibab squirrel (combined sexes).

Age x	Frequency Fx[1]	Survival Lx[2]	Mortality dx	Mort. Rate qx	Survival rate
0–1	58	1000	552	0.552	0.448
1–2	26	448	207	0.462	0.538
2–3	14	241	69	0.286	0.714
3–4	10	172	86	0.500	0.500
4–5	5	86	52	0.605	0.395
5–6	2	34	34	1.000	0.000

[1] Number of animals remaining in each age class from an initial population of 58.
[2] Number of animals converted to a base cohort of 1000.

Source: Author's original data.

Fluctuations

Fluctuation is a change in quantity in a short time. A common type of fluctuation is the yearly peaks brought about by high populations of small mammals born in the spring to the winter trough when animals die due to various environmental factors (Figure 6-6). While there is yearly change in most small mammal populations—for example, cottontails and quail—there is no great departure from the peaks and troughs between years, therefore, the population tends to oscillate over time giving the curve of annual change a flat appearance. Leopold (1933) has termed this type of annual change a stable population. The populations of some animal species fluctuate through peaks and troughs separated by several years in a predictable pattern; such fluctuations are known as **cycles**.

Two distinct multispecies cycles have been described: a short cycle of 3–4 years and a long cycle of 9–10 years. Lemmings and voles in the arctic and subarctic along with their predators, the arctic fox, red fox, and snowy owl, are short-cycle species. Elton (1942) was the first to discover and report on the short cycle. Short cycles are exhibited by animals having a limited breeding season or short life cycle and occur in simple ecosystems whose communities include only a few species. The long cycle became evident in the Hudson Bay Company fur records that indicate a 9–10 year cycle for the Canada lynx (Elton and Nicholson 1942). A snowshoe hare population increase is accompanied by an increase in the lynx population, which lags the snowshoe peaks and valleys. Other long-cycle species include ruffed grouse, muskrat, bobwhite, and pheasants (Errington 1945; Hewett 1954; Kozicky et al. 1955). A second type of fluctuation is an irruption (or eruption), which is marked by a sharp increase in a short time followed by a population crash with a return to stability. The Kaibab deer herd crash of 1940 is considered the classic example of an irruption (Leopold 1943; Leopold et al. 1947); irruptions have also been reported for jackrabbits (Wagner and Stoddart 1972), quail (Gullion 1960), and thar in New Zealand (Caughley 1970). Irruptions can occur when new habitat is created by disturbance as happened in California in the Devil's Garden deer herd (Salwasser 1979).

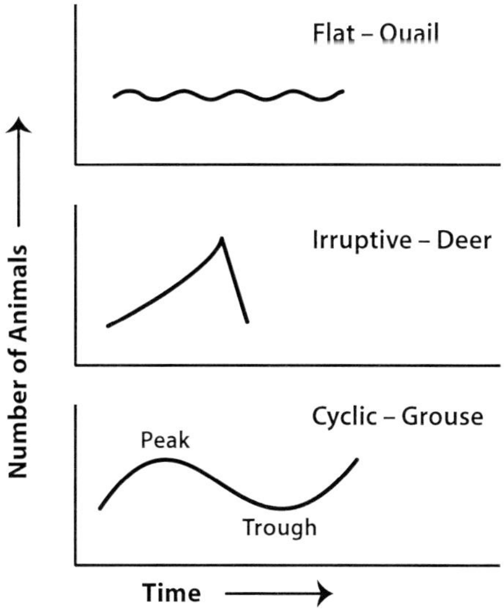

Figure 6-6 Population fluctuations.

Certain events or conditions seem to correlate with high populations and have been hypothesized, at least in part, to be the cause of cycles. Some of these theories are the ozone theory, sunspot theory, stress theory, predator-prey theory, and random theory (Christian 1950; Calhoun 1952; Cole 1954; Krebs et al. 1973). In the past, cycles were a subject of major concern and study, but the topic now seems to be of little interest. This is because most biologists accept the validity of crude cycles but also know that usually, cycles are not a factor that can be managed. This is not the case for irruptive species, such as deer, where increases can be detected and management action taken before a crash occurs.

6-3 The Environment of Wildlife

The environment of an animal was defined earlier as all the factors affecting its ability to survive and reproduce (Figure 6-4). When these environmental factors come together in a specific place and time, then that place becomes an animal's **habitat**. In discovering what the environmental factors are, focus is directed to an animal's **Umwelt,** which is the totality of the animal's surroundings (Klopfer 1969). Umwelt as perceived by humans may be entirely different from what an animal experiences. The challenge to managers is to understand and describe the environment of wild animals in a way that can be used for decision-making, for planning and management purposes to meet the requirements of state and federal policy, for mandates, and for the needs of private land owners.

Decimating Factors

Decision-making by managers is primarily focused on meeting an animal's essential **life needs** while being aware that **decimating factors,** or lack of life needs factors, can affect desired outcomes. An understanding of life needs specific to a particular animal species emerges from the accumulated knowledge of ecological relationships and the growing comprehension of food chains, food webs, and nutrient cycling responsible for ecosystem functioning. Managers spend most of their time working on food and cover because they account for a majority of the management problems. But an understanding of the influence of the other environment factors enhances a manager's ability to make decisions.

HAZARDS. A hazard is anything (excluding disease) capable of causing injury or death to an animal. Wildlife must cope with both natural and man-made hazards. Natural hazards include such things as water, snow, wind, and fire. Man-made hazards include roads and fences. In a broader context hazards also include mankind's hunting activity and habitat destruction. But these two activities involve an active agent while the concept of hazards we are attempting to clarify is the encounter of wildlife with the forces of nature and the artifacts of man, the intention of which is not the injury or killing of animals. Probably the most common and greatest threat to wildlife from natural and man-made hazards is from fire and man-made artifacts, like roads.

FIRE. Fire has both direct and indirect effects. The direct effect is death by burning as well as suffocation, particularly in the case of small mammals (Lawrence 1966). The air temperature at which animals are killed is 63°C (145°F) (Howard et al. 1959). Direct mortality has been documented in animals as varied as birds and voles (Bock and Lynch 1970; Hakala et al. 1971) deer, wood rats, and rabbits (Chew et al. 1958; Buech et al. 1977). In most cases, even in large fires, death by burning is rare (Vogl 1977). In general, the loss of vertebrates directly to the effects of fire is negligible. While there may be some loss of individuals, the effect on the population is insignificant.

The impact of fire upon an animal is a function of the animal's size and mobility and the extent, intensity, and speed of the fire. Adult deer and elk commonly avoid injury because they can easily move away from advancing fire but very young animals may not be so fortunate. Chipmunks, shrews, woodrats, and mice are generally reluctant to leave burning woodpiles and grass stands therefore they are probably killed directly (Bendell 1974). However, those species living underground can survive intense fire due to the great decrease in temperature below the ground surface (Howard et al. 1959). The indirect effects of fire on wildlife, though difficult to quantify, are substantially greater due to the loss of habitat that leads to starvation, predation, or decimation by weather. These indirect effects usually persist for 1–3 years (Cook 1959; Lawrence 1966) depending on fire severity and extent. Starvation following fires that burn large areas results from the loss of vegetation used for food, particularly in the case of species of limited mobility (home range). Even when starvation does not result in death, an animal's ability to reproduce is often severely impacted because when food is limited, nutrients are used for body maintenance rather than reproduction. Predation on small mammals following fire is generally increased since a survivor has to search for food at greater distances from safe cover. Fire may reduce vegetation to an extent that animals are frequently exposed in open areas thereby increasing their vulnerability to predators. It is not uncommon to see hawks and owls in recently burned areas, presumably because small mammals are easier to hunt in the open spaces where there is little or no vegetation—decimation for some is habitat improvement for others. Additional decimation of animals from weather exposure after fire results from a loss of cover that otherwise provided protection from wind, water, and heat/sun. Although general cover may still exist in a burned area, the special types of vegetative cover, such as for nesting, feeding, resting, and loafing may be too thin to protect both adults and young from the elements.

Little information is available on the effects of fire on amphibians, reptiles, and birds. Amphibians have been observed to remain in mud and water until a fire has passed. Reptiles may be able to escape due to their ability to detect fire with their heat-sensing pits used in locating prey (Komarek 1969). Birds are most vulnerable during nesting and fledging periods. Fire devastates ground nesting birds not only because it destroys nests, but also because it removes protective cover and eliminates insects used for food (Daubenmire 1968).

Roads. Roads and highways kill animals directly and indirectly. Each year the number of miles of paved highways and improved dirt roads in forested areas increases. About 18% of the nation's 3.8 million miles of roads are located within cities and towns while the remaining 82% are in rural areas where the chance of wildlife encounters is increased (CEQ 1981). Indirect road effects arise from the noise and disturbance caused by traffic that may prevent animals such as deer and elk from using an area. There are indications that many species have a tolerance level to roads so that flight does not lead to displacement (Lyon 1984). Data from a study in Wyoming on the effects of logging and recreational road traffic on elk indicate that the elk moved very little—a distance of about 1 km (0.6 mi)—during timber harvest operations and animals moved back into the harvested area after activity stopped (Ward 1976). The same study concluded elk were little affected by logging and recreation roads during the day within 0.40 km (0.25 mi) once the elk became used to the traffic. Elk seem to avoid areas of heavy vehicle use but the level of avoidance depends upon the extent of cover (Perry and Overly 1976; Lyon 1984).

Low-mobility species are affected by roads at the time they are constructed. A direct loss of habitat for small mammals, such as mice and rabbits, as well as a loss of animals in burrows, occurs where a road is constructed. This negative impact is somewhat offset by the increase in food and cover plants in the disturbed soil. The practice of seeding forbs and grasses along paved highways to stabilize berms attracts small mammals and increases their probability of being killed. The figures available on death of animals from vehicle accidents are for white-tailed deer. In Michigan, deaths exceed 20,000 a year (Hansen 1983). There is no estimate of the total number of wild animals killed each year by cars but the number must be high enough to justify the publication of *Flattened Fauna* (Knutson 1987), a text that provides keys to species of dead animals found on highways.

DISEASE. A disease is an impairment of the normal state of a living animal or plant that affects the performance of the vital functions (*Webster's Seventh New Collegiate Dictionary* 1966). Diseases affect individual animals by either killing them directly or so debilitating their physiology or behavior that they die from starvation or become easy prey for predators. Diseases are caused by bacteria (tularemia, botulism, Lyme), fungi (aspergillosis), viruses (foot and mouth, rabies), and toxic chemical substances including pesticides. However, diseases are rarely a simple one cause-effect relationship but are generally the sum of changes in the environment (Karstad 1979). Diseases transmitted from one animal to another either by direct or indirect contact are called "contagious." Diseases capable of invading body tissue are termed "infectious" and disease-producing agents are called "pathogens."

Managers are concerned about diseases in wild animals because there are interactions between humans and wildlife, livestock and wildlife, and between different species of wildlife.

Wildlife to Humans. An example of a disease transmitted from wildlife to humans and also to domestic animals from bites by skunks, foxes, and small mammals is rabies (*Rhabdovirus*). Rabies quickly overcomes an animal's immunological system because the virus multiplies rapidly; such a disease is described as virulent. Another particularly feared disease is bubonic plague (*Yersinia pestis*) or "Black Death." The fear arises from our knowledge of the large number of deaths that occurred in Europe in the fourteenth century (Tuchman 1978) and the fact that death from plague still occurs in the southwestern United States where there are large rodent populations. For the period 1950–1989, 46 cases of plague were reported in humans in Arizona. The disease is not caused by rodents but by the bites of fleas and ticks (disease vectors) inhabiting rodent burrows and dens.

Another disease transmitted by ticks is Lyme disease. Ticks as intermediate hosts get bacterium (*Borrelia burgdorferi*) from the blood of infected animals, particularly rodents, and then transmit the bacterium to humans or other animals when they suck blood (Davidson and Nettles 1988). The disease was formerly called "Lyme arthritis" and was named for the town of Old Lyme in Connecticut where it was first recognized. Lyme disease has become more prevalent in recent years and is often difficult to detect and treat.

Livestock to Wildlife. The second form of disease transmission, from livestock to wildlife and vise versa, led at least in one case to the destruction of a deer herd on the Stanislaus National Forest in the 1920s (Allen 1974). Foot and mouth disease (a virus) was devastating livestock in California and could only be controlled by killing infected deer. After eliminating over 34,000 deer the disease was brought under control. Many diseases are transmitted from one wildlife species to another. An example is tularemia caused by the bacterium *Francisella tularensis*. The disease is transmitted by ticks and fleas and occurs in over 45 species of vertebrates but is present most often in rodents and rabbits (Davidson and Nettles 1988).

Naturally Produced Toxins. Another form of disease arises out of naturally produced toxins. A good example is botulism, well known in the wildlife profession because it occurs with such frequency. The disease is caused by the bacterium *Clostridium botulinum*, which produces a toxin so powerful that when ingested by waterfowl it results in death. This bacterium typically develops in warm, stagnant water low in oxygen. Noninfectious diseases include such ailments as tumors (virus) and warts (virus). Also included in the disease category are parasites of roundworms, tapeworms, and flukes occurring in a host-parasite relationship. The immature parasite requires an intermediate host, animals, for a "habitat."

Lead. In recent years mankind has added to the disease problem by introducing lead into the environment in hunting waterfowl. Waterfowl feed on lake and pond bottoms where they pick up shot along with food items. The shot is retained in the gizzard, lead is absorbed into the blood stream, and the waterfowl develop lead poisoning, which causes them to become emaciated and their ability to walk or fly is impaired (Davidson and Nettles 1988). Raptors feeding on waterfowl with lead shot in their gizzard can also be infected with lead poisoning (Pattee and Hennes 1983). Pesticides are chemical compounds used to control plants and animals such as weeds, insects, rodents, and predatory animals. Those designed to kill insects are designated as insecticides and those designed to control vegetation are herbicides. For an account of the use of herbicides in the United States and how their once widespread application directly and indirectly affected the ability of plants and animals to survive and reproduce, the book *Silent Spring* by Rachel Carson (1962) is an accurate presentation. One of the best known pesticides is DDT, a synthetic chlorinated hydrocarbon that was made available for public use in 1945 and later banned for most uses by the Environmental Protection Agency in 1972. It is the most persistent chlorinated hydrocarbon followed by dieldrin, toxaphene, lindane, chlordane, heptachlor, and aldrin. Aldrin and dieldrin were banned in 1974 and heptachlor and chlordane were banned in 1975. Other pesticides include the organic phosphorus compounds of parathion and malathion used to kill parasites on plants and animals.

Certain animals are more sensitive to pesticides than others. As a general rule, crustaceans (lobsters, shrimp, crabs), shellfish (clams and oysters), and fish are most sensitive, followed by amphibians (frogs, salamanders), reptiles, birds, and mammals (USDI 1966). Pesticides may kill animals directly but most exert indirect effects through the food chain either by depletion of a food supply or reduction of reproductive capacity after an animal consumes contaminated food species. Fish and birds are good indicators of pesticide contamination. Birds feed on insects likely to have been sprayed and fish come into contact with run-off water containing the contaminant. Pesticides not only affect wildlife but also humans through the food chain or by direct contact. Most people in North America have measurable amounts of pesticides in their bodies (CEQ 1981).

Predators. A predator is an animal that kills and feeds on other animals. Hawks eat mice, mountain lions eat deer, and foxes eat rabbits. Because of this "kill and eat" trait probably no other topic in wildlife management elicits as much emotion and debate as the subject of predators, particularly as they affect animals that humans want to hunt or view, such as deer, quail, squirrels, and rabbits. "All living things are destined to die and be recycled as a part of the flow of energy through the life community or stated another way, a creature must feed, and sooner or later it will be fed upon" (Allen 1979).

Over millions of years predators have evolved together with their prey. If predators destroy their prey completely, then the former must necessarily die. As a result a delicate balance between prey and predator has evolved. In ecological systems, predators are believed to perform the function of removing excess numbers and culling the prey population by removing weak, diseased, or abnormal individuals; however, this function of predators has not been well authenticated in the literature. One of the first studies on predator–prey relationships was in Iowa on mink predation of muskrats (Errington 1967). The results generally confirmed the findings that when a muskrat population was high, predation by mink was also high, but predation was not even across all age classes. Paul Errington also discovered that social interactions within muskrat populations contributed to limiting their populations. In Errington's study the muskrats selected were those that were suffering from disease and starvation so the predation was compensatory not additive. This simply means that the predator replaces disease and starvation as the decimating factor. If predation were additive then the loss of animals would be added to that caused by disease and starvation. However, all predation is not compensatory; for example, bears killing elk calves or their own young, or raccoons eating waterfowl eggs. The idea that predators take only the sick and weak is intuitive but oversimplifies the predator–prey relationship. Some general propositions of predator–prey relationships that can be formulated from ecological studies are:

- The density of a predator population is always much less than that of its prey.
- Predators feed on a variety of prey, not a single species.
- Reproduction rates of prey species are far greater than that of predator species.
- The prey species is almost always smaller in size than the predator but when small predators prey on larger species it is usually on the young.

One study of predator control often cited as the reason for an increase in deer populations centered on the Kaibab Plateau in Arizona. Before predator control began in 1906, the North Kaibab had an estimated 4,000 deer (Rasmussen 1941). From 1906 to 1923 mountain lions, wolves, coyotes, and bobcats were removed by hunters and trappers. Twenty years later, just before a die-off, the deer population was estimated to be 100,000 on the Plateau, however, the accuracy of this estimate is questionable because there is no supporting field data. Also the results are confounded by the fact that 200,000 sheep and 20,000 cattle were removed from the study area during the same period (Caughley 1970).

A more recent accounting of the environmental events and conditions refutes Caughley's analysis that a reduction in the livestock/deer food competition caused an eruption in the Kaibab deer population (Binkley et al. 2006). John Mitchell and Duane Freeman (1993) suggest that there was a deer irruption followed by a die-off and the causes included a reduction of predators, livestock competition, hunting, over-browsing, and a change in the long-term fire regime. A true accounting of the **ecological events** causing a deer die-off on the North Kaibab will probably never be known.

An interesting idea that developed from predator–prey studies is that predators can maintain species diversity (Spight 1967). Diversity is enhanced because predators keep abundant species in check so that they cannot out-compete less abundant species for food, shelter, and other necessities. Predators are an important component of an animal's environment, but the effects of predation on a population must be considered in connection with the other environmental influences.

HUMANS. In many parts of the world humans are now in conflict with and will increasingly be so with other animal species for the use of the earth's limited resources. An understanding of the conflicts between humans and wildlife is one of the keys to the well-being of both. There is no doubt that humans directly (legal and illegal hunting) and indirectly (habitat destruction) affect the reproduction and survival of a large number of animals. Also, there is something about the presence of humans that causes undomesticated animals to flee or to maintain their distance. Animals can adapt to human presence after several generations of being in a local population where the interaction occurs. These interactions often lead to problems as animals, for example deer, first become present and then a nuisance to people.

Man as a hunter directly affects an animal's welfare but man as a destroyer of habitat also has an indirect effect on the habitat's ability to survive. While hunting legally and illegally affects animals directly, we can assume that legal hunting, where it is part of a continuing system, is under control. Illegal hunting is a different situation. The direct effect from habitat destruction is a loss of resources, but man remains indirectly one cause of the effect on wildlife. Care must be taken in describing destruction because the loss of habitat for one animal is often a gain for another. So destruction must be considered in terms of the species affected. However, there are three forms of habitat destruction that adversely affect all wild species:

1. Urbanization and the wildland-urban interface (Chapter 3)
2. An increase in clean agriculture (removal of all plant material)
3. Soil loss

Large losses in habitat for big game species have been documented in the West. For example, in Wyoming, a state of wide open spaces, it is estimated that from 1967 to 1977 approximately 150 square miles of forest land were urbanized and lost as deer and elk habitat (Rippe and Rayburn 1981). In the Southwest, riparian vegetation consisting of large deciduous trees is dwindling due to the activities of woodcutters, recreationists, grazing, and an increase in home construction. Riparian areas are important and critical habitat for many species, but the loss of riparian habitat has not been documented adequately due to lack of long-term inventory data. As human populations increase there must be a corresponding increase in living, work, and commercial space as well as in food growing areas, railroads, and highways for transport. Since the land base is fixed, such increase must necessarily be appropriated from land used by wildlife species. The loss of habitat to urbanization will inevitably proceed as long as human populations continue to increase.

Life Needs

The life needs of wildlife in a habitat depend heavily upon the adequacy of **food**, **cover**, **water**, and **space** and the desires of people (bio-politics). Habitat management for a given species involves the manipulation of vegetation as the most important factor contributing to maintaining a population that can survive and reproduce (see Chapter 5). Even carnivores are indirectly dependent on plants for energy since most of the animals they eat are herbivores. Many habitat conservation measures require the seeding, planting, or protection of areas containing certain trees, shrubs, and herbaceous plants as wildlife food or cover plants. Plant species possess nutritional elements in differing quantities and their value for one wildlife species may be different than that for another. On the other hand, a single plant species may possess food or cover value for a number of wildlife species. Clearly such plants are those that managers should preferentially employ in habitat improvement projects.

Single live **trees** provide the physical structure needed for a nest by a squirrel or a hawk, for overnight roosting by wild turkeys, or for perching and resting by doves and pigeons. When single trees accumulate to make a small group and can be identified as a stand, then additional protection is provided to many forms of wildlife. In their early developmental stage, sprouts, limbs, or twigs of trees are food for browsing animals, such as deer and elk. The importance of live trees in providing wildlife food and cover cannot be overemphasized. Approximately 28 genera make up the groups of tree species across North America that are commonly used by some wildlife species (Table 6-7). A large number of birds and several mammals, as a functional group, have trees with cavities as part of their habitat components (Table 6-8).

Cavity nesting birds are a particularly important class of forest birds because the majority are insectivorous and help control endemic forest insects that can damage valuable timber trees and impact forest health/beauty. Across North America, snags are used by 85 species of cavity nesting birds (Scott et al. 1977). In the Southwest, soft snags larger than 15 inches are used by 53 species of primary and secondary cavity nesting birds. A soft snag is a tree in an advanced stage of decay and deterioration. **Primary cavity nesters** (Pcn) (e.g., woodpeckers) excavate holes in the soft decaying wood. They make new holes each year, leaving their old ones to be used again by other wildlife. Holes and fresh chips at the base of trees are signs that the tree is used by cavity nesters. **Secondary cavity nesters** (Scn) (e.g., owls, kestrels, bluebirds, wood ducks) cannot excavate their own holes and so make their homes in those abandoned by Pcn and natural cavities created by broken tops, lost branches, and lightning strikes.

Snags larger than 20 inches in diameter are used more frequently by cavity nesters. There are several reasons why large snags are the most beneficial: they provide more space for birds (particularly larger birds needing larger cavities), better nest insulation, more longevity against the forces that fell snags (ice and wind), and more substrate for insects that are a major source of food. Large snags 28–30 inches in diameter are essential for species such as the pygmy nuthatch, which roosts communally to conserve heat energy. In one instance over 160 birds simultaneously used one tree cavity (Cunningham et al. 1980). Brown creepers and Williamson's sapsuckers also require large snags.

Table 6-7 Trees used by a number of wildlife species for food and cover.

Common Name	Genus	Common Name	Genus
Alder	*Alnus*	Hickory	*Carya*
Ash	*Fraxinus*	Incense cedar	*Libocedrus*
Aspen	*Populus*	Juniper	*Juniperus*
Beech	*Fagus*	Larch	*Larix*
Birch	*Betula*	Magnolia	*Magnolia*
Blackgum	*Nyssa*	Maple	*Acer*
Baldcypress	*Taxodium*	Mulberry	*Morus*
Cedar	*Thuja*	Oak	*Quercus*
Cherry	*Prunus*	Pine	*Pinus*
Chinaberry	*Melia*	Sassafras	*Sassafras*
Dogwood	*Cornus*	Sourwood	*Oxydendrum*
Douglas fir	*Pseudotsuga*	Spruce	*Picea*
Fir	*Abies*	Tuliptree	*Liriodendron*
Hemlock	*Tsuga*	Willow	*Salix*

The number of dead trees, snags, logs, and stumps to leave as a component of wildlife habitat is variable depending on the species of trees and wildlife, forest region, and management objectives. For example, in the Arizona and New Mexico ponderosa pine forests, the U.S. Forest Service policy is to leave three good snags per acre within 500 feet of forest openings and water, with two good snags per acre over the remaining forest. In the Southwest good snags are considered to be dead more than six years, more than 18 inches dbh, and with more than 40% bark cover. In the Pacific Northwest the suggested snag density is six to seven per acre for ponderosa pine (Thomas et al. 1979a).

Table 6-8 Wildlife species that use dead trees and snags for food and cover.

Common Name	Life-form	Use	Common Name	Life-form	Use
Abert's squirrel	Ma	Den	Northern flicker	Bi	Pcn
Acorn woodpecker	Bi	Pcn	Northern Flying squirrel	Ma	Den
Ash throated flycatcher	Bi	Scn	Nuttall's woodpecker	Bi	Pcn
Barn owl	Bi	Scn	Pileated woodpecker	Bi	Pcn
Barn swallow	Bi	Scn	Plain titmouse	Bi	Scn
Black bear	Ma	Den	Pygmy nuthatch	Bi	Scn
Black backed woodpecker	Bi	Pcn	Porcupine	Ma	Den
Black capped chickadee	Bi	Scn	Prothonotary warbler	Bi	Scn
Boreal chickadee	Bi	Scn	Raccoon	Ma	Den
Bridled titmouse	Bi	Scn	Red squirrel	Ma	Den
Brown-crested flycatcher	Bi	Scn	Red bellied woodpecker	Bi	Pcn
Brown headed nuthatch	Bi	Scn	Red breasted nuthatch	Bi	Scn
Carolina chickadee	Bi	Scn	Red cockaded woodpecker	Bi	Pcn
Carolina wren	Bi	Scn	Red headed woodpecker	Bi	Pcn
Chestnut backed chickadee	Bi	Scn	Ringtail	Ma	Den
Chimney swift	Bi	Scn	Siberian tit	Bi	Scn
Douglas' squirrel	Ma	Den	Southern flying squirrel	Ma	Den
Downy woodpecker	Bi	Pcn	Strickland's woodpecker	Bi	Pcn
Dusky-capped flycatcher	Bi	Scn	Three toed woodpecker	Bi	Pcn
Eastern screech-owl	Bi	Scn	Tree swallow	Bi	Scn
Fox squirrel	Ma	Den	Tufted titmouse	Bi	Scn
Gila woodpecker	Bi	Pcn	Vaux's swift	Bi	Scn
Golden fronted woodpecker	Bi	Pcn	Western bluebird	Bi	Scn
Gray squirrel	Ma	Den	Western gray squirrel	Ma	Den
Great crested flycatcher	Bi	Scn	Western screech-owl	Bi	Scn
Hairy woodpecker	Bi	Pcn	White breasted nuthatch	Bi	Scn
House wren	Bi	Scn	White headed woodpecker	Bi	Pcn
Ladder-backer woodpecker	Bi	Pcn	Williamson's sapsucker	Bi	Pcn
Lewis' woodpecker	Bi	Pcn	Winter wren	Bi	Scn
Mountain bluebird	Bi	Scn	Yellow bellied sapsucker	Bi	Pcn

Pcn = Primary cavity nester (excavates cavity)
Scn = Secondary cavity nester (uses cavity excavated by Pcns)
Den = A cavity used by mammals
Bi = Bird
Ma = Mammal

Live trees produce **mast,** which is the name for nuts, seeds, buds, or fruits of trees and shrubs that is food for livestock and certain species of wildlife. Hard mast is nuts from oak, beech, walnut, chinquapin, and hickories. Soft mast comes from the fruits and berries of dogwood, viburnum, elderberry, hackberry, hawthorn, grape, raspberry, and blackberry. In the West, gymnosperms produce large numbers of seeds that fall to the ground and never germinate. Most of these seeds become food for birds and small mammals.

Trees such as mature stands of ponderosa pine in a good year can produce over 340,000 seeds per acre per year (ac/yr) or about 28 pounds ac/yr. In the East, the common and widely distributed white oak can produce over 200,000 acorns ac/yr or approximately 1,300 pounds ac/yr (Burns and Honkala 1990). In the East, acorns are a major food source for wild turkey and deer. The yearly production of mast can be highly variable because of the influence of temperature and rainfall.

The planting and production of **shrubs** was emphasized in the past, particularly in the West, due to their importance in maintaining big game populations. More recent ecological thinking has encouraged their cultivation for a variety of other life-forms as well. Many shrubs are browse food plants for deer and elk but also provide seeds and fruits for birds and small mammals and in many cases, provide cover for the same groups of species as well (Table 6-9).

Table 6-9 Shrubs used by a number of wildlife species for food and cover.

Common Name	Genus	Common Name	Genus
Apache-plume	*Fallugia*	Manzanita	*Arctostaphylos*
Azalea	*Rhododendron*	Menziesia	*Menziesia*
Bearmat	*Chamaebatia*	Mountainmahogany	*Cercocarpus*
Barberry	*Berberis*	Oak	*Quercus*
Blueberry	*Vaccinium*	Palmetto	*Sabal*
Ceanothus	*Ceanothus*	Paperflower	*Psilostrophe*
Buckthorn	*Rhamnus*	Porliera	*Porliera*
Bush-honeysuckle	*Diervilla*	Rabbitbrush	*Chrysothamnus*
Buttonbrush	*Cephalanthus*	Raspberry	*Rubus*
Calliandra	*Calliandra*	Rose	*Rosa*
Chokecherry	*Aronia*	Sagebrush	*Artemesia*
Cinquefoil	*Potentilla*	Saw palmetto	*Serenoa*
Cliffrose	*Cowania*	Serviceberry	*Amelanchier*
Condalia	*Condalia*	Silktassel	*Garrya*
Cyrilla	*Cyrilla*	Snowberry	*Symphoricarpos*
Dalea	*Dalea*	Cherry	*Prunus*
Elder	*Sambucus*	Euonymus	*Euonymus*
Gooseberry	*Ribes*	Sumac	*Rhus*
Hackberry	*Celtis*	Viburnum	*Viburnum*
Hawthorn	*Crataegus*	Waxmyrtle	*Myrica*
Hazelnut	*Corylus*	Willow	*Salix*
Holly	*Ilex*	Winterfat	*Eurotia*
Huckelberry	*Gaylussacia*	Witchhazel	*Hamamelis*
Juniper	*Juniperus*	Wolfberry	*Lycium*
Kalmia	*Kalmia*		

Approximately 1,000 species of shrubs, semi-shrubs, and woody vines grow in the U.S. National Forests and adjacent lands (Dayton 1931). Alexander Martin, Herbert Zim, and Arnold Nelson (1951), in their classic book, account for over 100 wildlife species using shrubs for food. It is probably not an exaggeration to assert that every shrub species provides food or cover for some animal species.

The association of specific animal species with specific shrub species is recognized in their respective common names, such as bearberry, rabbitbrush, deerbrush, and antelope bitterbrush. Shrubs grow in pure stands in the absence of an overstory or in scattered single or small clumps as understory plants. Production of shrubs follows the well-documented inverse relationship of high percent tree canopy/low production or low percent tree canopy/high production. However, total productivity is modified by soil and rainfall so there is great variation from location to location. Furthermore, production is commonly lower in shaded stands.

Plants include the many species of grasses—such as wheatgrasses, bluegrasses, and bromes—and the low, wide-leaved flowering plants—such as buttercups, cinquefoils, and clovers—often referred to as forbs. Grasses, grass-like plants, and **herbaceous** plants are important sources of wildlife food and provide cover for a range of animals from small mammals—for example, chipmunks, mice, and ground squirrels—to large grazing animals including elk, deer, and pronghorn.

Given limits of space, it is not possible here to list and describe all the grass or forb species of value to all wildlife species, but an appreciation of their importance can be obtained by reading current issues of the *Wildlife Society Bulletin* or *Journal of Wildlife Management.* Articles in these journals contain information on herbaceous plants used by various wildlife species either for food or cover. Some of the most common genera are listed in Tables 6-10 and 6-11.

FOOD. The food resource involves not only the availability of food necessary to obtain energy, but the conservation of this energy once it has been obtained. The consumption of food is determined by **availability** and **palatability**. Availability refers to what is obtainable during the various seasons of the year. Availability also relates to the physical location of the food, such as browse out of reach to deer and elk. Food palatability includes qualities of preference, nutritive content, and digestibility. Preference describes the hierarchical ordering of food; the highest being the item selected more frequently than any other. This hierarchical priority can generally be ordered as:

- High Value (Preferred): Eaten most frequently and in large amounts when available from an array of food plants.
- Moderate Value (Staple): Second choice but still provides nutrition. Eaten in moderate amounts when an array of food plants is available.
- Low Value (Stuffing-emergency): Fulfills short-term needs but generally low in nutritional value. Generally eaten in small amounts unless it is the only food plant available.

Nutrients are needed by all living organisms for body maintenance and reproduction but requirements vary by species. More information is available on the food requirements of wildlife species than any other topic in the wildlife pro-

Table 6-10 Forbs used by a number of wildlife species for food and cover.

Common Name	Genus	Common Name	Genus	Common Name	Genus
Agoseris	*Agoseris*	Groundcherry	*Physalis*	Royal fern	*Osmunda*
Alumroot	*Heuchera*	Groundsmoke	*Gayophytum*	Senna	*Cassia*
Arnica	*Arnica*	Hawkweed	*Hieracium*	Snapweed	*Impatiens*
Balsamroot	*Balsamorhiza*	Hollyfern	*Polystichum*	Strawberry	*Fragaria*
Brodiaea	*Dichelostemma*	Knotweed	*Polygonum*	Sweetclover	*Melilotus*
Bundleflower	*Desmanthus*	Lespedeza	*Lespedeza*	Turkey-mullein	*Eremocarpus*
Buttercup	*Ranunculus*	Lettuce	*Lactuca*	Tansymustard	*Descurainia*
Cinquefoil	*Potentilla*	Lupine	Lupinus	Thistle	*Cirsium*
Clover	*Trifolium*	Marshpurslane	*Ludwigia*	Tickclover	*Desmodium*
Croton	*Croton*	Medic	*Medicago*	Trailing-arbutus	*Epigaea*
Dandelion	*Taraxacum*	Oxalis	*Oxalis*	Vetch	*Vicia*
Deervetch	*Lotus*	Parthenium	*Parthenium*	Violet	*Viola*
Dock	*Rumex*	Partridgeberry	*Mitchella*	Waterlily	*Nymphaea*
Eriophyllum	*Eriophyllum*	Peavine	*Lathyrus*	Wintergreen	*Gaultheria*
Eupatorium	*Eupatorium*	Penstemon	*Penstemon*	Wild-buckwheat	*Eriogonum*
Filaree	*Erodium*	Plantain	*Plantago*	Willow-herb	*Epilobium*
Fleabane	*Erigeron*	Pussytoes	*Antennaria*	Wyethia	*Wyethia*
Galax	*Galax*	Ragweed	*Ambrosia*	Yarrow	*Achillea*
Goldenrod	*Solidago*				

Table 6-11 Grass and grass-like plants used by a number of wildlife species for food and cover.

Grass	Genus	Grass	Genus
Bluegrass	*Poa*	Muhly	*Muhlenbergia*
Bristlegrass	*Setaria*	Needlegrass	*Stipa*
Brome	*Bromus*	Panicum	*Panicum*
Cane	*Arundinaria*	Paspalum	*Paspalum*
Cupscale	*Sacciolepis*	Saltgrass	*Distichlis*
Cutgrass	*Leersia*	Squirreltail	*Sitanion*
Fescue	*Festuca*	Wheatgrass	*Agropyron*
Junegrass	*Koleria*	Wildrye	*Elymus*
Grass-like	**Genus**	**Grass-like**	**Genus**
Bulrush	*Scirpus*	Sedge	*Carex*
Rush	*Juncus*	Spikerush	*Eleocharis*

fession. One has only to read the *Journal of Wildlife Management* from the early 1940s to current issues to obtain an appreciation of the research efforts expended by biologists in determining food requirements. Comprehensive publications are available that describe food habits at the broad level (Martin et al. 1951), regional level (Patton and Ertle 1982), and local level (Stephenson 1974). The nutritional requirements of game animals have been most extensively studied because the wildlife profession generally began by managing hunted species such as deer, elk, turkey, rabbits, squirrels, and quail. Due to the great variation in the nutritive content of plants, wildlife biologists are still seeking a single index of plant nutritional value.

Table 6-12 Protein requirements of domestic and wild animals.

Species	Protein (%)
Beef cattle	8.3–9.3
Sheep	7.7–8.2
White-tailed deer	6.7–16
Wild turkey	15–20
Bobwite quail	12–28
Fox squirrel	15–20

Sources: Nat. Acad. Sc. 1963, Nat. Acad. Sc. 1964, French et al. 1956, Halls 1970, Nestler 1949, Nat. Acad. Sc. 1962.

Animals need vitamins, fats, proteins, minerals, and carbohydrates. But protein seems to be the most likely indicator of plant nutritional quality. Some of the factors affecting the crude protein content of plants are soil moisture, canopy closure, soil nutrients, grazing intensity, and burning. While protein is a good indicator of plant quality, the complete nutritive value of a plant must be compared with the requirements of a species. Table 6-12 presents examples of the daily protein requirements for a few animals, including domestic livestock, for a comparison.

Chemist Justus von Liebig's (1840) study of plant nutrition lead him to formulate the principle of limiting factors or the **law of the minimum**. Essentially the law states that the nutrient in least supply is the **limiting factor** for plant growth. Liebig's law was extended to wildlife when research began to show that any one of the resource factors—food, cover, space, or water—limits the survival of wildlife species. Furthermore, each species possesses a range of tolerance to its environment, meaning there can be too much as well as too little of a specific factor such as temperature or succession (Tables 2-1, 2-2). The inability to cope with any factor that exceeds the limits of such a range is often referred to as the law of tolerance. Any resource factor exceeding the limits of tolerance for a particular organism becomes a limiting factor for that organism. In short, all organisms require a complex of resources and environmental conditions and have a range of tolerance to any one of them (Odum 1971).

COVER. Once energy has been obtained through a proper diet, an animal must then conserve this energy through some form of cover. Cover is the physical habitat component or landscape feature providing protection from hazards and predators. The cover requirements of individual species are as variable as species' food requirements. Cover is furnished by vegetative **structure** (trees, shrubs, grasses, and forbs) as well as by stages within life-form (seedling, sapling, etc.), topographic features (aspect, hills, valleys, soil, etc.), and water. Although cover provides protection, there are special types of cover associated with specific functions for each wildlife species. Some of the cover types that have been identified in literature are thermal (heat and cold) cover for deer and elk, roosting cover for turkeys, nesting cover for grouse, resting cover for quail, and escape cover for squirrels. For example, thermal cover for elk in Oregon and Washington is defined as any stand of coniferous trees 40 feet or more in height with an average canopy closure exceeding 70%; and for deer, coniferous trees either pole or sapling size, at least five feet tall with 75% canopy closure (Thomas 1979). These various forms of cover are not mutually exclusive; for example, hiding cover for deer and elk often serves as thermal cover also. Like food, cover is subject to annual and long-term changes (ecological time), as in the change in location of forest stands. Care must be taken in describing cover in general terms. For example, thermal cover for elk in the summer may not exist in the same stand in the winter if the stand is composed of deciduous trees. In addition to general types of cover, for some species there is a subset that can be identified as shelter. Shelter is associated with the rearing of young, such as in a nest. For example, the Abert's squirrel builds a nest in a single ponderosa pine where young are born. This nest occurs in a tree surrounded by a group of trees. The tree group and individual nest tree provides cover from weather and predators, but a nest (shelter) provides additional protection for the young (Patton 1984).

Not all species have specific shelter requirements but some types of shelter used are rock crevices, tree cavities, hollow logs, ground dens, and burrows. When discussing cover and shelter requirements of specific species, caution is urged in not transferring the requirements across geographical boundaries or vegetation types unless studies or experience have shown that the requirements are similar in the hierarchical levels being compared.

WATER. Water—the one abundant liquid on earth—is, when pure, tasteless, odorless, and transparent. Water is a solvent of a great variety of substances, both solid and gaseous. Not only does it dissolve more substances than any other liquid but, what is more important, it dissolves those substances that are most needed for the maintenance of life. Water is the greatest medium of exchange in the world. It brings down the gases from the atmosphere; it transfers ammonia from the air into the soil for plant food; it leaches out the soluble constituents of the soil; and it acts as a chemical agent in nutrition and also in the processes of putrefaction and decay, which keep the world's available food supply in circulation.

Water is needed by different species for metabolism (drinking) or as a medium for reproduction and rearing of young. Species needing free water for metabolism must find it in their home range. Water is a resource necessary for animals to survive and reproduce and like nutritional requirements, water requirements vary by species (Table 6-13). Examples of research to determine water requirements of wild species are provided by Wesley et al. (1970) for antelope and Turner (1973) for bighorn sheep. Species can adapt to regions with low precipitation by conserving water; going underground (rodents) or seeking shade to avoid heat (deer and elk); undergoing morphological adaptations like the body size and shape (jackrabbit); or using metabolic water (kangaroo rat).

Research has shown that precipitation in the form of rainfall has an effect on the distribution and

Table 6-13 Water requirements for some common species.

Animal	Water (gallons/day)	Spacing (miles)
Range cattle	7.4–10	—
Sheep	0.2–1.5	—
Mule deer	0.6	1–3
Antelope	1–2	2–3
Elk	5–8	1–3
Quail	750/covey/yr	0.5–1
Turkey	500/flock/sum	1–2
Songbirds	1–2/group	0.25–0.5
Mourning dove	2–5/flock	3–5
Pheasant	2–5/flock	0.5–1

Sources: Stoddart et al. 1975, Bissell et al. 1955, Payne and Copes 1986.

reproduction of some species; for example, in Arizona the number of young quail is related to precipitation (Gallizioli 1965); in South Carolina drought reduced the number of bobwhite young per female (Rosene 1969); in California, the number of young quail are reduced when rainfall is less than 6.3 inches (McMillan 1964); and deer densities in Texas are related to the precipitation of the previous year (Teer et al. 1965). Many factors affect the distribution, availability, and potability of water for use by wildlife and, excluding rainfall patterns, most result from mankind's activities such as dams and reservoirs, livestock use, wetland drainage, pollution in many forms (sewage, acid rain, chemicals, oil spills, etc.). All these are part of the web of indirect factors. In studies to determine the food and water requirements of a species, investigators must consider the following questions:

- What is available by season?
- What is accessible (height of browse plants, plants in rocky areas, waterholes, lakes, etc.)?
- What is palatable and potable?
- What nutritional levels of plants and amounts of water are needed for physiological processes at each season?
- What is the individual adapted to eating (physical adaptations)?

SPACE. All wildlife species have a living space. Each area contains varying densities of the species but in some areas, patches are unoccupied. Occupied and unoccupied areas make up a **geographical range**. The geographical range provides an indication of the adaptability of a species to various climatic, topographic, and vegetative factors. For example, black bears, coyotes, and bald eagles are found in every forest region in the continental United States. Within a geographic range an individual has a **home range**.

Home range delineates the space where an animal finds all its life needs to survive and reproduce. The size of area used by mammals correlates with body weight and is roughly related to reproduction characteristics (Emlen 1973). Species of low mobility are more at risk from habitat destruction (e.g., severe wildfire and timber harvesting) than highly mobile species because of the scale of their space needs; highly mobile species are able to find a space that fits their needs more easily than low-mobility species when they are displaced.

Home ranges can be large (square miles) as for the mountain lion or small (a few acres) as for the cottontail. One part of the home range is the daily cruising radius or the distance an animal travels in one day in search of food and cover. Home ranges can be any shape extending from approximately circular to elliptical. A home range and cruising radius roughly define an animal's mobility. In recent years, attention has been given to the size of an area that an animal uses, and this has developed into a generalization of relative distinction between two animal groups.

- Species with low mobility, high fecundity, small in size, and having all their requirements met in identifiable vegetation communities are often described as **habitat specialists** (Harris 1988).
- Species large in size, low fecundity, high mobility, and requiring large landscapes are referred to as **habitat generalists**.

Based on the need to consider home ranges of many species, the need to preserve areas for interior species, and the desires of society for areas with old forest conditions, it is realistic that 50,000 to 100,000 acres might be appropriate area sizes for landscape-level planning that considers a variety of wildlife species (Table 6-14). Species with large home ranges are also subject to habitat fragmentation because of management activities. The minimum arrangement of food and cover in close proximity to one another for a species is termed **juxtaposition** (King 1938).

Table 6-14 Approximate home ranges of mammals that use forest ecosystems.

Species	Acres	Size Class*
Ground squirrels	2	2
Cottontail	5	2
Raccoon	370	4
Gray fox	741	4
White-tailed deer	865	4
Red fox	3,112	5
Coyote	3,458	5
Mule deer	6,175	5
Black bear	6,422	5
Elk	6,669	5
Moose	13,350	6
Bobcat	32,604	6
Brown bear	73,371	6
Mountain lion	74,100	6

*Size Class is a base 10 order of magnitude.
Class 1 < 1 ac, Class 2 = 1–10 ac, Class 3= 11–100 ac, Class 4 = 101–1,000 ac, Class 5 =1,001–10,000 ac, Class 6 = 10,001–100,000 ac.

Sources: Yensen and Sherman; Chapman and Litvaitis; Gehrt; Cypher; Bekoff and Gese; Mackie et al.; Pelton; Peek; Bowyer et al.; Anderson and Lovallo; Schwartz et al.; Pierce and Bleich; all in Feldhamer et al. 2003.

Food available in one area (opening, meadow, etc.) and cover (tree stand, shrubs, etc.) nearby clearly reduce the energy required for and the natural hazards associated with the search for food away from cover. Equally clear is that the expenditure of energy to search for food and cover cannot be greater than that obtained from the food source; if this is the case, the animal loses weight. Another term related to juxtaposition is **interspersion**, which is the distribution of food and cover for a species over the landscape. The arrangement of stands to provide food and cover in a checkerboard pattern provides landscape diversity to meet the habitat requirements for a number of species.

6-4 Habitat—A Special Place

Vegetation that is the primary component of habitat is controlled by nature but can be influenced by humans. In the absence of human interference, vegetation will naturally develop in a way that will always contain habitat components for a number of species. A balanced system is what one would expect in undisturbed wilderness areas, but vegetation is always changing even though the rate of change may be difficult to detect. Each successional stage provides different types of food and cover that satisfy the minimum requirements of some species but not all species, and that provides optimum conditions for others. Thus, several species may have all their requirements met in a single stage while some require one stage for food and another for cover. Because seral stages progress over time the vegetation cycle of food and cover for different wildlife species occurs on the same tract of land but at different times.

As presented in Chapter 5, land management can be strategically planned and implemented to maintain diversity of vegetation conditions (structure and composition) across a landscape for multiple species or groups of species with overlapping habitat components. Using this ecosystem management approach, vegetation can be kept in or directed toward a seral stage producing the combinations of food and cover required by one or more wildlife species. This is done by directly manipulating succession to return stages to a lower level of secondary succession or to promote the development of later stages of succession. Another alternative if the environmental factors are suitable, is to convert the area to a completely different vegetation type (e.g., forest to grassland or hardwoods to conifers).

Who Is Responsible?

Control of wildlife populations directly is the responsibility of game and fish agencies and is usually accomplished by regulating hunting seasons and bag limits. However, wildlife populations can be controlled indirectly through habitat manipulation—activities that are the responsibility of state, federal, or private land management agencies. Direct techniques used to modify succession in forested areas are: prescribed fire, controlled grazing, timber harvesting, herbicides, and seeding and planting.

In addition to such techniques employed by land managers, nature has from time immemorial controlled succession by volcanic eruptions such as Mount St. Helens in 1980, wildfires such as those in Yellowstone National Park in 1988 and the growing occurrence of mega-fires, or by large insect and disease outbreaks (endemic and introduced). While humans may not manipulate vegetation for brief periods of time, natural forces are always at work throughout the world.

Initially wildlife managers were largely concerned with population regulation of a small number of game species by controlling the length of hunting seasons and the bag limit. But with the rapid increase in understanding of ecological processes and a growing public and professional concern for maintaining the integrity of ecological systems, wildlife managers are increasingly turning to cooperative efforts with other natural resource managers to meet the complex planning and management requirements of a large number of species. Wildlife managers can no longer focus exclusively on only animals that hunters find desirable or others find undesirable. They increasingly must plan for and manage all wildlife species, many of which are a matter of indifference to a segment of our human population.

The opportunity to improve habitat conditions for wildlife can often be accomplished at reduced costs through indirect management practices by coordinating with other management activities such as timber, range, engineering (roads, etc.), watershed, and recreation projects. Conservation practices differ from direct habitat management practices only in the ways they are financed and brought to completion. These measures generally use funds from several functions to accomplish the job. Indirect habitat management is one level of wildlife management and should not be underestimated even though results in terms of increased wildlife populations may not be immediately evident.

General Conservation Practices

Most specific wildlife management practices listed below are the result of research or many years of practical experience in all ecosystems. The list is not all-inclusive, but it does contain a number of generic measures suitable for a variety of forest types. Detailed descriptions of each practice can be found in many state and federal publications dating from the 1930s to the present time. The practices listed are for the most part purposely undertaken to benefit wildlife and so all the costs are usually paid from wildlife funds. In discussing management, we use the term habitat and wildlife in a generic way with an understanding that any habitat is species specific and wildlife is more than just birds and mammals (the focus of this textbook). There are many practices that can be used to improve wildlife habitat conditions for species in any defined forested ecosystem, but caution is urged before transporting technology from one forest region to another to first document that the practice actually benefits the targeted species or group of species. All the practices listed in the sections below are part of ecosystem management.

- Maintain (plant, cultivate, and fertilize) permanent openings (food plots) with wildlife food plants: Benefits many species but generally is used to attract deer and turkey to specific areas.
- Cut, burn, or disc concentrations of shrub species to stimulate growth of browse: Used primarily to produce food for deer and elk.
- Create snags by directly killing trees: Benefits many species of cavity nesting birds and provides hunting perches for raptors.
- Fence riparian areas to exclude cattle: Benefits many species but cost is high.
- Restrict access to critical habitat of threatened and endangered species: Proven technique but cost may exclude its use in many areas. Good public involvement project for volunteers.
- Plant food and cover species along streams to stabilize bank erosion and provide shade: Generally good conservation technique. Benefits aquatic species.
- Create water sources by building small ponds, guzzlers, and trick tanks: Benefits many species. Cost can be high and maintenance is required.
- Improve existing water sources such as seeps and springs to create permanent water sources: May need to be fenced to exclude livestock.
- Build dams to create shallow lakes for waterfowl habitat: Good technique but can be very expensive. Best situation is for a multi-resource structure as in flood control.
- Build small check dams in eroded wet meadows: Generally gets quick results. Provides conditions to establish riparian vegetation.
- Remove invading trees from wet meadows: Expensive but effective in maintaining high forage producing areas for deer and elk, and "bugging areas" for turkey poults.

Timber Management. Timber management activities affect a large number of wildlife species either positively or negatively depending on the requirements of the individual species. The removal of tree overstory and alteration of structure and composition can be used as a management tool to meet specific wildlife objectives in addition to broader management goals (Chapter 5). The coordination measures listed here are intuitive but are based on sound ecological principles.

- Locate timber roads and skid trails to eliminate or minimize siltation of streams consistent with modern riparian protection policies.
- Reserve, in connection with tree planting and regeneration, natural openings, access ways, and brush areas for food and cover retention or planting (e.g., for deer, elk, and turkey).
- Plant or reserve groups of trees for cover (thermal, escape, hiding, etc.) for targeted species.
- Reserve and release fruit and nut trees and shrubs to increase food production for many species.
- Retain den and roost trees for squirrels, raccoons, turkeys, band-tailed pigeons, and some raptors.
- Create openings by harvesting in dense timber stands to benefit deer, elk, and turkey, as well as many species of small mammals and birds.
- Plan timber harvest roads to provide stable and seasonal access for hunters and fishermen.
- Retain tree buffer areas or riparian vegetation along streams and lakeshores to maintain cold water temperatures.
- Include provisions in timber sale contracts to prevent entry of large amounts of small-sized debris into perennial streams; retaining large-sized debris may be desirable in some streams.
- Favor aspen reproduction in timber sales where it is a seral stage. Management objectives must be clearly defined especially where beaver or deer could cause damage to a watershed.
- Bridge water courses on haul roads where feasible, instead of culverts; utilize native bottom culvert systems.
- Limit the number of stream crossings in road design for logging operations.
- Adjust slash disposal plans to include requirements of wildlife species; slash piles benefit small mammals and birds.
- Prevent log skidding across meadows and seeps, along streambanks, or through food, escape, nesting, or roosting cover; such areas must be identified before the sale.
- In releasing coniferous timber species, leave hardwoods in strips or patches to provide a mast crop for squirrel, deer, and turkey.
- Suspend logging during breeding and spawning seasons to avoid interference with reproduction cycles.
- Scatter the location of small timber sales to break up large areas of a single age class, and otherwise provide for a diversity of food and cover areas for many species.

- Seed skidways, roadsides, and landings on sale areas with wildlife food mixtures of forbs, grasses, and shrubs to provide forage for many species.
- Provide for juxtaposition of stands in different vegetation types.
- Provide for interspersion of habitat units throughout the vegetation type.
- Thin some dense stands of trees to stimulate understory growth for deer and elk forage.
- Close permanent and seasonal logging roads to protect a species during all or part of a year.
- Mark and save a given number of snags per acre for cavity nesting birds.

Range Management. Conservation measures undertaken in range management programs are largely directed toward including wildlife requirements for grazing and browsing species such as deer, elk, and bighorn sheep. The following measures are those that range conservationists and wildlife biologists have generally agreed upon, due to their wide application. They are practical, beneficial, and can be accomplished with little difficulty.

- Allocate a percentage of forage to wildlife species in range allotment plans to keep total livestock/wildlife numbers within carrying capacity.
- Manage livestock to protect key wildlife areas. This is especially important in overlapping winter-use areas.
- In placing salt on the range, distribute it to allow use by big game animals. While this technique was used extensively in the past, there is some question about its value for wildlife.
- Reseed depleted grass ranges with palatable species to reduce the use of shrubs by livestock; plant according to local guidelines.
- Reduce shrub use by livestock through grazing of moderate intensity for shorter periods.
- Install escape ramps for small game in livestock watering tanks to keep water from being polluted by dead animals.
- Include legumes, browse, and other important wildlife foods in range reseeding mixtures. Plant species adapted to local conditions and palatable to target species such as deer, elk, and turkey.
- Leave browse for wildlife in reseeding projects involving brush removal. Make certain that browse species are not decadent.
- Defer spring livestock grazing until young wildlife are no longer dependent on green feed. This is very effective in important gamebird habitat and also benefits deer and elk.
- Control, fence, or eliminate livestock driveways across winter game ranges. Be sure to note that fencing is expensive.
- When initiating brush control on winter ranges, consider wildlife needs in the early planning stages of a project in relation to the surrounding area.

- Consider the possible competition effects on wildlife when changing the kind of livestock in an allotment. Competition between livestock and wildlife must be considered early in the planning process.
- Reserve forage at heavily used hunting and fishing camps for recreational pack and saddle stock.
- Do not build range improvements that concentrate or encourage use of key wildlife habitat by domestic livestock. Plan for a good distribution of improvements.
- Locate stock watering improvements to benefit wildlife. Water improvements have proved an important factor in increasing deer, elk, and turkey where food and cover were available (but not water).
- Fence springs at water developments to provide cover and nesting sites for upland birds. This is an effective technique in heavily stocked livestock allotments.
- Consider the effects on other wildlife when initiating predator control, particularly if poison is used.
- Where there is direct competition for forage between livestock and big game, select a reasonable balance. This must be provided for in the planning stage.

Engineering Practices. Engineering concerns deal primarily with cultural and structural features including roads. The placement of buildings, fences, recreational areas, electrical lines, and water structures can have highly detrimental effects on many wildlife species. As a consequence, conservation measures undertaken with engineering projects are largely targeted at mitigating negative impacts on habitat and populations. The following measures are among those wildlife biologists have found to be most acceptable and successful:

- Limit borrow (remove without replacing) from gravel bars to above water level. Borrow pits below high-water stage should be leveled and provided with drainage to avoid trapping fish when high water recedes.
- Install culverts (an encased opening that allows water to flow under a road) in streams at the grade of the stream and below stream level. Drop-offs at outlet are impossible for fish to navigate. Drop-offs may undercut culverts and require expensive maintenance procedures.
- Eliminate or minimize the use of heavy equipment in streambeds to prevent silting of the stream.
- Consider the access needs for hunters and fishermen when planning roads and trails.
- In locating roads, avoid streambanks, meadows, water courses, and other areas used by wildlife. Locating roads away from water or wet areas is also a good soil conservation technique.
- Establish wildlife food plants on cut-and-fill slopes. A cut area removes material from one area to fill in another area. This is a good conservation practice to reduce erosion.

- In constructing roads along a stream, leave fringes of timber and brush to provide shade and bank protection.
- Keep construction debris out of streams to reduce siltation of spawning areas.
- Provide for spreading of drainage water from roads to prevent silting of streams. This reduces siltation and pollution.
- Consider designing road fills to serve as dams, which can provide additional water. This must be planned carefully so that the dirt dam is stable.

Recreation Management. Recreation measures are aimed at controlling the behavior of people; not just hikers, picnickers, etc., but hunters and fishermen as well. Considerable effort is required to mitigate the effects of recreational activities on both habitat and populations particularly on sites having fragile soils or containing sensitive plants and animals. Most recreation control measures require long-range planning involving public review and acceptance.

- Restrict cross-country use of four-wheel drive vehicles or all-terrain vehicles to designated roads in sensitive soil areas. This is a controversial technique that always requires public review.
- Locate hunting and fishing camps and enforce camping regulations so that stream pollution or damage to habitat is held to a minimum.
- Corral horses used at hunting camps located within deer and elk winter ranges. Enforcement may be difficult in remote areas.
- Restrict camping and grazing of pack and saddle animals at heavily used hunting and fishing sites. The terrain immediately adjacent to these areas may need to be closed to grazing.
- Restrict speed boating in high-quality fishing waters.
- Consider camping needs of hunters and fishermen when selecting areas for campground development.
- Through hunting and fishing clubs and other organizations, develop programs of cooperative camp cleanup, sanitation, and fire prevention. This is an effective way to educate sports enthusiasts on just how much trash is left by their activities.
- Plan for the needs of noncommercial hunters and fishermen for campsite areas before issuing special permits to commercial guides.

Watershed Management. Within the large land areas classified as a forest a great many smaller communities and ecosystems can be identified. Among them are marshes, riparian zones, and aquatic habitats of emergent vegetation, all of which add to the total diversity and wildlife productivity of a forest type. The management and protection of watersheds as the source of water for streams and lakes have always been important functions of forest management. Watersheds control water quality. When maintained in the proper condition they prevent soil erosion and serve as filters for the streams within the watershed boundary. Foresters and wildlife biologists are concerned with the effects of various watershed uses and treatments on the natural character of the stream environment providing fish habitat.

The key to protecting forest streams is to harvest small areas over longer time intervals to even out water flows. However, if larger, permanent water flows are needed and would benefit trout waters, then heavier harvesting, if properly done, may improve water temperature and depth. Livestock, deer, elk, and moose tend to concentrate much of their use close to or in riparian zones. Overuse of streamside vegetation by grazing animals contributes to the destruction of stream banks and streamside vegetation, thereby increasing erosion. Fencing has been used to restrict livestock access to streams, but deer and elk can easily jump a fence. Successful management under these circumstances requires a balanced multiple-use approach, which includes limiting livestock and wildlife numbers together with routine watershed maintenance.

Recreational campground use is often the cause of erosion and streambank deterioration. In addition, water quality in the absence of sanitation facilities is adversely affected. Considerable progress has been made in monitoring water quality as a result of the National Environmental Policy Act. However, an increasing human population, resulting in more use of outdoor areas, necessitates constant monitoring of water quality in heavily used areas. In some situations, erosion can be controlled by building structures to stabilize the banks or by plantings aimed at the same goal. Other control measures include limits on campground use. Considerable watershed restoration is done under state and federal programs that can be programmed for specific projects in small watersheds with small work crews that involve the following activities:

- Use browse and legumes in erosion-control plantings to benefit wildlife species.
- Include wildlife food plants in bank stabilization projects.
- Integrate watershed plans and big game hunting to provide for game management in municipal watersheds to increase recreational opportunities for a local area.
- Include the protection and improvement of wildlife habitat in water impoundment programs. If planned properly, this step can benefit many locally important species.
- Safeguard or improve aquatic habitats when using structures to alter stream courses, streambanks, or for channel and gully control.
- In gully and other erosion control plantings, use fruit-bearing and thicket-forming species to provide food and cover for wildlife.

Riparian Zone

Vegetation immediately adjacent to streams or along the edges of lakes and ponds is characterized by plant species and life-forms differing from those of the surrounding forest and is termed **riparian.** Riparian areas are of particular concern because their existence depends on trees. Riparian tree composition depends on elevation but typically consists of deciduous trees from the genera Populus, Acer, Salix, and Alnus. The marked contrasts between riparian and upland vegetation produces structural diversity and edge characteristics enhancing its utility for wildlife. Because of their long, narrow shape, stream riparian areas contribute few acres to the total available habitat; however, they are highly productive so their value to wildlife is well out of proportion to their small area (Thomas et al. 1979b). The highest density of birds ever recorded in North America was found in riparian vegetation in Arizona (Johnson 1970). The density of 1,324 pairs per 100 acres of noncolonial birds is also among the highest densities in the world, including tropical forests.

Not only are riparian areas important to terrestrial animals, but they also control the associated lotic habitat for amphibians and fish. Canopies provide shade, root systems stabilize banks, and plant detritus and insects provide nutrients for stream organisms (Meehan et al. 1977). Riparian areas create an oasis effect in dry lands and because of their cooler microclimate and free water they are major resting places for many north-south migrating birds. In addition, tall trees along a stream or watercourse at times create a flue-like condition causing an updraft, which brings in air underneath the vegetation and over the water.

Riparian vegetation provides a natural travel corridor for local species such as deer, raccoons, turkey, and bear. In the eastern United States the riparian zone is not as distinct as it is in the West where higher rainfall east of the 100th meridian results in more vegetation perpendicular to the aquatic region of the riparian zones. The increased moisture and deeper soils integrate the vegetation types, lessening the distinction between where riparian starts and ends.

Soil—A Medium for Growth

One of the most important resources that managers must always consider in a plan is **soil**, a critical component that facilitates nutrient flows in an ecosystem in the Up-circuit and Down-circuit (Figure 1-1). While food and cover directly affect an animal's ability to survive and reproduce, soil nutrients directly affect the ability of plants to survive and reproduce. Therefore, soil is an indirect factor affecting the ability of an animal to survive and reproduce. It is well documented that agricultural production is directly related to the level of nutrients in the soil. Nutrients in the soil are transferred to nutrients in plants that are in turn transferred to animals when consumed. If the nutrients in plants are in sufficient quantity and proper relationships, the consuming animal remains healthy. This is as true for wildlife as it is for domesticated animals. Studies in Missouri have shown that the highest density of game animals (rabbits, raccoons, etc.) is in areas of the highest soil fertility (Crawford 1950).

When mule deer are removed from low-fertility California chaparral and fed optimum diets in confinement they reach a larger size after the first year of feeding than some adults in the wild attain in a lifetime (Dasmann 1981). Results of research in Pennsylvania on white-tailed deer followed the same trend as in California (French et al. 1956). Other research has shown that the number of young produced by cottontails (Hill 1972), and white-tailed deer (Cheatum and Severinghaus 1950) is correlated to increasing soil fertility. In other studies on white-tailed deer body weight has been shown to be related to soil fertility (Smith et al. 1975).

Allen (1974) emphasized that in terms of fertility, soil and water are equivalent since water is needed to make the soil nutrients available. Managers must remember that soil is a major factor influencing vegetation and that vegetation is the driving force for wildlife habitat, which translates into food and cover. Perhaps the most important soil–wildlife relationship is that wildlife production, like any other crop of the land, is directly proportional to fertility of the soil and that land that is not sufficiently fertile to produce vegetation for food and cover is marginal for wildlife (Denney 1944).

Habitat Extensions

The definition of habitat stated that habitat is the sum of all the factors affecting an animal's ability to survive and reproduce in a specific place. However, a scale factor of space must be considered as one part of the habitat definition (Chapter 5). There is **primary habitat**, that is all the combined habitat areas and environmental factors necessary to support a viable population of a species, and there is **secondary habitat**, an area where an organism may spend part of its time but that does not meet all its life requirements (Harris 1984). Secondary habitat seems to apply mostly to highly mobile species at the landscape scale. These animals may have to move because of weather conditions.

The innate factor includes behavior traits of migration and reproduction activities. These innate behavioral characteristics have a distinct bearing on primary habitat. For example, in the southwestern United States elk migrate from summer use of high-elevation, mixed conifer forest habitat to a low-elevation, pinyon-juniper woodland habitat during the winter. Winter and summer habitat together must be delineated as primary elk habitat, but individually they are secondary habitats. Because there is confusion in the use of the general term habitat, researchers have defined subsets of the term to provide a common understanding of its meaning to provide consistency and to eliminate ambiguity (Block 1993). The following definitions were developed for ornithological use but they will provide a common understanding for discussing issues for all species.

- **Habitat use**: The manner in which a species uses a collection of environmental components to meet life requisites. Habitat use can be regarded in a general sense, or separated into specific acts or needs such as foraging, nesting, or roosting.
- **Habitat selection (preference):** Innate and learned behavioral responses that allow animals to distinguish among various components of the environment resulting in the disproportional use of environmental conditions to influence survival and reproduction.
- **Habitat suitability (quality):** The ability of the environment to provide conditions appropriate for survival, reproduction, and viability. Suitability is a continuous variable measured by the intrinsic rate of population increase.
- **Macrohabitat:** Landscape-scaled features that are correlated with the distribution and abundance of populations. Often used to describe seral stages or discrete arrays of specific vegetation types.
- **Microhabitat**: Specific, recognizable features of the environment that act as proximal cues to elicit a settling response from an individual animal.

- **Critical habitat**: Physical or biological features that are essential to the conservation of the species and which may require special management considerations or protection.

Wildlife biologists are continually having to create new ways to manage wildlife populations to survive and reproduce in their specific habitats in areas being managed under the concept of ecosystem management for multiple-use purposes. The existence of wildlife transcends private, urban, state, and federal boundaries. As a result, different approaches to management and planning are in use according to who has the authority for making decisions. For federal lands, legislation often introduces new terms and concepts into a structured planning process. In the process there can be several ways to manage wildlife populations, and planners have to select from a variety of approaches to accomplish a set of objectives.

6-5 Wildlife Management Strategies

The management of wildlife depends on who owns wild animals and who owns the animal's habitat. Under English Common Law all wild animals are public property whose management agencies have authority to enforce wildlife regulations. This common law is the basis for the legal system governing wildlife ownership in the United States. The legal status of who owns resident wildlife was settled by the Supreme Court when the Court ruled in a case involving Connecticut that the state had the right to control and regulate the common property of game (Bean 1977). However migratory birds, endangered species, and exotic imports are under federal jurisdiction.

Single Species Management

For years, game and fish departments have been using **single species management** for hunted or trapped species. The experience gained from single species management has resulted in an accumulation of a large amount of data for deer, elk, turkey, squirrels, quail, and rabbits and has generally led to successful programs by management agencies. In addition, the public has made an impact on federal and state agencies to do research on and provide more information for managing non-game species such as birds of prey and passerines for bird watching. William Block, Deborah Finch, and Leonard Brennan (1995) provide a good discussion on single versus multispecies approaches for management.

GAME AND NON-GAME. In the early stages, information on a single species tends to be at a broad scale concerned with distribution and general food and cover requirements in a **species home range**. As information accumulates, data is refined and becomes more specific pertaining to, for example, a particular mountain range or management unit. Single species management has largely been guided by obtaining information on food and cover requirements to use in developing written habitat guidelines. The data are often compiled into a rating system for habitat, as has been done in the past for big game in Montana (Cole 1958), California (Dasmann 1948), and New Mexico (New Mexico Game and Fish Dept. 1973).

These guidelines have proved to be a real asset to field biologists. Their success has led to recognition of the need to quantify requirements to better predict the effects of land treatments on wildlife habitat for specific species especially for those classified as game or non-game species. The simulation of forage as the critical component of habitat for elk in Montana (Giles and Snyder 1970) and for deer in Colorado (Wallmo et al. 1977) are examples of techniques used for quantifying data for management purposes.

THREATENED AND ENDANGERED. A species designated as threatened or endangered requires a different approach to management. One section of the law establishes the criteria of critical habitat to include the need for space and population growth, food, water, nutritional requirements, breeding sites, and lack of disturbance. All federal agencies are prohibited from authorizing, funding, or carrying out actions that destroy or adversely affect critical habitats. Threat to a species from habitat destruction, pollution, over-harvesting, disease, predation, and other natural or man-made factors must be reviewed and evaluated before an animal or plant can be placed on the federal endangered or threatened list. The Fish and Wildlife Service and National Oceanic and Atmospheric Administration National Marine Fisheries Service administers the Act.

MANAGEMENT INDICATOR SPECIES. The term **management indicator species** (MIS) was introduced to the wildlife profession by the National Forest Management Act (NFMA) but research shows that this may not be a valid use of indicators (Mannan et al. 1984; Verner et al. 1984; Szaro 1985). The National Forest Management Act of 1976 (36 Code of Federal Regulations [CFR] 219.19(a)(1)) defines MIS as plant and animal species, communities, or special habitats selected for emphasis in planning, and which are monitored during forest plan implementation in order to assess the effects of management activities on their populations and the populations of other species with similar habitat needs that they may represent. MIS is no longer required in the planning rules for NFMA. MIS as an ecological indicator is a plant or animal so closely associated with a particular environmental factor that the presence of the species is clearly indicative of the presence or absence of the factor (Clements 1916, 1949; Patton 1987). In most cases a carefully selected plant indicator identifies a cause-effect relationship. Indicators were first used in resource management by plant ecologists working with soil productivity and agricultural crops (Schantz 1911). In the 1930s, vegetation condition and trend techniques were developed for range management using plant indicators (Talbott 1937). Modified forms of these techniques are still in use today.

FEATURED SPECIES. A featured species approach has been in use for many years in the eastern forest types. One species is selected for an area and management efforts are concentrated to fulfill its needs through coordination of timber and wildlife habitat activities (Holbrook 1974). Selected attributes within the habitat of the featured species are managed to meet the needs of other species (Figure 6-7). If properly planned and implemented, featured species management provides a variety of habitat conditions for other species that overlap the requirements of the featured species.

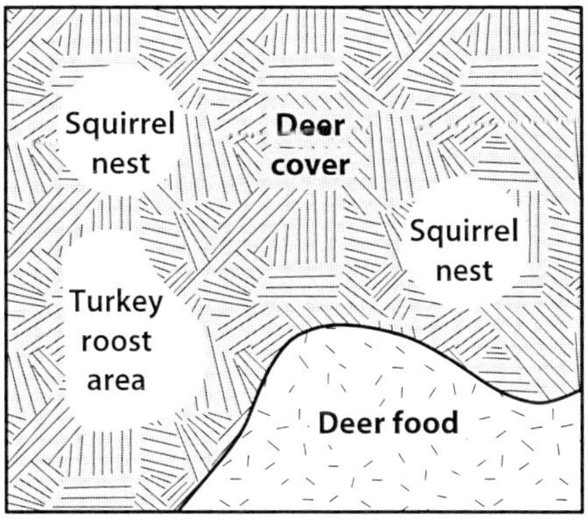

Figure 6-7 Featured species.

FOCAL SPECIES. An approach, similar to featured species, is to select a small subset of species as **focal species** (Rowland and Vojta 2013). Species are selected whose status reflects the integrity of a larger ecological system to which it belongs. Focal species can be selected on the basis of their functional role in ecosystems. Under the National Forest Management Act (1976) focal species is now included in forest management plans. Single species management as practiced to maintain game and non-game species is a form of focal species management.

Multispecies Management

Multispecies management involves **grouping species** with common habitat requirements and then managing for identified overlapping critical habitat components. Attempts at a multispecies classification for species was developed in Texas as early as the 1940s, resulting in a scheme classifying 49 mammals and 171 birds into nine categories from "conspicuously insectivorous birds" to "tree protectors" (Taylor 1940). The intent of the classification was to show how habitat of nongame species was ecologically interrelated to the habitat of game animals. A modification of Taylor's efforts resulted in a classification for North American mammals and birds by forest habitat preference (Yeager 1961). In this scheme, birds and mammals were placed into three groups according to whether they were primarily of forest or brushland, whether they associated with forest or brushland, and lastly whether they only rarely, if ever, used woody cover.

GUILDS. Groups of species exploiting the same class of environmental resources in a similar way is defined as a **guild** (Root 1967). A guild groups species together for habitat reasons, without regard to taxonomic criteria. The important consideration when developing guilds or related systems of species grouping is whether the result provides information that can be used in decision-making. Typically, guilds are contained in a matrix table with one axis representing the feeding habitat and the other axis, the reproducing habitat. Useful information from research on the theory of guilds with an emphasis on birds, has been provided by Severinghaus 1981; Short and Burnham 1982; and Mannan et al. 1984. An example of a guild matrix for birds was developed for the pine-oak woodlands of California

(Verner 1984). The primary feeding and nesting zones form the two axes of the matrix. Within the matrix, birds are assigned to each cell according to their feeding and nesting habits. The matrix has a potential of 30 guilds if all cells are filled.

LIFE-FORMS. Thomas (1979) used a **life-form** approach to multispecies management in the Blue Mountains of Oregon, and the concept has been extended to other areas (Brown 1985). A life-form is a single category synthesizing information on where an animal feeds and breeds (Haapanen 1965). Although the life-form is a specific category, it is represented by a matrix like the guild. Each life-form combination has an identification number and forms an easy-to-use table. All the vertebrates (26 amphibians and reptiles, 263 birds, and 90 mammals) known to occur in the Blue Mountains were ultimately grouped into 16 life-forms. The most obvious advantage of both the guild and life-form concepts is that a large amount of information can be condensed into tabular form for quick reference.

FUNCTIONAL GROUPS. The **phenomenon of life** creates **functional groups** that facilitate energy flow and nutrient cycling through an ecosystem. These functional groups can be selected to form a sub-set of **focal species** to meet monitoring requirements for wildlife in forest plans. These species are the biological engines that make an ecosystem work (Table 6-15). Each of the forest regions described in Chapter 2 has a listing of species developed from a checklist. The regional checklists will need to be authenticated for use at a local level. In most cases where a forest plan has been approved, there is already a list of species documented for the area of concern. If needed for local areas, additional functional groups can be developed to account for other groups, especially for perching birds and rodents.

Table 6-15 Examples of functional groups that facilitate ecosystem functioning in forest regions and forest types.

Functional Group	Life-form	Example
Birds of prey	bi	Bald eagle
Birds of prey–cavity using	bi	Flammulated owl
Birds–cavity creator	bi	Northern flicker
Birds–cavity using	bi	Brown creeper
Birds–ground nesting	bi	Wild turkey
Birds–perching	bi	American robin
Carnivore–large	ma	Black bear
Carnivore–small	ma	Gray fox
Insectivore	ma	Pygmy shrew
Herbivore–large	ma	Elk
Rodent–large	ma	Beaver
Rodent–small	ma	Deer mouse
Rodent–tree squirrel	ma	Eastern gray squirrel

Wildlife Habitat Relationships

Wildlife habitat relationships (WHR) is a systems approach to comprehend all the associations with and uses of habitat by wildlife. It is a way to synthesize the environmental factors that affect an animal's ability to survive and reproduce in a particular place (Figure 1-2). It is also a way of analyzing the complexities of plant–animal interactions defined by **relations** and **attributes** in a database that can respond to questions using **structured query language.** The process depends upon linkages (relationships) and interactions between plant and animal species in a given ecosystem.

The ecological approach to using WHR in a systematic way has been documented by Patton (1978), Thomas (1979), Verner and Boss (1980), and Hoover and Wills (1984) and this approach was adopted as policy by the U.S. Forest Service (Nelson and Salwasser 1982). The ultimate goal of WHR is to manage groups of species with common habitat components and this was the goal of the first relational data model (RUN WILD) developed for National Forests in Arizona and New Mexico (Patton 1978). A similar approach using management assemblages is suggested by Block et al. (1995). A starting point for a basic WHR relational data model for forest ecosystems is outlined in Chapter 8.

Laissez-Faire

All of the strategies considered for managing habitats for wildlife assume funds are available to support and maintain their implementation over time as part of an administrative unit. However, given the shortage of personnel and finances, management may not be available but the resource still exists. In that case the only way forward is a **laissez-faire policy** to let nature take its own course. This is the course of action that was in place for managing public lands before the Fish and Wildlife Coordination Act was passed in 1934. Given all the environmental factors that affect an animal's ability to survive and reproduce in a particular place, there is an amount of uncertainty that exists in selecting a management strategy. Uncertainty exists because of the unknowns that can affect both habitats and wildlife. Former secretary of defense Donald Rumsfeld tried to explain uncertainty in this way:

> There are known knowns. These are things we know that we know. There are known unknowns. That is to say, there are things that we know we don't know. But there are also unknown unknowns. There are things we don't know we don't know. (Rumsfeld 2002)

Some of the things about resource management **we know we don't know** are:

- the effects of climate change on animals and plants at different scales of management,
- when major events such as fire, wind, and water will happen,
- what laws will be passed by congress and how they will affect the management of natural resources,
- when a major disease outbreak will eliminate a specific wildlife or plant species,
- when a threatened or endangered species will become extinct, or
- when management plans will be fully funded to meet the requirements of laws enacted by congress.

Adaptive Management for Wildlife. The ideas of adaptive management specifically for wildlife populations have been in place for many years. In the

1940s and 1950s state resource agencies were struggling with setting hunting and fishing seasons and bag limits. That process involved using the limited biological knowledge that was available at that time along with experience of "game wardens" and "conservation officers" to arrive at a set of regulations. These regulations were then reviewed by the public—in sometimes hotly debated sessions—with the result that public pressure dictated the next year's seasons. Once regulations were established, then the affected agency had a year to observe the outcomes of the various seasons and limits to repeat the process for the next year. Ultimately the process resulted in an acceptable set of regulations with only minor changes from year to year.

Today the process is about the same but with more concern by the public about forest resources and the effects of management activities on animals and habitats such as endangered and threatened species, old forest conditions, forest health and sustainability, riparian vegetation, timber harvesting, and livestock grazing. Adaptive resource management will be used regularly in the future as resource agencies have to adjust to constrained budgets and reduced personnel levels.

The final question has to be considered: Can all of the known unknowns cause a management policy of laissez-faire to be in effect or can adaptive management keep the planning process in place?

6-6 Summary

Population dynamics deal with changes in populations over time. This change is brought about by such individual and population factors as birth rate, mortality, immigration, and emigration. Populations possess characteristics of natality and density, which are not properties of individuals. Individuals produce young that are part of a population and under ideal conditions the population will grow, initially exponentially, until the decimating factors come into play to keep the population at carrying capacity. Understanding population dynamics involves rates of change from one year to the next.

Two theoretical models, inhibited and uninhibited, help to explain how a population grows and then levels off. Several examples for pheasant and deer indicate that populations of wild animals do increase in keeping with theoretical models. These models, while not useful for decision-making, do provide the conceptual basis for understanding what is possible when animals are introduced into favorable environments or reduced to low numbers through some decimating factor. The opposite of natality is mortality and this latter population characteristic can be modeled with the use of life tables. Life tables indicate the loss of animals by age class until the last animal in a cohort has died. Life tables are also useful in identifying age classes that might be subject to differential mortality.

Two systems have evolved for managing wildlife: single species and multispecies. The development of both systems started with single species largely because information on hunted or fished species was needed. The featured species system, which directs management to one species but in the process maintains habitat for other species, grew out of single species management. Environmental laws enacted in the 1970s required management information on all species of wildlife, not simply game animals. This requirement led to the development of new ways of combining single species information into a multispecies format. As a consequence of new ways of thinking about wildlife resulting from the rapid accumulation of scientific information and legislatively driven changes in planning for and management of wildlife, ecological factors became the fourth approach to multispecies management. Ecological factors emphasize indicators, diversity, edge, and habitat extent to set management direction. Both single species and multispecies management systems are of value and provide a range of management tools from which to select in making decisions.

REFERENCES

Allen, D.L. 1974. *Our wildlife legacy.* Funk and Wagnalls, New York.

Allen, D. L. 1979. *Wolves of Minong: Their vital role in a wild community.* Houghton Mifflin Co., Boston, MA.

Anderson, E.M., and M.J. Lovalio. 2003. Bobcat and lynx. P. 758–786 in *Wild mammals of North America,* G.A. Feldhamer, B.C. Thompson, and J.A. Chapman (eds.). The Johns Hopkins University Press, Baltimore, MD.

Andrewartha, H.G., and L.C. Birch. 1984. *The ecological web: More on the distribution and abundance of animals.* University of Chicago Press, IL.

Bean, M.J. 1977. *The evolution of national wildlife law.* Praeger, New York.

Bekoff, M., and E.M. Gese. 2003. Coyote. P. 467–481 in *Wild mammals of North America,* G.A. Feldhamer, B.C. Thompson, and J.A. Chapman (eds.). The Johns Hopkins University Press, Baltimore, MD.

Bendell, J.F. 1974. Effects of fire on birds and mammals. P. 73–138 in *Fire and ecosystems*, T.T. Kozlowski and C.E. Ahlgren (eds.). Academic Press, New York.

Binkley, D., M.M. Moore, W.H. Romme, and P.M. Brown. 2006. Was Aldo Leopold right about the Kaibab deer herd? *Ecosystems* 9:227–241.

Bissell, H.D., B. Harris, H. Strong, et al. 1955. The digestibility of certain natural and artificial foods eaten by deer in California. *Calif. Fish and Game* 41:57–78.

Block, W.M. 1993. The habitat concept in ornithology: Theory and applications. P. 35–91 in *Current Ornithology,* Vol 11, D.M. Power (ed.). Plenum Press, New York.

Block, W.M., D.M. Finch, and L.E. Brennan. 1995. Single-species versus multiple-species approaches for management. P. 461–476 in *Ecology and management of neotropical migratory birds,* T.E. Martin and M. Finch (eds.). Oxford University Press, New York.

Bock, C.E., and J.F. Lynch. 1970. Breeding bird populations of burned and unburned conifer forest in the Sierra Nevada. *Condor* 72:182–189.

Bowyer, R.T., V.V. Ballenberghe, and J.C. Kie. 2003. Moose. P. 931–964 in *Wild mammals of North America,* G.A. Feldhamer, B.C. Thompson, and J.A. Chapman (eds.). The Johns Hopkins University Press, Baltimore, MD.

Brown, E.R. (ed.). 1985. Management of wildlife and fish habitats in forests of western Oregon and Washington. USDA Forest Service, Pacific Northwest Region, Portland, OR.

Brown, J.L. 1969. Territorial behavior and population regulation in birds: A review and reevaluation. *Wilson Bull.* 81:293–329.

Buech, R.R., K. Siderits, R.E. Radtke, et al. 1977. Small mammal populations after a wildfire in northeast Minnesota. Res. Pap. NC-151. USDA Forest Service, St. Paul, MN.

Burns, R. M., and B. H. Honkala. 1990. *Silvics of trees of North America.* Vol 1, conifers, Vol 2, hardwoods. Agric. Hdbk. 654. USDA Forest Service, Washington, DC.

Calhoun, J. B. 1952. The social aspects of population dynamics. *J. Mammal.* 33:139–159.

Carson, R. 1962. *Silent spring*. Fawcett Publications, Greenwich, CT.
Caughley, G. 1966. Mortality patterns in mammals. *Ecology* 47:906–918.
Caughley, G. 1970. Eruption of ungulate populations with emphasis on Himalayan Thar in New Zealand. *Ecology* 51:53–72.
Caughley, G. 1979. What is this thing called carrying capacity? P. 2–8 in *North American elk: Ecology behavior, and management*, M.S. Boyce and L.D. Hayden Wing (eds.). University of Wyoming, Laramie.
CEQ. 1981. Environmental trends. Council on Environmental Quality. Exec. Office of the President, Washington, DC.
Chapman, J.A., and J.A. Litvaitus. 2003. Eastern cottontail. P.101–125 in *Wild mammals of North America,* G.A. Feldhamer, B.C. Thompson, and J.A. Chapman (eds.). The Johns Hopkins University Press, Baltimore, MD.
Chapman, R.N. 1931. *Animal ecology.* McGraw-Hill, New York.
Cheatum, E.L., and C.W. Severinghaus. 1950. Variations in fertility of white-tailed deer related to range conditions. *Trans. N. Amer. Wildl. Conf.* 15:170–189.
Chew, R.M., B.B. Butterworth, and R. Grechman. 1958. The effects of fire on the small mammal populations of chaparral. *J. Mammal.* 40:253.
Christian, J.J. 1950. The adreno-pituitary system and population cycles in mammals. *J. Mammal.* 31:246–260.
Clements, F.E. 1916. Plant succession: An analysis of the development of vegetation. *Carnegie Inst. Pub.* 242:1–112. Washington, DC.
Clements, F.E. 1949. *Dynamics of vegetation.* H.W. Wilson Co., New York.
Cole, G.F. 1958. Range survey guide. Federal Aid Project W-37-R. Montana Dept. of Game and Fish, Missoula.
Cole, L.C. 1954. Some features of random population cycles. *J. Wildl. Manage.* 18:2–24.
Cook, S.F. 1959. The effects of fire on a population of small rodents. *Ecology* 40:102–108.
Crawford, W.T. 1950. Some specific relationships between soils and wildlife. *J. Wildl. Manage.* 30:396–398.
Cunningham, J., R. Balda, and W. Gaud. 1980. Selection and use of range by secondary cavity nesting birds of the ponderosa pine forest. USDA Forest Service Research Paper RM-222, Rocky Mountain Forest and Range Experiment Station, Fort Collins, CO.
Cypher, B.L. 2003. Foxes. P. 511–546 in *Wild mammals of North America,* G.A. Feldhamer, B.C. Thompson, and J.A. Chapman (eds.). The Johns Hopkins University Press, Baltimore, MD.
Dasmann, R.F. 1981. *Wildlife biology.* John Wiley and Sons, New York.
Dasmann, W.P. 1948. A critical review of range survey methods and their application to deer range management. *Calif. Fish and Game* 34:189–207.
Daubenmire, R. 1968. Ecology of fire in grasslands. P. 209–273 in *Advances in ecological research,* Vol 5. Academic Press, New York.
Davidson, W.R., and V.F. Nettles. 1988. Field manual of wildlife diseases in the southeastern United States. SE Coop. Disease Study, University of Georgia, Athens.
Davies, N.B. 1978. Ecological questions about territorial behavior. P. 317–335 in *Behavioral ecology: An evolutionary approach,* J.R. Krebs and N.B. Davies (eds.). Sinauer Associates, Sunderland, MA.
Dayton, W.A. 1931. Important western browse plants. Misc. Pub. 101. U.S. Dept. Agric., Washington, DC.
Denney, A.H. 1944. Wildlife relationships to soil types. *Trans. N. Amer. Wildl. Conf.* 9:316–323.
Edwards, R.Y., and C.D. Fowle. 1955. The concept of carrying capacity. *Trans. N. Amer. Wildl. Conf.* 20:589–602.
Einarsen, A.S. 1945. Some factors affecting ring-necked pheasant population density. *Murrelet* 26(3):39–44.

Elton, C.S. 1942. *Voles, mice, and lemmings.* Clarendon Press, London.
Elton, C.S., and M. Nicholson. 1942. The ten-year cycle in numbers of lynx in Canada. *J. Animal Ecology* 11:215–244.
Emlen, J.N. 1973. *Ecology: An evolutionary approach.* Addison Wesley, Reading, MA.
Errington, P.L. 1945. Some contributions of a 15-year study of the northern bobwhite to a knowledge of a population phenomena. *Ecol. Mono.* 15:1–34.
Errington, P.L. 1967. *Of predation and life.* Iowa State University Press, Ames.
Feldhamer, G.A., B.C. Thompson, and J.A. Chapman (eds.) 2003. *Wild mammals of North America: Biology, management, and conservation,* Second Edition. The Johns Hopkins University Press, Baltimore, MD.
Fox, M.W. 1980. *The soul of the wolf.* Little, Brown and Co., Boston.
French, C.E., L.C. McEwen, N.D. MacGruder, et al. 1956. Nutrient requirements for growth and antler development in the white-tailed deer. *J. Wildl. Manage.* 20:221–232.
Gallizioli, S. 1965. *Quail research in Arizona.* Arizona Game and Fish Department, Phoenix.
Gehrt, S.D. 2003. Raccoon. P. 611–634 in *Wild mammals of North America,* G.A. Feldhamer, B.C. Thompson, and J.A. Chapman (eds.). The Johns Hopkins University Press, Baltimore, MD.
Giles, R.H., Jr., and N. Snyder. 1970. Simulation techniques in wildlife habitat management. P. 23–49 in *Modeling and systems analysis in range science,* D. Jamison (ed.). Range Science Dept. Series No. 5. Colorado State University, Fort Collins.
Grinnell, J. 1943. *Philosophy of nature.* University of California Press, Berkeley.
Gullion, G.W. 1960. The ecology of Gambel's quail in Nevada and the arid Southwest. *Ecology* 14:518–536.
Haapanen, A. 1965. Bird fauna of the Finnish forests in relation to forest succession. *I. Annu. Zoo. Fenn.* 2:153–196.
Hakala, J.B., R.K. Seemel, R.A. Richey, et al. 1971. Fires effects and rehabilitation methods: Swanson-Russian River fires. P. 87–89 in *Proceedings: A symposium on fire in the northern environment,* C.W. Slaughter, R.J. Burns, and G.M. Hansen (eds.). USDA Forest Service, Portland, OR.
Halls, L.K. 1970. Nutrient requirements of livestock and game. P. 10–24 in *Range and wildlife habitat evaluation,* Misc. Pub. No. 1147, A. Paulsen, Jr. and E.H. Reid (eds). USDA Forest Service, Washington, DC.
Hansen, C.S. 1983. Costs of deer-vehicle accidents in Michigan. *Wildl. Soc. Bull.* 11:161–164.
Harris, L.D. 1984. *The fragmented forest: Island biogeography theory and the preservation of biological diversity.* University of Chicago Press, IL.
Harris, L.D. 1988. Reconsideration of the habitat concept. *Trans. N. Amer. Wildl. and Nat. Resour. Conf.* 53:137–144.
Hewett, O. (ed.). 1954. A symposium on cycles in animal populations. *J. Wildl. Manage.* 18:1–112.
Hill, E.P. 1972. *The cottontail in Alabama.* Bull. No. 440, Alabama Agric. Exp. Sta. Auburn University.
Holbrook, H.L. 1974. A system for wildlife habitat management on southern National Forests. *Wildl. Soc. Bull.* 2:119–123.
Hoover, R.L., and D.L. Wills (eds.). 1984. *Managing forested lands for wildlife.* Colorado Div. of Wildl. in cooperation with USDA Forest Service, Rocky Mountain Region, Denver, CO.
Hornocker, M.G. 1969. Winter territoriality in mountain lions. *J. Wildl. Manage.* 33:457–464.
Howard, H.E. 1920. *Territory in bird life.* Dutton, New York.
Howard, W.E., R.L. Fenner, and H.E. Childs, Jr. 1959. Wildlife survival in brush burns. *J. Range Manage.* 12:230–234.
Johnson, R. 1970. *Tree removal along southwestern rivers and effects on associated organisms.* American Philosophical Society Yearbook. Philadelphia, PA. 431–432 p.

Karstad, L. 1979. Diseases of wildlife. P. 123–127 in *Wildlife conservation*, R.D. Teague and E. Decker (eds.). The Wildlife Society, Washington, DC.

Kendeigh, S.C. 1941. Territorial and mating behavior of the house wren. *Illinois Biol. Mono.* 10:1–20.

King, R.T. 1938. The essentials of a wildlife range. *J. For.* 36:457–464.

Klein, D.R. 1968. The introduction, increase, and crash of reindeer on St. Matthew Island. *J. Wildl. Manage.* 32:350–367.

Klopfer, P.H. 1969. *Habitats and territories: A study of the use of space by animals.* Basic Books, New York.

Knutson, R.M. 1987. *Flattened Fauna: A field guide to common animals of roads, streets, and highways.* Ten Speed Press, Berkeley, CA.

Komarek, E.V., Sr. 1969. Fire and animal behavior. *Tall Timbers Fire Ecol. Conf.* 9:161–207.

Kozicky, E.L., G.O. Hendrickson, and P.G. Homer. 1955. Weather and fall pheasant populations. *J. Wildl. Manage.* 19:136–142.

Krebs, C.J., M.S. Gaines, B.L. Keler, et al. 1973. Population cycles in small rodents. *Science* 179:35–41.

Lawrence, G.E. 1966. Ecology of vertebrate animals in relation to chaparral fire in the Sierra Nevada foothills. *Ecology* 47:278–291.

Leopold, A. 1933. *Game management.* Charles Scribner's Sons, New York.

Leopold, A. 1943. Wildlife in American culture. *J. Wildl. Manage.* 7:1–6.

Leopold, A., L.K. Sowls, and D.L. Spencer. 1947. A survey of over-populated deer ranges in the United States. *J. Wildl. Manage.* 11:162–177.

Linduska, J.P. (ed.). 1964. *Waterfowl tomorrow.* U.S. Fish and Wildl. Serv. Washington, DC.

Lyon, L.J. 1984. Road effects and impacts on wildlife and fisheries. Paper presented at Symposium on Forest Trans., 11–13 December, Casper, WY. USDA Forest Service, Region 2, Denver, CO.

MacArthur, R.H., and E.O. Wilson. 1967. *The theory of island biogeography.* Princeton University Press, Princeton, NJ.

Mackie, R.J., J.G. Kie, D.F. Pac, and K.L. Hamlin. 2003. Mule deer. P. 889–905 in *Wild mammals of North America,* G.A. Feldhamer, B.C. Thompson, and J.A. Chapman (eds.). The Johns Hopkins University Press, Baltimore, MD.

Mannan, R.W., M.L. Morrison, and E.C. Meslow. 1984. The use of guilds in forest bird management. *Wildl. Soc. Bull.* 12:426–430.

Martin, A.C., H.S. Zim, and A.L. Nelson. 1951. *American wildlife and plants: A guide to wildlife food habits.* Dover Publications, New York.

McCullough, D.R. 1979. *The George Reserve deer herd: population ecology of a K-selected species.* University of Mich. Press, Ann Arbor.

McMillan, I.I. 1964. Annual population changes in California quail. *J. Wildl. Manage.* 28:702–711.

McWhirter, N. 1981. *Guinness book of world records.* Sterling Publishing Co., New York.

Mech, L.D., L. Frenzel, Jr., R. Meam, and J. Winship. 1971. Ecological studies of the timber wolf in northeastern Minnesota. Res. Pap. NC 52. USDA Forest Service, St. Paul, MN.

Meehan, W. R., F. J. Swanson, and J. R. Sedell. 1977. Influences of riparian vegetation on aquatic ecosystems with particular reference to salmonid fishes and their food supply. P. 137–145 in *Importance, preservation, and management of riparian habitat,* R. Johnson and D.A. Jones (tech. cords.). Gen. Tech. Rep. RM-43, USDA Forest Service, Rocky Mountain For. and Range Exp. Sta., Fort Collins, CO.

Mitchell J.E., and D.R. Freeman. 1993. Wildlife-livestock-fire interactions on the North Kaibab: A historical overview. USDA Forest Service Gen. Tech. Report RM-222. Fort Collins, CO.

National Academy of Science. 1962. Nutrient requirements of laboratory animals. P. 990 in *Nutrient requirements of domestic animals.* Natl. Res. Council Pub., Washington, DC.
National Academy of Science. 1963. Nutrient requirements of beef cattle. P. 1137 in *Nutrient requirements of domestic animals.* Natl. Res. Council Pub., Washington, DC.
National Academy of Science. 1964. Nutrient requirements of sheep. P. 1193 in *Nutrient requirements of domestic animals.* Natl. Res. Council Pub., Washington, DC.
Nelson, R.D., and H. Salwasser. 1982. The Forest Service wildlife and fish habitat relationships program. *Trans. N. Amer. Wildl. and Nat. Res. Conf.* 47:174–182.
Nestler, R.B. 1949. Nutrition of bobwhite quail. *J. Wildl. Manage.* 13:342–358.
New Mexico Game and Fish Dept. 1973. *Big game habitat analysis for New Mexico.* N.M. Game and Fish Dept., Mimeo, Santa Fe.
Nice, M. 1941. The role of territory in bird life. *Amer. Mid. Nat.* 26:441–487.
Odum, E.P. 1971. *Fundamentals of ecology.* W.B. Sanders Co., Philadelphia, PA.
Pattee, O.H., and S.K. Hennes. 1983. Bald eagles and waterfowl, the lead shot connection. *Trans. N. Amer. Wildl. and Nat. Resour. Conf.* 48:230–237.
Patton, D.R. 1978. *Run Wild: A storage and retrieval system for wildlife habitat information.* USDA Forest Service Gen. Tech. Rep. RM-51, Fort Collins, CO.
Patton, D.R. 1984. A model to evaluate Abert squirrel habitat in uneven aged ponderosa pine. *Wildl. Soc. Bull.* 12:409-414.
Patton, D.R. 1987. Is the use of management indicator species feasible? *West. J. Applied For.* 2:33–34.
Patton, D.R., and M. Ertle. 1982. *Wildlife food plants of the Southwest.* Wildl. Unit Tech. Rep. USDA Forest Service, SW Region, Albuquerque, NM.
Payne, N.F., and F. Copes. 1986. *Wildlife and fisheries habitat improvement handbook.* Wildlife and Fisheries Report, USDA Forest Service, Washington, DC.
Peek, J.M. 2003. Wapiti. P. 877–888 in *Wild mammals of North America,* G.A. Feldhamer, B.C. Thompson, and J.A. Chapman (eds.). The Johns Hopkins University Press, Baltimore, MD.
Pelton, M.R. 2003. Black bear. P. 547–555 in *Wild mammals of North America,* G.A. Feldhamer, B.C. Thompson, and J.A. Chapman (eds.). The Johns Hopkins University Press, Baltimore, MD.
Perry, C., and R. Overly. 1976. Impact of roads on big game distribution in portions of the Blue Mountains of Washington. P. 62–68 in *Proceedings: A symposium on elk, logging, and roads,* J. M. Peek (chair). University of Idaho, Moscow.
Pierce, B.M., and V.C. Bleich. 2003. Mountain lion. P. 744–757 in *Wild mammals of North America,* G.A. Feldhamer, B.C. Thompson, and J.A. Chapman (eds.). The Johns Hopkins University Press, Baltimore, MD.
Rasmussen, D.I. 1941. Biotic communities of the Kaibab plateau. *Ecol. Mono.* 3:229–275.
Rippe, D.J., and R.L. Rayburn. 1981. Land use and big game population trends in Wyoming. W/CRAM-81-W22. U.S. Fish and Wildlife Serv., Washington, DC.
Root, R.B. 1967. The niche exploitation pattern of the blue-gray gnatcatcher. *Ecol. Monogr.* 37:317–349.
Rosene, W. 1969. *The bobwhite quail: Its life and management.* Rutgers University Press, New Brunswick, NJ.
Rowland, M.M., and C.D. Vojta (eds.). 2013. *A technical guide for monitoring wildlife habitat.* USDA Forest Service Gen. Tech. Report WO-89, Washington, DC.
Rumsfeld, D. 2002, February 12. *Pentagon news briefing.* U.S. Department of Defense, Washington, DC.
Salwasser, H. 1979. The ecology and management of the Devil's Garden interstate deer herd and its range. Ph.D. Dissertation. University of California, Berkeley.

Schantz, H.L. 1911. Natural vegetation as an indicator of the capabilities of land crop production in the Great Plains Area. Bul. 210, Bureau of Plant Industry, U.S. Department of Agriculture, Washington, DC.

Schoener, T.W. 1968. Sizes of feeding territories among birds. *Ecology* 49:123–141.

Schwartz, C.C., S.D. Miller, and M. Haroldson. 2003. Grizzly bear in *Wild mammals of North America,* G.A. Feldhamer, B.C. Thompson, and J.A. Chapman (eds.). The Johns Hopkins University Press, Baltimore, MD.

Scott, V.E., K.E. Evans, D.R. Patton., et al. 1977. Cavity-nesting birds of North American forests. Agric. Hdbk. 511. USDA Forest Service, Washington, DC.

Severinghaus, C.W. 1972. Weather and deer populations. *The Conservationist* 27:28–31.

Severinghaus, W.D. 1981. Guild theory development as a mechanism for assessing environmental impacts. *Envir. Manage.* 5:187–190.

Short, H. L., and K. P. Burnham. 1982. Technique for structuring wildlife guilds to evaluate impacts on wildlife communities. Special Sci. Rep. Wildl. 244, U.S. Fish and Wildl. Serv., Fort Collins, CO.

Silvy, N. J. (ed.). 2012. *The wildlife techniques manual,* Seventh edition, Vol 1. The Wildlife Society, Washington, DC.

Smith, C.M., E.D. Michael, and H.V. Wiant, Jr. 1975. Size of West Virginia deer as related to soil fertility. *W.Va. Agric. and For.* 6:12–13.

Spight, T.M. 1967. Species diversity: A comment on the role of the predator. *Amer. Nat.* 101:467–474.

Stephenson, R.L. 1974. Reproduction biology and food habits of Abert's squirrels in central Arizona. M.S. thesis, Arizona State University, Tempe.

Stoddart, L.A., A.D. Smith, and T.W. Box. 1975. *Range management.* McGraw-Hill, New York.

Szaro, R.C. 1985. Guild theory: Fact or fiction. *Environ. Manage.* 10:681–688.

Talbott, M.W. 1937. Indicators of southwestern range conditions. Farmers Bul. 1782, U.S. Dept. Agric., Washington, DC.

Taylor, W.P. 1940. Ecological classification of the mammals and birds of Walker County, Texas and some adjoining areas. *Trans. N. Amer. Wildl. Conf.* 5:170–176.

Teer, J.G., J.W. Thomas, and E.A. Walker. 1965. *Ecology and management of white-tailed deer in the Llano Basin of Texas.* The Wildlife Society, Washington, DC.

Thomas, J.W. (tech. ed.). 1979. Wildlife habitats in managed forests: The Blue Mountains of Oregon and Washington. Agric. Hdbk. 553. USDA Forest Service, Washington, DC.

Thomas, J.W., C. Maser, and J.E. Rodiek. 1979a. Snags. P. 60–77 in *Wildlife habitats in managed forests: The Blue Mountains of Oregon and Washington,* J.W. Thomas (ed.). Agric. Hdbk. 553. USDA Forest Service, Washington, DC.

Thomas, J.W., C. Maser, and J.E. Rodiek. 1979b. Riparian zones. P. 40–47 in *Wildlife habitats in managed forests: The Blue Mountains of Oregon and Washington,* J.W. Thomas (ed.). Agric. Hdbk. 553. USDA Forest Service, Washington, DC.

Tuchman, B.W. 1978. *A distant mirror: The calamitous 14th century.* Alfred A. Knopf, New York.

Turner, J.C., Jr. 1973. Water, energy and electrolyte balance in the desert bighorn sheep, Ovis canadensis. Ph.D. dissertation. University of California, Riverside.

USDA. 2000. Forest cover types (map). National Atlas. USDA Forest Service, Washington, DC and U.S. Geological Survey, Denver, CO.

USDI. 1966. *Fish, wildlife and pesticides.* U.S. Fish and Wildlife Service, Washington, DC.

Verner, J. 1984. The guild concept applied to management of bird populations. *Envir. Manage.* 8:1–14.

Verner, J., and A. Boss. 1980. *California wildlife and their habitats: Western Sierra Nevada.* USDA Forest Service Gen. Tech. Rep. PSW-37, USDA Forest Service, Berkeley, CA.

Verner, J., M.L. Morrison, and C.J. Ralph. 1984. *Wildlife 2000: Modeling habitat relationships of terrestrial vertebrates*. University of Wisconsin Press, Madison.
Vogl, R. J. 1977. Fire: A destructive menace or natural process. P. 261–289 in *Recovery and restoration of damaged ecosystems*, J. Cairns, Jr., K.L. Dickson, and E.E. Herricks (eds.). University of Va. Press, Charlottesville.
von Liebig, J. 1840. *Chemistry in its application to agriculture and physiology.* Taylor and Walton, London.
Wagner, F.H., and L.C. Stoddart. 1972. Influence of coyote predation on black-tailed jackrabbit populations in Utah. *J. Wildl. Manage.* 36:329–343.
Wallmo, O.C., L.H. Carpenter, W.L. Regelin, et al. 1977. Evaluation of deer habitat on a nutritional basis. *J. Range Manage.* 30:122–127.
Ward, A.L. 1976. Elk behavior in relation to timber harvest operations and traffic on the Medicine Bow Range in South Central Wyoming. P. 32–43 in *Elk, logging, roads*, J.M. Peek (chair). University of Idaho, Moscow.
Webster's Seventh New Collegiate Dictionary. 1966. *Webster's seventh new collegiate dictionary,* Third edition. Houghton Mifflin Company, Boston, MA.
Wesley, D.E., K.L. Knox, and J.G. Nagy. 1970. Energy flux and water kinetics in young pronghorn antelope. *J. Wildl. Manage.* 34:908–912.
Wilson, E.O. 1975. *Sociobiology: The new synthesis.* Belknap Press, Cambridge, MA.
Wynne-Edwards, V.C. 1962. *Animal dispersion in relation to social behavior.* Hafner, New York.
Yeager, L.E. 1961. Classification of North American mammals and birds according to forest habitat preference. *J. For.* 9(4):671–674.
Yensen, E., and P.W. Sherman. 2003. Ground squirrels. P. 211–231 in *Wild mammals of North America,* G.A. Feldhamer, B.C. Thompson, and J.A. Chapman (eds.). The Johns Hopkins University Press, Baltimore, MD.

All Things Wild and Wonderful

7

Robin Hood had some things right! Not only do the Robin Hood legends address the theme of economic equality ("stealing from the rich to give to the poor"), they also focus on the rights of the ordinary citizen to use the forests, including hunting game animals. This latter component of the legend derives from reality—the Magna Carta, signed in 1215 by King John of England. Although most often held out as a fundamental pillar of our legal system (which is trial by a jury of peers), some 75–80% of the original document addressed grievances about the use of the king's forests; in other words, forest and wildlife policy and management.

The magnitude of the forestry component of the Magna Carta led to the creation of a separate "Charter of the Forest" in 1218 and revised in 1225. Thus, the Forest Charter served as a foundational document that helped define the relationship between the management of habitat (the landowner, generally) and the management of the animals, a right generally reserved to the states, as described in more detail in Chapter 6 and later in this chapter. The fundamental concept is that of management—the use of means to accomplish some end goal. Effective management requires effective planning—"the process of establishing goals, defining general criteria for measuring success, formulating a structure for the final deliverable, and creating a plan for the subsequent phases" (Kister 2016, pg. 202)—the focus of this chapter.

7-1 Forest Planning: The Big Picture

Forest planning involves a complicated and unique interaction between law, policies, processes, and scale. Many other industries possess one or more of these elements. For example, the pharmaceutical and energy industries are subject to very strict and complicated regulations through federal and sometimes state laws. Agriculture occurs on a vast scale across the country. However, three characteristics of forests frame the larger forest planning process.

Distinctive Aspects of Forests

Sound planning for forest resources requires a clear understanding of some of their distinctive aspects as a natural resource. According to G. Robinson Gregory (1987) six of these aspects are:

- **Trees in a forest are both the "factory" and "product."** In contrast to other industrial complexes, cutting trees results in the destruction of some of the factory's productive capacity.
- **Immobility of standing timber.** Although eventually delivered to a converting facility (e.g., a saw or pulp mill) forests are inextricably tied to the land in a certain location. Moving an automobile factory would cost perhaps billions of dollars; but like agriculture, essentially no amount of money can move the forest as a production unit.
- **Length of production period.** It takes a long time for trees to reach a point where they qualify as a product. Some hardwood plantations may reach merchantability in 5–10 years but generally, a first product from thinning a forest may not occur for 15 years with a final harvest in 25 years. Thus, time figures into planning of forests in a big way.
- **Production flexibility.** Forests tend to have long "shelf lives." Delaying the harvest of a stand by 1 or 2 or 5 years will have relatively little effect on product quality or volume. In addition, once a tree reaches certain size thresholds it can be used for multiple products.
- **Externalities.** Not only do forests produce trees as a salable product but the production of these trees come with other positive and negative benefits that effect other uses of forests—a phenomenon known as "externalities." Examples of a positive externality is growing trees in a forest thereby creating recreation opportunities and the construction of roads to harvest timber creates access to these opportunities with little or no cost to recreationists. Of course, cutting trees can have negative effects on the quality of the recreation experience a forest produces.
- **Land ownership patterns.** No other industry in the United States has such a varied ownership as does forests, especially in terms of ownership of the productive capacity of forests. As noted in Chapter 3, public agencies own some 50% of the forest land in this country.

When combined, these distinctive aspects greatly complicate the entire planning process.

Desired Future Conditions and Dynamics

Over the past 15–20 years, a shift in emphasis in planning goals has occurred, especially in public agencies. Instead of focusing on the production aspects of forest management—traditionally measured in terms of the amount of timber cut—increasingly, management goals have changed to emphasize stand structure with timber harvests to achieve an end, rather than the end itself. Foresters have long known that the conditions of forests, especially in the West, bear little resemblance in many cases to the conditions that existed before European-American settlement.

A combination of unrestrained harvesting, massive fires, and aggressive fire suppression efforts, resulted in forest conditions with tree densities far exceeding historic conditions, substantial changes in species mixes, and smaller trees (Chapter 3; Covington and Moore 1994; Bonnicksen 2000). Although a discussion of the causes for these extends beyond the scope of our topic here, the Forest Service has embraced a planning approach that includes the explicit recognition of desired future conditions in its 2012 Planning Rule process (USDA 2018). The rule provides the guidelines and standards that frame revisions of forest plans to include the use of the best available science. And though the term implies a static condition, the philosophy behind the concept is to identify the natural processes and management regimes that maintain desired future conditions over time.

Sustainability

Foresters have identified and embraced the concept of sustainability for hundreds of years. Forests in Europe had experienced systematic harvesting for centuries in support of a stable society that had relatively little technological change for hundreds of years. The foresters had fears of a timber shortage. Given that forests provided the prime source of energy and building materials, such a shortage could prove devastating to the local and national economies. Therefore, foresters developed the concept of sustained yield. This concept called for an equal annual or periodic flow of timber harvest in perpetuity (Davis et al. 2001).

Until the 1900 establishment of the first forestry school in the United States at Yale University, foresters in the United States received their education in Europe. They brought back with them the concept of sustained yield and the framework of forest regulation to help achieve it (Chapter 5). This concept, codified for the Forest Service in the Multiple Use Sustained Yield Act, has served as a first principal for the management of both public and private forests.

Interest in the sustainability of forest resources has increased dramatically in the past 10 years. The Sustainable Forestry Initiative (SFI) under the auspices of the Program for the Endorsement of Forest Certification (PEFC) has developed strong standards and practices to certify forest management activities based on 13 Principles, 15 Objectives, 37 Performance Measures, and 101 Indicators. These requirements include measures to protect water quality, biodiversity, wildlife habitat, species at risk and forests with exceptional conservation value and include auditing (SFI 2018).

The American Tree Farm System has developed similar standards for private landowners that also include third-party audits and certification (ATFS 2018). In addition, various analytical approaches have been developed to help plan and evaluate sustainability projects (Heink et al. 2016; Diaz-Balteiro et al. 2017b). The work in this area has progressed to the point that Diaz-Balteiro et al. (2017a) conducted a systematic review of 271 scientific articles on the use of multi-criteria methods to help determine the sustainability of systems.

Scale and landownership have major effects on sustainability. The original concept of sustained yield derived from concerns about a potential shortage of timber accessible from villages and towns, given the limited transportation systems of the time. As transportation has improved, accessible areas have expanded. This creates a more robust planning problem. Imagine trying to achieve a sustained yield of timber from a five-acre forest. Given that it takes a long time for trees to grow, many years could pass between harvests—a very difficult situation if wood is the primary building material. On the other hand, creating a sustained yield plan for 500,000 acres, although computationally much more complex, provides the opportunities to create a system with an annually available harvest. Of course, this scenario includes the hidden assumption of identical management goals for all the 500,000 acres. Considering landownership patterns, this situation seldom occurs. As a result, management plans and activities for a given area tend towards being both piecemeal and uncoordinated.

7-2 Resource Planning

Previous chapters have discussed trees, forests, environment, habitat, wildlife, and ecosystem management. Individually, none of these provides the components for effective management, but they require integration into a scheme that gives all the elements direction and purpose. Today's effective resource manager must devote time and energy to devise efficient ways to complete inventories and projects within the systematic context of the planning process. This process requires the proper allocation of time, personnel, and physical resources.

Why Plan

Although the overall forest land acreage in the United States has remained stable since the 1930s (Chapter 3), the land base for resource management has decreased due to the increasing number of cultural landscape features (e.g., increased wilderness areas and the expansion of human settlements and the wildland-urban interface) needed by an increasing human population and changing wants and needs. Present conflicts in allocating land to wildlife habitat can only intensify with population growth. The problems of allocating resources increase in complexity because different wildlife species have different food and cover requirements; this requires setting aside a variety of habitats to accommodate these species. The distribution of forest ownership, as detailed in Chapter 3, further complicates and enhances the need for planning.

Resource planning is a special case of the generalized planning process where the focus of attention is on some natural resource variable—e.g., wildlife habitat, timber production, recreation use, etc. But before discussing the resource planning process (Section 7-3), it is important to answer the question of "Why?" as in "Why plan?" Simply stated, we plan for two reasons:

- we want to determine if we can get to where we want to go from where we are and, if so, how to get there and
- in many cases, especially in natural resources, we have legal requirements to adhere to.

Public Sector Planning

Most state and federal natural resource agencies plan at some level because of laws. Not only does planning discharge the agencies' legal responsibilities but it also responds to a world of change and competition for financial resources and space. Change can be orderly to meet specific goals and objectives or it can be random and non-directed. Orderly human-induced change requires some level of control. The planning process establishes criteria, principles, and procedures for control and documents the sequence in which events should happen. The planning process and the resultant plan serve as an end or goal.

Wildlife and forest managers must compete with the natural resource management goals of other agencies for resources to manage for wildlife food and cover, but because wildlife is part of an interdependent biological community, forest wildlife managers must also cooperate "so that there is a place to compete for" (Leopold 1933, pg. 239). Given the statutes and regulations governing the actions of managers; long-range planning, in terms of the natural development of life communities, is one of the realistic and efficient ways to bring land into condition for maximum production of wildlife (Allen 1973).

The planning process serves as the forum for the integration of competing demands and objectives to marshal all the available resources into a practical working tool for management and staff. While neither well documented nor thoroughly understood, in a general way, wild animals are crucial in the functioning of ecosystems, including nutrient cycling and energy transfer. Because of the ecological roles wild animals have as an ecosystem component, the wildlife management function in an agency deserves a funding level equal to that which other natural resource functions, such as timber, water, and livestock, have had for many years.

To adequately assemble all plant and animal populations at any unit level requires a detailed plan of action. Employing the knowledge and insights of a variety of professional personnel, foresters, ecologists, and wildlife biologists in the planning process is a necessity. The trend away from single species or single resource management to ecosystem management is a positive approach to insuring that few if any animal species go below viable population levels because of human activities, unless done as a planned objective. Exceptions to multispecies management and planning exist: when single species management is biologically sound and legally justifiable or required, such as with endangered species and wild horses and burros.

To compete for dollars and space, to cooperate in the formulation and selection of alternatives, and to integrate wildlife objectives into ecosystem management plans, forest biologists must have access to the kind of information outlined in the preceding chapters and later in this chapter. Now more than ever, biologists as planners must investigate and understand how wildlife interacts with and affects other ecosystem components. Ecosystem managers must ensure the full consideration and inclusion of all resources in the planning process. At the federal level, the allocation of habitat to forest wildlife takes place in the planning process at the forest level based on the mandated principles of multiple use of resources. A further complication in the allocation process results from legislation requiring the maintenance of viable populations of all native vertebrate species in the planning area. This legislative mandate is a consequence of the biological fact that virtually all animal species are included in and vital to the functioning of an ecosystem.

The balance among resource use, timber, water, livestock, wildlife and ecosystem functioning, energy flow, and nutrient cycling reflects the system of political and biological decisions that result in forest plans. A political decision determines the acreage designated for a given habitat and how many animal species will be maintained in the same habitat. A biological decision determines the effect of the political decision on animal and plant species in the management unit and provides alternatives whenever conflicts arise.

Land management planning in and of itself does not create, protect, or enhance the habitat of any animal species since planning does not directly affect an animal's ability to survive and reproduce. Land-use planning depends in the first instance on legislative authority empowering a federal or state agency to act and in the second instance on the policies and procedures formulated by that agency to use its authority. A long list of legislative statues provides the authority for various U.S. agencies—the Fish and Wildlife Service, the Bureau of Land Management, and the Forest Service—to manage natural resources, including fish and wildlife on public lands, and to cooperate with state agencies in conducting these management activities. While federal agencies do have the authority in a few cases to manage populations (e.g., migratory species) on public lands, the generally accepted division of responsibility is that habitat management on federal land is the responsibility of the federal agency, but state agencies control and enforce population regulation under state law—a situation derived from provisions in the Magna Carta.

Because this book is particularly concerned with ecosystem management across the United States, it is appropriate to summarize a few of the major laws controlling federal land management agencies. The Authority and Direction section of Chapter 6 includes a discussion of several important Acts that we will not repeat here. But, the three following Acts are important because of the direction they set for lands administered by the Secretary of Agriculture and the Secretary of the Department of Interior.

SIKES ACT (1967). Sikes Act directs the Secretary of Agriculture and the Secretary of the Interior, in cooperation with state agencies, to plan, develop, maintain, and coordinate programs for the conservation and rehabilitation of wildlife, fish, and game. It also provides for fish and wildlife conservation programs on federal lands, including authority for cooperative state/federal plans and authority to enter into agreements with states to collect fees for funding programs identified in those plans.

WILD AND FREE ROAMING HORSE AND BURRO ACT (1970). This act requires the Bureau of Land Management and the Forest Service to manage wild (actually feral) horses and burros on lands under their jurisdiction. Migratory birds and threatened and endangered species, where federal agencies have the responsibility to manage animal populations, are also managed under these two agencies as laid out in this Act.

FEDERAL LAND AND POLICY MANAGEMENT ACT (1976). The Federal Land and Policy Management Act is an organic act for the Bureau of Land Management in the Department of Interior, roughly comparable to the National Forest Management Act for the Forest Service. It authorizes the management of public lands to provide habitat for wildlife and domestic animals.

Legislative authority to manage does not automatically result in a management strategy by a state or federal entity. The agency must develop a policy to guide management in the allocation of staff, finances, and other resources to achieve their goals. Policy, in short, is the setting of a course of direction. The first step in formulating policy is the preparation of a statement that summarizes the objectives to be realized and the process by which they will be reached. Policy statements must be clear and easily communicated through a chain of command.

An example of a policy statement is found in a Forest Service Memorandum (USDA 2008, par. 1), which states:

> The purpose of this memorandum is to state the policies of the U.S. Department of Agriculture with respect to management of fish and wildlife and their habitats and to prescribe specific actions to implement the policies. The Department's prime responsibility is to help maintain sufficient and efficient production capability of farm, forest, water, and rangeland resources for the public benefit, now and in the future, and to encourage and support proper use, management, and conservation of those natural resources. Programs to meet this mission are carried out through research, education, technical and financial assistance to landowners, managers, producers and consumers, and through the management of public land for which the Department is responsible, in cooperation with State and local agencies.

The laws listed here focus on the wildlife management aspects of the mission of the Forest Service. *The Principal Laws Relating to Forest Service* (USDA 1993) contains a comprehensive identification and description of all the laws that effect Forest Service management.

Some states and local governments have their own management planning requirements. For example, the Michigan Department of Natural Resources manages 4.5 million acres of forest land, including 700,000 acres specifically designated as "State Wildlife Areas" (MDNR 2018). Their plan details a strategic framework for management, including the identification of management priorities as well as processes for coordination with other organizations (Derosier et al. 2015). The Wisconsin Department of Natural Resources has created a similar plan for its state forests (WDNR 2015).

Private Sector Planning

Private forest lands contain over 57% of the total forests in the United States (Chapter 3). Although some federal laws (e.g., the Endangered Species Act) regulate management practices on these lands, private landowners do not face the same planning requirements outlined for the public sector, especially for the U.S. Forest Service and the Bureau of Land Management. However, many private landowners have voluntarily created management plans for their forest holdings specifically designed to enhance wildlife habitat. For some owners, the enhancement serves to increase the economic value of their lands as individuals and groups will pay for the exclusive access to lands for hunting. Other landowners have non-consumptive use of wildlife—e.g., viewing—as a key management goal. Regardless of intent, considerable time, effort, and resources go into the creation and implementation of wildlife habitat management plans.

Owners of small tracts of forest lands have unique challenges when trying to manage their lands. Legal title to the land may be spread across multiple individuals as the land was passed along over time or that may have resulted from the investment of multiple individuals. In such cases, landowner goals may differ. This can occur even if the original title rested in a set of individuals who had similar interests. Over generations, interests and goals change and such changes can increase the difficulty of management. In addition, the owner(s) may have neither the expertise nor time to manage their own lands.

Land parcel size can also increase management difficulties. Sustained yield considerations are difficult to maintain over time with small tract size when harvests occur at very long intervals. A landowner may also face financial difficulties that force a timber sale that interrupts the overall forest management plan. Property taxes, usually due annually, may also force the early cut of timber so that the property does not go tax delinquent. Pressures to sell may also come from increasing urban/suburbanization and development within the wildland-urban interface and resulting increases in property values.

Managing for wildlife habitat adds yet another layer of complexity to the planning process. As noted earlier in this chapter, wildlife, especially larger species (e.g., brown bear and cougar), can range over many property boundaries and may have habitat needs that spread into the hundreds of square miles. In such case, management of one square mile (640 acres) may have relatively little wildlife living within the area unless these acres include key habitat features of den/roost (e.g., a cave for bats) or water.

Private forest land holders have multiple sources for help if they have an interest in creating forest and wildlife management plans for their property. Forestry programs associated with Land-Grant Universities (LGU) will have extension services that can help small private forest landholders develop management plans. The State and Private Forestry division of the U.S. Forest Service also assists non-industrial private landowners, often working in cooperation with the University Extension Service.

Many state forestry agencies have their own small landowner assistance units that may also work with their counterparts from the Forest Service and universities. Larger industrial private landowners may also offer management assistance. Finally, numerous forestry and wildlife management consultants exist. The size of their firms and range of services they offer vary from individual practitioners to multinational corporations. The Association of Consulting Foresters provides one avenue to help find assistance in management planning. Regardless of intent, considerable time, effort, and resources go into the creation and implementation of forest and wildlife habitat management plans. Some states also have strict environmental planning laws that regulate private planning. For example, the California Environmental Quality Act of 1970 (CEQA) states:

> CEQA applies to certain activities of state and local public agencies. A public agency must comply with CEQA when it undertakes an activity defined by CEQA as a "project." A project is an activity undertaken by a public agency or a private activity which must receive some discretionary approval (meaning that the agency has the authority to deny the requested permit or approval)

> from a government agency which may cause either a direct physical change in the environment or a reasonably foreseeable indirect change in the environment. (CNRA 2014, "References")

In addition, the state's Z'Berg-Nejedly Forest Practices Act (1973) regulates timber harvesting on state and private lands in California. This Act requires a state-approved Timber Harvest Plan (THP) prior to any harvesting operations. The Forest Practices Act requires that all THPs consider wildlife and fisheries (CNRA 2012). Other states with NEPA-like laws include Connecticut, Georgia, Hawaii, Minnesota, and Washington—both the state and the District of Columbia.

Clean Water Act of 1972

Transcending both public and private land management planning, the Federal Clean Water Act provided the regulatory framework for regulating pollutants into the waters of the United States. Of importance to the management of forest lands are the regulations that apply to non-point source pollutants. Non-point source pollutants are those that cannot be attributed to a single location, such as a factory or a sewage treatment plant, but instead result from larger scale activities, such as agricultural practices. In forest management, such activities include road construction and maintenance, timber harvesting, and site preparation, which are some of the key management activities used to manipulate wildlife habitat (Chapter 6). Thus, land managers, both public and private, must create plans that meet Clean Water Act requirements.

7-3 Planning Process

Planning is a process that sets a management direction for the future. It is necessary not only to discharge a responsibility of an organization but to properly respond to a world of change to meet specific goals given an expected level of funding and personnel.

Description

Given that we know why we plan, at least in a general sense, leads us to the topic of how we plan. Good planning involves a continuing cycle of development, implementation, and evaluation that recognizes and incorporates the iterative nature of the process (Figure 7-1).

Unfortunately, these latter two steps—implementation and monitoring—often receive little or no attention in the planning process. The process stops with the plan of "what to do" with insufficient (or non-existent) treatment of how to achieve the results (implementation), or to monitor and assess the efficacy of the planning and implementation (evaluation). Ignoring these latter two steps all but ensures failure.

The iterative nature of planning frames the details of the multi-step planning process. Establishing goals serves as a convenient and traditional starting place for planning, although the process has multiple potential starting places. From goals, we move to identifying desired outcomes, data collection, the generation of alternatives, and eventually to revising plans based on monitoring (Figure 7-2).

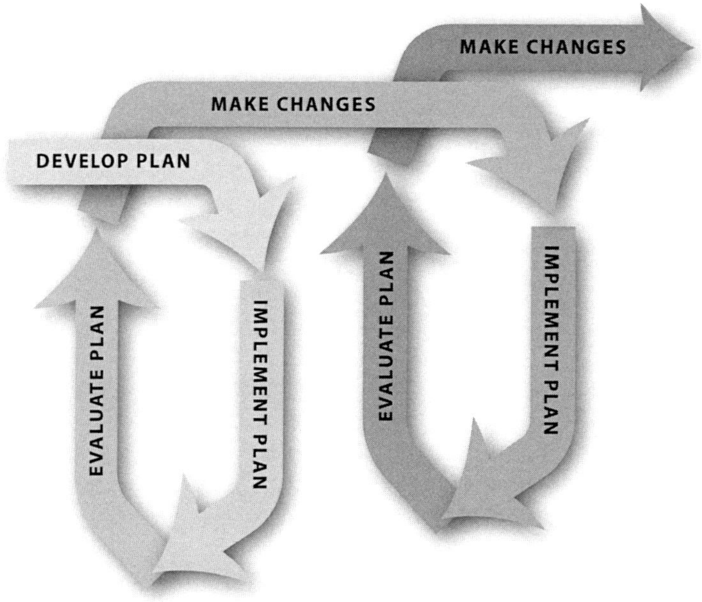

Figure 7-1 Iterative planning process.

Adapted from Environmental Protection Agency 2008.

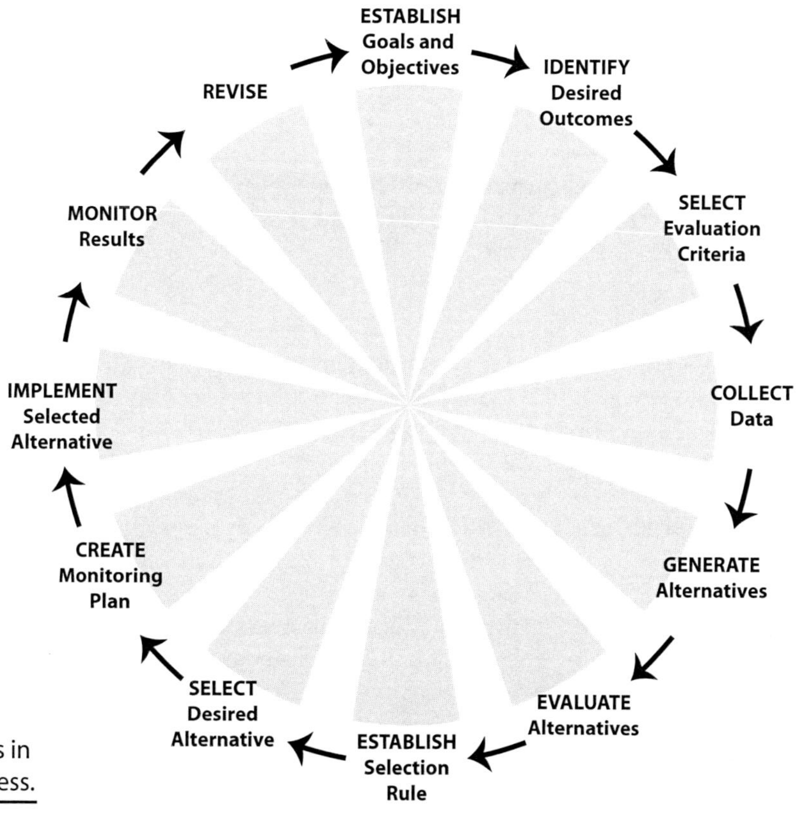

Figure 7-2 Steps in the planning process.

Each of these steps has its own process, creating a "double iterative" process of circles within circles. For example, evaluating alternatives may require statistical analysis (Chapter 4) and the use of a decision support system. The data needs for the evaluation tools will then drive the needed data, which in turn shapes inventory design (and costs). Alternatively, a decision support system (Section 7-4) may be created or an existing one modified to use existing data. The overall planning goals and evaluative criteria define the larger discussion about data needs and evaluation methods. However structured, feedback loops exist between and among the different steps in the planning process.

Planning Hierarchy

A characteristic of planning is its hierarchical nature. Planning exists at various levels in scale, complexity, and specificity. A classic three-level hierarchy describes planning at the strategic, operational, and tactical levels (Figure 7-3).

For example, a timber harvest plan may contain extreme detail with respect to specific activities, their timing, and the required resources—all of which are elements of tactical planning. However, the sale area may only total a few hundred acres. On the other hand, an organization's operational plan may encompass millions of acres but will generally not have the level of specifics or detail of a project-level plan. Strategic plans guide an agency or organization in meeting long-term goals and are stated in economic, technical, and political terms. Strategic plans are often referred to as "long-term plans."

Figure 7-3
Planning hierarchy.

Source: U.S. Department of Homeland Security, Federal Emergency Management Agency 2010.

The scale of the plan will dictate not only the kind and level of detail of information included but the process and format as well. Plans at the national and regional levels must be presented in a standardized format to facilitate understanding by management agencies, conservation organizations, and interested constituencies (aka "stakeholders") because these plans relate to the same laws and sets of conditions.

Operational plans guide specific actions and allocation of resources on a shorter time and smaller area scale. Such plans may be based on function, such as a transportation plan, or be linked to designated land areas—or an overlay of function and location. An operational plan focuses on specific land management units (or parts of them), specific ecosystems and certain organisms therein, the relationships between the environmental characteristics of the unit and their organisms, and the ways to maintain or allocate these relationships. Operational plans are known in natural resource professions as management plans. While national and regional plans are important, field foresters and biologists tend to focus their efforts in preparing the local plan or management unit plan, so the focus here is on this kind of planning. Tactical plans disaggregate management plans to a scale that allows implementation.

The Forest Service employs a three-level planning model. At the broadest level, the agency creates a national strategic plan that sets the direction for the entire organization across all areas of its mission. Land management (operational) planning, takes place at the forest level. Project or activity (tactical) planning occurs within a national forest. All plans have certain elements in common, which must be included to have a useful working document. Two of these common elements, both of high priority in the process, are the setting of goals and objectives. **Goals** are broad statements of central themes that set management direction. Seldom quantifiable, they generally have no time limit. **Objectives**, specific statements of activities undertaken to realize the goals for a management unit, can be measured and monitored over time.

One approach, known as Management by Objectives (MBO), fits well with a planning methodology widely adopted by many federal and state agencies. The technique depends upon adopting specific objectives and determining how they will be measured. Agency personnel are then expected to complete each objective in terms of the measurement stipulated.

Natural resource management must have at its basis ecological principles and biological laws derived from the accumulated knowledge arising out of research and experience. These principles and laws are seldom explicitly stated or identified in support of actual project objectives. A clear statement of the ideas, principles, and laws that underpin decision-making will strengthen the biologist's or natural resource manager's position in proposing specific goals and objectives in a plan.

Earlier chapters included discussions of time in relation to succession (Chapter 2), forest rotation, and cutting cycles (Chapter 5). All these factors are common to management plans and must be included; another time factor, the planning period, is important and must also be part of the plan. Building time checkpoints at critical junctures into management plans provide structured opportunities to evaluate accomplishments and set new directions if required. Such checkpoints serve as a basis for a strong monitoring plan (Figure 7-2).

Managers clearly need to know the starting date on which a plan comes into operation, but equally important is to understand the planning horizon—the period that the plan addresses—and the date for plan revision. An outdated plan is as useless as no plan at all, and indeed may cause damage. State and federal laws often dictate the planning period. Revisions can result from advancing technology, to correcting errors or incorporating new data, or to addressing public concerns about a management goal or activity. While the life of a plan should be clearly stated, it should not be so rigid that it precludes revisions and updates as conditions change. Ensuring that plans at the different levels remain consistent over time, especially during plan revisions, poses a key challenge to planners and decision makers.

Information Needs

Management unit plans require considerable data, much of which may already be available (hopefully); such data may have one or more of the following characteristics, however: too little, too old, in the wrong format, or at inappropriate level of detail (Chapter 4). Data needs also vary depending on hierarchical level. But regardless of level, all planning requires at least some data in each of four major domains: biophysical, sociopolitical, spatial, and temporal.

Biophysical

- Soil surveys: Used to establish vegetation potential and productivity as well as operability constraints.
- Allotment analysis reports: Establish past and present livestock use of an area.
- Timber inventory records: Stocking and stand tables to estimate current and future growth and yield. Stand structure data can also serve as inputs for wildlife habitat models.
- Exclosure location and data: Helps to establish plant successional trend and use patterns.
- Wildlife census data: Information on populations both current and historical.
- Vegetation type maps: Keyed to identify food and cover areas.
- Range condition and trend surveys: Provides information on herbage production and utilization.
- Hydrology inventory: The location of seeps, springs, streams, and small catchments helps determine the need for water development projects for habitat improvement.
- Recreation facilities: To help identify human impacts on the planning unit.
- Roads and trails: Identify access routes that affect the planning unit.
- Utility corridors and rights of way: Identify areas with dedicated and restricted use and condition.
- Climate and weather: Help determine potential vegetation, site productivity, access, and operability.

Sociopolitical

- Land ownership: Identify potential objectives and constraints (Chapter 3).
- Special use areas: Pre-defined or designated areas with specified activities that restrict available management options.
- Political boundaries: Set limits on planning area.
- Cultural and historical sites: Officially identified and designated areas of cultural or historical significance. Management activities are often restricted in such areas.
- Zoning: Legal designations from state and/or local governments that can restrict or prohibit certain management activities on private land.
- Human population locations and magnitude: Help identify current and potential uses of land.
- Current land uses: Frame planning process.

Spatial

- Topography: Needed to locate boundaries, forest vegetation types, erosion potential, and operability.
- Exclosure location and data: Helps to establish plant successional trend and use patterns.
- Vegetation type maps: Serves to help identify food and cover areas.
- Hydrology inventory: The location of seeps, springs, streams, small catchments, and lakes helps determine the need for water development projects for habitat improvement.
- Recreation facilities: To determine human impact on the management unit.
- Political boundaries: Set limits on planning area.
- Roads and trails: Identify access routes and condition that impact the management unit.
- Current land uses: Frame planning process.
- Human population locations and magnitude: Help identify current and potential uses of land.

Temporal

- Historical data: Fire history, farming, logging, mining activity, etc.
- Wildlife census data: Information on populations both current and historical.
- Game kill records: Often used to supplement historical census data.
- Range condition and trend surveys: Provide information on herbage production and utilization.

Clearly, these domains overlap. For example, a list of recreation sites (biophysical domain) has much more value if linked to the location of these sites (spatial domain). Miles of road (biophysical domain) provides some information about potential maintenance needs, especially at the long-term planning level, but without roads identified and tracked on the landscape (i.e., spatial domain of the data) such data have relatively little value for operational or tactical planning.

The organization conducting the planning may already have some of the information listed or can acquire it relatively easily. However, over time, planning efforts will require new or updated data. In such cases, the organization may need to conduct inventories, using the various tools discussed in Chapter 4. Such inventories can be extremely expensive in time, personnel, and cost. Therefore, taking great care to clearly and precisely identify data needs early in the planning process will help to define and schedule any necessary data acquisition efforts. The overall planning process should also include protocols for data storage, preferably electronically in a GIS and/or database, with these protocols implemented and monitored to ensure ready data access and data integrity over time.

7-4 Decision Support Systems

Prior to the mid-1970s the public and private sector natural resources planning goals and objectives were relatively simple, requiring only minimal amounts, types, and complexity of data and analysis. Therefore, the analysis and presentation of data generally did not require the aid of complicated automated systems. Pencils, paper, mechanical calculators, aerial photographs, mylar overlays, and sufficient analytical planning rules address planning requirements. Today, the situation has changed dramatically. In the public sector, because the types and amounts of information needed to meet the requirements of the natural resource laws and agency regulations implementing these laws are sufficiently complex, new ways of analyzing data have been developed and incorporated into the planning process. In the private sector, corporations developed increasingly sophisticated timber growth and yield, and harvest scheduling models to address much more complex and rapidly changing social and economic systems. These new analytical tools are generically called "Decision Support Systems."

Description

Simply stated, "decision support systems (DSSs) are defined as interactive computer-based systems that aid users in judgment and choice activities" (Druzdzel and Flynn 2010, pg. 6). DSSs have been adopted across a wide range of organizations, both public and private, in a variety of industries from wind power plants in Romania (Bâra et al. 2013) to the stone and marble industry in Palestine (Hanieh et al. 2016). Other industries using DSSs include health care, airline maintenance, law enforcement, and financial planning (Power 2018).

DSSs vary widely in complexity and capabilities. For natural resource management planning, DSSs generally include some type of database linked to a simulation model to predict the future, an optimization program to help allocate resources, a geographic information system to incorporate spatial characteristics of the problem, and a capacity to display results; all are managed thru the user interface (Figure 7-4).

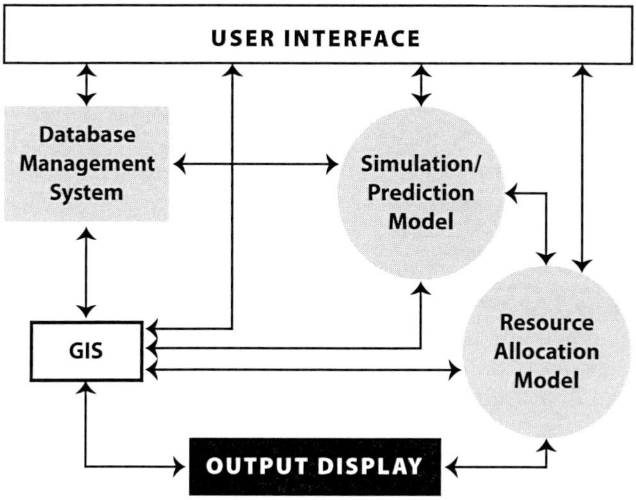

Figure 7-4
Generalized structure of a decision support system.

Some information flows through the system will occur automatically, such as the outputs from the simulation model moving into the allocation model. Others, such as specifying desired outputs or constraints, will include direct user input. The value of DSSs lies in the amount of data they can process, the complexity of the problems they can tackle, and especially in the replicability of the outcomes they produce. In other words, DSSs can reduce bias in decision-making by proving consistent rule-based results. The rules themselves may be biased but once established, a DSS will use these rules consistently.

Examples

Two notes of caution. First, GiGo or "Garbage in, Garbage out." The most sophisticated and complex DSS has little or no value without good data. Second, a decision support system never "makes" a decision. Instead, the value of the DSS rests with the information it provides to decision makers.

PUBLIC SECTOR. In the mid-1970s the Forest Service adopted a linear programming model called "FORPLAN" (Mitchell and Kent 1987) as its response to the needs for a national and regional system of planning, specifically those associated with the National Forest Management Act (NFMA). FORPLAN (FORest PLANning) was designed to provide management alternatives by presenting various levels and combinations of goods and services that a forest can provide. FORPLAN is the primary tool used in developing the forest plans required under the NFMA.

FORPLAN software uses data supplied for each analysis area. The data must be developed for management prescriptions and estimated costs, together with the outputs, effects, and benefits for each prescription or combination of goods and services. This information is then used to generate matrices of managerial options and associated benefits, costs, and so forth. The product is a listing of the optimum combinations of prescriptions and associated benefits. FORPLAN received criticism based on its complexity, structure, and lack of transparency (Kent et al. 1991). However, over time the model has increased in sophistication and scope, aided in part by the increased computing power.

PRIVATE SECTOR. Forest decision support systems date back to at least the mid-1960s when linear programming was used to plan forest roads (Boughton 1967) but expanded in the 1970s with applications to scheduling timber harvests (Johnson and Scheurman 1977). Other early applications include reforestation planning (Buongiorno and Teeguarden 1973) and planning for an integrated forest industry (Barros and Weintrab 1982). These early models tended to focus on single outputs but as in the public sector, forest planning in the private sector also increased in complexity. As a result, single objective models grew into much more sophisticated multi-objective models.

A major leap in capability came with the ability to include spatial modeling into the DSSs. The power and availability of GIS made such DSSs possible. Including spatial constraints is particularly important when planning forest operations as fragmentation of the landscape is of major concern in providing high quality wildlife habitat. The capability to incorporate spatial modeling allows individual areas (e.g., a patch or stand) to be treated individually rather than as part of a larger forest type. In addition, spatial models recognize the relative positions of each patch to other patches (Baskent and Keles 2005).

Data Needs

DSSs have become extremely powerful and valuable tools in both public and private forest management planning. Their value is constrained, however, by data needs (Segura et al. 2013). Although new data acquisition methods continue to emerge, (Chapter 4) the modeling needs have tended to increase faster than acquisition capabilities. In addition, DSSs have tended to focus more on market (i.e., goods and services that can have some direct monetary value assigned to them and can be sold/rented/leased) rather than on non-market values (i.e., goods and services that do not have direct monetary values assigned to them and thus cannot be sold/rented/leased). This focus results in part, no doubt, due to more data being available on market rather than on non-market values! The DSSs high data needs justifies repeating the warning about "GiGo"—Garbage in, Garbage out. The requirements to plan using data-needy DSSs offers up the real temptation of using poor—but available—data. Resisting such temptation is the mark of the true natural resource professional. The ultimate decision support system is the human brain.

7-5 Planning and the National Environmental Policy Act

A full and detailed description of implementing the National Environmental Policy Act (NEPA) (Section 6-1) and NEPA-like planning processes in the private and public sectors is far beyond the scope of our work here. A substantial body of literature has focused on this topic. Indeed, a quick web search of the words

"NEPA" and "forest planning" generated over 40,000 search results. Multiple textbooks, commercially available workshops, in-house training, and university courses exist that focus on the topic of NEPA planning in general with many of these specific to forest planning.

In the simplest of terms, however, NEPA requires a detailed analysis of the effects, both immediate and cumulative over space and time, of any project having a significant impact on the environment. In addition, NEPA requires that the planning include multiple alternatives to accomplish the proposed action, including a "no action" alternative. For clarification, the no-action alternative represents the continuation, and not the cessation, of current activities. NEPA also mandates multiple opportunities for public comment during the process and that the agency conducting the planning directly respond to the comments.

To accomplish this goal, the Forest Service and other Federal agencies, have created Interdisciplinary Teams (IDTs or ID Teams). As the name indicates, a variety of professionals populate these teams—foresters, wildlife biologist, geologists, engineers, etc.—based on the exact nature of the proposed project. These teams work together to create alternatives and analyze the environmental effects of each alternative proposed. The planning process results in an environmental analysis, a document written when, based on an initial analysis, a proposed project is found to have no significant impact on the environment. If this initial analysis indicates that a project warrants further analysis, then a draft Environmental Impact Statement (EIS) is created, which details the different alternatives and the estimated effects of each alternative on the environment and how the alternative meets project goals; these goals are known as the project's "purpose and need." Public comment is solicited on the draft EIS as one source of information to generate the final EIS. The final EIS includes a recommended alternative that is forwarded to the decision maker. The process is documented through a "Record of Decision." Lawsuits often accompany the EIS process because the public does not agree with the decision.

As this even very simple explanation indicates, the EIS process can consume large amounts of resources and the time needed to complete the planning can range from months to years. Multiple critiques of the process exist. The Government Accounting Office reported that "little information exists on the costs and benefits of completing NEPA analyses. Agencies do not routinely track the cost of completing NEPA analyses, and there is no government-wide mechanism to do so" (USGAO 2014, pg. 10).

Ronald Bjorkland (2013) noted that the lack of project monitoring requirements is a major flaw in the law. Daniel R. Mandelker (2010) cited NEPA's cumbersome decision-making process, created essentially in the absence of any guidance in the law itself or in the record of legislative intent, as its largest flaw. In addition, Mandelker also notes NEPA's failure to address large-scale projects, such as forest planning, and its exclusion of activities, such as transportation plans, as a weakness.

On the other hand, he notes that the project review process results in minimizing negative ecological effects. Kevin DeGood (2018 "Conclusion") stated that NEPA serves a larger purpose of transparent decision making that results in "more just, democratic, productive, and sustainable U.S. society." Regardless of opinion, NEPA will continue as a reality in Forest Service and Bureau of Land Management planning for the foreseeable future. Only major changes in federal law, such as the repeal of NEPA, could change this reality.

7-6 Lessons Learned

Based on the literature (e.g., Doppelt and Onsgaard 2006; Klosterman 2013) and the authors' combined planning experience in both the public and private sectors, the authors have identified the following lessons learned about forest planning:

1. Planning is expensive in terms of time, money, and energy.
2. Planning in the absence of clearly defined goals and objectives is doomed to fail.
3. Plans tend to start failing as soon as implementation begins ("Humans plan, the deities laugh").
4. Plans tend to become increasing complex.
5. Complex plans become difficult to understand by both the people in the organizations doing the planning and the public who have interest in the plan.
6. Involving the public in planning greatly increases the time and resources needed to arrive at a decision, but can make for a better and more politically acceptable plan.
7. Planning without strong communication, implementation, and monitoring creates a high potential for failure.
8. Plans always contain errors.
9. There's no such thing as a perfect plan.
10. Planners with field experience generally are better planners than planners without field experience.
11. Without a good plan, management is doomed to fail.
12. With a bad plan, management is doomed to fail.
13. Without good leadership, planning efforts are doomed to fail.
14. Garbage in, garbage out.
15. The better you can identify your planning assumptions, the better your plan.
16. Predicting the future is difficult, at best.
17. Don't over-rely on technology.
18. KIASAP: Keep It As Simple As Possible.

Given all the insights identified as part of the planning process we know that we know (the Rumsfeld effect, Section 6-6) that we still have a lot to learn about planning.

7-7 Summary

The context that surrounds forestry—land ownership patterns, previous uses, laws, regulations, the unique attributes of forests—combine to create an extremely complex planning environment. Just as with our physical environment, this planning environment has multiple interacting components whose relationships we do not fully understand. In some cases, it is not known that some of these interactions exist. The frequent occurrence of unintended consequences in forest management attests to this ignorance of interactions.

Such complexity could lead to despair and resignation. But the need for management of forest resources to meet the wants and needs of humans, including desires to maintain and enhance wildlife habitat and to conserve and protect wildlife populations, leads to development approaches and aids to planning. For all its flaws, NEPA helps identify potential negative effects of proposed actions and decision support systems and other computer models help to cope with at least some of the complexity inherent in forest systems.

However, the task of helping move towards forest and wildlife management goals—including desired future conditions for forest structure/composition and the stand dynamics that maintain them—belongs to humans and their ability to innovate. Identifying needs, as opposed to wants, serves as the first step in the planning process. Very simply, if you do not know where you are going, it does not matter what course of action you pursue; therefore, planning becomes irrelevant. Natural resource managers have many tools to help achieve management goals. Laws and the ability to plan form the foundation of the needed tools, along with subject knowledge, of course. The key is combining these elements to create a path to achieve management goals.

References

Allen, D.L. (chair). 1973. *North American wildlife policy.* Wildl. Manage. Inst., Washington, DC.

ATFS. 2018. *Standards of sustainability 2015–2020.* Available at https://www.treefarmsystem.org/standards-review; accessed July 5, 2018.

Bâra, A., I. Botha, I. Lungu, and S.V. Oprea. 2013. Decision support system in national power companies. A practical example (Part I). *Database Systems Journal* IV(1):37–46.

Barros, O., and A. Weintraub. 1982. Planning for a vertically integrated forest industry. *Operations Research* 30(6): 1168–1182. https://doi.org/10.1287/opre.30.6.1168

Baskent, E.Z., and S. Keles. 2005. Spatial forest planning: A review. *Ecological Modelling* 188(2/4):145–173.

Bjorkland, R. 2013. Monitoring: The missing piece: A critique of NEPA monitoring. *Environmental Impact Assessment Review* 43:129–134.

Bonnicksen, T.M. 2000. *America's ancient forests from the ice age to the age of discovery.* John Wiley & Sons, New York. 594 p.

Boughton, W.C. 1967. Planning the construction of forest roads by linear programming. *Australian Forestry* 3:111–120.

Buongiorno, J., and D.E. Teeguarden. 1973. An economic model for selecting Douglas-fir reforestation projects. *Hilgardia* 42(3):35–120.

CNRA. 2012. Resource management. *CALFIRE. California Natural Resources Agency.* http://calfire.ca.gov/resource_mgt/resource_mgt; accessed July 6, 2018.

CNRA. 2014. The California Environmental Quality Act CEQA. *California Natural Resources Agency.* Available at http://resources.ca.gov/ceqa; accessed July 6, 2018.

Covington, W.W., and M.M. Moore. 1994. Southwestern ponderosa pine forest structure: Changes since Euro-American settlement. *J. For.* 92(1):39–47.

Davis, L.S., N.K. Johnson, P. Bettinger, and H. Theodore. 2001. *Forest management,* Fourth edition. McGraw-Hill Higher Education, New York.

DeGood, K. 2018. The benefits of NEPA: How environmental review empowers communities and produces better projects. *Center for American Progress.* Available at https://www.americanprogress.org/issues/green/reports/2018/01/16/444899/benefits-nepa-environmental-review-empowers-communities-produces-better-projects; accessed July 6, 2018.

Derosier, A.L., S.K. Hanshue, K.E. Wehrly, J.K. Farkas, and M.J. Nichols. 2015. *Michigan's wildlife action plan.* Michigan Department of Natural Resources, Lansing. Available at www.michigan.gov/dnrwildlifeactionplan; accessed June 28, 2019.

Diaz-Balteiro, L., R. Alonso, M. Martínez-Jaúregui, and M. Pardos. 2017b. Selecting the best forest management alternative by aggregating ecosystem services indicators over time: A case study in central Spain. *Ecological Indicators* 72:322–329.

Diaz-Balteiro, L., J. González-Pachón, and C. Romero. 2017a. Measuring systems sustainability with multi-criteria methods: A critical review. *European Journal of Operational Research.* 258(2):607–616. Available at http://www.sciencedirect.com/science/article/pii/S0377221716307147. https://doi.org/10.1016/j.ejor.2016.08.075; accessed June 28, 2019.

Doppelt, B., and G. Onsgaard. 2006. *Lessons learned from the USDA Forest Service community-based watershed restoration partnerships: An assessment of the accomplishments, limitations, and transferable knowledge.* Factor 10, Inc., Eugene, OR.

Druzdzel, M.J., and R.R. Flynn. 2010. Decision support systems. P. 9 in *Encyclopedia of Library and Information Sciences,* Third edition, M.J. Bates and M.N. Maack (eds.). Taylor and Francis, Boca Raton, FL. DOI: 10.1081/E-ELIS3-120043887

Gregory, G.R. 1987. *Resource economics for foresters,* Second edition. John Wiley and Sons, Inc., New York.

Hanieh, A.A., S. AbdElall, and A. Hasan. 2016. Decision support system for industrial social performance. *Procedia CIRP* 40:329–34. DOI: 10.1016/j.procir.2016.01.052

Heink, U., J. Hauck, K. Jax, and U. Sukop. 2016. Requirements for the selection of ecosystem service indicators: The case of MAES indicators. *Ecological Indicators* 61:10–26. Available at https://doi.org/10.1016/j.ecolind.2015.09.031

Johnson, K.N., and H.I. Scheurman. 1977. Techniques for describing optimal timber harvest and investment under different objectives—Discussion and synthesis. *Forest Science Monograph* 18:31.

Kent, B.B., B. Bare, R.C. Field, and G.A. Bradley. 1991. Natural resource land management planning using large-scale linear programs: The USDA Forest Service experience with FORPLAN. *Operations Research 39*(1):13–27.

Kister, T.M. 2016. Improving the information development process: A refined iterative development model. *Technical Communications Online* 63:186–211. Available at https://www.stc.org/techcomm/2016/08/01/improving-the-information-development-process-a-refined-iterative-development-model/; accessed June 28, 2019.

Klosterman, R.E. 2013. Lessons learned about planning. *Journal of the American Planning Association* 79(2):161–169. DOI: 10.1080/01944363.2013.882647

Leopold, A. 1933. *Game management.* Charles Scribner's Sons, New York.

Mandelker, D.R. 2010. The National Environmental Policy Act: A review of its experience and problems. *Washington University Journal of Law & Policy* 32:292–312. Available at http://openscholarship.wustl.edu/law_journal_law_policy/vol32/iss1/9

MDNR. 2018. *About the Michigan DNR.* Available at https://www.michigan.gov/dnr/0,4570,7-350-79137---,00.html; accessed June 28, 2019.

Mitchell, T. R., and B. M. Kent. 1987. Characterization of the FORPLAN analysis system. P. 3–14 in *FORPLAN: An evaluation of a forest planning tool,* T.W. Hoekstra, A.A. Dyer, and D.C. Le Master (tech. eds.). Gen. Tech. Rep. RM-140, USDA Forest Service, Fort Collins, CO.

Power, D. 2018. *Who uses computerized decision support?* Available at http://dssresources.com/faq/index.php?action=artikel&cat=&id=118&artlang=en; accessed July 3, 2018.

Segura, M., D. Ray, and C. Maroto. 2013. Decision support systems for forest management: A comparative analysis and assessment. *Computers and Electronics in Agriculture* 101:55–67.

SFI. 2018. *Forest management standard.* Available at http://www.sfiprogram.org/sfi-standards/forest-management-standard; accessed July 5, 2018.

USDA. 1993. The principal laws relating to Forest Service activities. Agri. Hdbk. 453. U.S. Department of Agriculture, Washington, DC. Available at http://ir.library.oregonstate.edu/concern/defaults/bv73c177x

USDA. 2008. Forest Service memorandum. Policy statement. Office of the Chief, Washington, DC.

USDA. 2018. *Forest Service planning rule.* Available at https://www.fs.usda.gov/planningrule; accessed July 5, 2018.

USGAO. 2014. National Environmental Policy Act: Little information exists on NEPA analyses. Government Accountability Office Report to Congressional Requesters. GAO-14-370. 38 p.

WDNR. 2015. Fish, wildlife and habitat management plan, October 1, 2015–September 30, 2025. Wisconsin Department of Natural Resources. 41 p.

Knowledge from Literature

8

Often overlooked in the continuing education of professional foresters and wildlife biologists is the experience that can be obtained from literature. Two great ways for professionals to increase their knowledge is to stay current through the literature and by learning from the experiences of others. Literature is a valuable tool for managers to have in maintaining their competency for making decisions. However, library services or other sources of literature may not always be convenient or readily available in a work environment, even with the access provided through the internet. No matter what the situation, questions might just be answered by making a telephone call or by clicking a mouse.

Knowing who to contact and who might have the needed information is helpful if a person has some basic knowledge of who owns the land and who has the authority to make decisions (Chapter 3). This chapter presents contacts for a number of agencies that can answer questions or suggest publications that meet a variety of management needs. In addition to searching for literature through the channels provided, it is also a good practice for managers or their administrative unit to have a working library of their own. Recommended publications for a working library are provided to assist in finding information for managing both public and private forest lands.

8-1 U.S. Department of Agriculture

Most of the federal forest land in the 50 states is managed by the U.S. Forest Service in the Department of Agriculture, which has research and development programs headquartered at the following research station locations:

- Northern Research Station, Newtown Square, PA;
- Pacific Northwest Research Station, Portland, OR;
- Pacific Southwest Research Station, Albany, CA;
- Rocky Mountain Research Station, Fort Collins, CO; and
- Southern Research Station, Asheville, NC.

These five research stations have 67 laboratories, 80 experimental forests located across all 50 states along with The Forest Products Laboratory (Madison, WI) and The International Institute of Tropical Forestry (Rio Piedras, PR). All of these administrative units have scientists in all areas of natural resource topics who produce hundreds of publications available free and online to download. Each of the administrative units listed has a website with a variety of choices and ways to contact topic specialists. Recent innovations include short science briefs to aid in the understanding of complex scientific and management issues, as well as user-friendly interfaces and search engines.

Readers looking for information on research and development activities concerning natural resources on federal and private forest land can visit each station's website to access their publication list. In addition, if readers have a title and publication number from a research station, they can enter this information into the search function on the web page to obtain a document directly. For example, Table 1-1 on species richness can be downloaded by just entering: *General Technical Report RMRS-GTR-332* in the search area and a printable document in PDF format will appear on their screen.

The Forest Service also has 154 National Forests in 39 states (Table 8-1). Each forest has employees in natural resource disciplines and they can be a source for management information for surrounding state and private land owners. In fact, each forest region has an administrative unit that focuses on what is called "State and Private Forestry." The focus of the state and private forestry program is on:

- conservation education,
- cooperative forestry,
- fire and aviation management,
- forest health protection,
- Office of Tribal Relations, and
- urban and community forestry.

Requests for information can be sent directly by email or written requests to the nine regional offices. The following regional offices are the contacts:

- Northern Region, Missoula, MT;
- Rocky Mountain Region, Golden, CO;
- Southwestern Region, Albuquerque, NM;
- Intermountain Region, Ogden, UT;
- Pacific Southwest Region, Vallejo, CA;
- Pacific Northwest Region, Portland, OR;
- Southern Region, Atlanta, GA;
- Eastern Region, Milwaukee, WI; and
- Alaska Region, Juneau, AK.

In 1935 The Soil Conservation Service (SCS) was established under Public Law 74–46. The Act was passed to help farmers and ranchers reduce soil loss on farm, grazing, and forest lands. This Act followed the days of the "dust bowl" when production of farms, grassland, and forest land was decimated by severe drought conditions and soil loss from strong winds. In 1994 Congress reorganized the Department of Agriculture and renamed the SCS to the **Natural Resources**

Table 8-1 National Forests in 39 states.

AK	Chugach, Tongass
AL	Conecuh, Talladega, Bankhead, Tuskegee
AR	Ouachita, Ozark-St. Francis
AZ	Apache-Sitgreaves, Coconino, Coronado, Kaibab, Prescott, Tonto
CA	Angeles, Cleveland, Eldorado, Inyo, Klamath, Lassen, Los Padres, Mendocino, Modoc, Plumas, San Bernardino, Sequoia, Shasta-Trinity, Sierra, Six Rivers, Stanislaus, Tahoe
CO	Arapaho, Grand Mesa, Gunnison, Pike, Rio Grande, Roosevelt, Routt, San Isabel, San Juan, Uncompahgre
FL	Apalachicola, Ocala, Osceola
GA	Oconee, Chattahoochee
ID	Panhandle, Boise, Caribou, Challis, Clearwater, Nez Perce, Payette, Salmon, Sawtooth, Targhee
IL	Shawnee
IN	Hoosier
KY	Daniel Boone
LA	Kisatchie
ME	White Mountains
MI	Hiawatha, Huron-Mainistee, Ottawa
MN	Chippewa, Superior
MO	Mark Twain
MS	Bienville, Delta, Desoto, Holly Springs, Homochitto, Tombigee
MT	Beaverhead-Deerlodge, Bitterroot, Custer, Flathead, Gallatin, Helena, Kootenai, Lewis and Clark, Lolo
NC	Croatan, Nantahala, Pisgah, Uwharrie
NE	Nebraska
NH	White Mountain
NM	Carson, Cibola, Gila, Lincoln, Santa Fe
NV	Humboldt-Toiyabe
NY	Finger Lakes
OH	Wayne
OR	Deschutes, Fremont, Malheur, Mount Hood, Ochoco, Rouge River, Siskiyou, Siuslaw, Umatilla, Umpqua, Wallowa-Whitman, Willamette, Winema
PA	Allegheny
SC	Francis Marion-Sumter
SD	Black Hills
TN	Cherokee
TX	Angelina, Davy Crockett, Sabine, Sam Houston
UT	Ashley, Dixie, Fishlake, Manti-LaSal, Uinta, Wasatch-Cache
VA	George Washington, Jefferson
VT	Green Mountain
WA	Colville, Gifford Pinchot, Mount Baker-Snoqualmie, Okanogan, Olympic, Wenatchee
WI	Chequamegon, Nicolet
WV	Monongahela
WY	Bighorn, Bridger-Teton, Medicine Bow, Shoshone

Conservation Service (NRCS). This new title reflected a broader assignment to provide technical assistance to farmers, ranchers, and land owners who wanted to make conservation improvements on their land. NRCS has state offices and districts within a state, which have cooperative projects involving private land owners. By typing NRCS and a state name into a computer browser, private land owners can find links to technical support, cooperative projects, and publications that cover a range of natural resource topics.

8-2 U.S. Department of Interior (USDI)

The **Bureau of Land Management (BLM)** is an agency in the Department of Interior and responsible for managing about six percent of America's forest land, mostly in the western United States (Table 3-2). In 1946, the BLM was created by combining the General Land Office and the Grazing Service. With the passage of the Federal Land and Policy Management Act (1976) the BLM was authorized to manage public lands to provide habitat for wildlife and domestic livestock. The public has access to a library maintained in The National Operations Center in Denver, CO. The national office is located in Washington, DC with national centers in Arizona and Colorado. In addition, the following states have BLM offices where landowners can get information about land management practices that might relate to private land adjoining BLM land: Arizona, Alaska, California, Colorado, Idaho, Montana, North Dakota, South Dakota, Nevada, New Mexico, Oregon, Washington, Utah, and Wyoming.

The USDI has other administrative units, such as the National Park Service, U.S. Fish and Wildlife Service, and U.S. Geological Survey (USGS), established to protect natural resources in dedicated areas of land in national parks and a system of wildlife refuges. These administrative units have educational materials and research publications available to the public in their respective regional or individual websites. The USGS is involved in research and public outreach programs through its system of 40 Cooperative Research Units in 38 states. A state agency and a university with a unit leader supervising research and graduate education programs support each unit. Research results are generally made available through publications in a professional journal or state agency, such as a game and fish department. The USGS also has research and education programs in designated core science areas such as biology and ecosystems in its seven regions that can be accessed through its website.

8-3 Land-Grant Colleges and Universities

In 1862 Congress passed the Morrill Act that authorized a land donation for each state that could be sold to fund public colleges whose focus was on agriculture and mechanical arts (engineering). Every state has a land-grant college and some have more than one. Several Acts have passed since 1862 that authorized the establishment of agricultural experiment stations, which receive funds to support research activities in a cooperative extension services program. Many forestry, wildlife, and other natural resource degree programs are associated with land-grant colleges and their faculty contribute information to the extension specialists. Extension specialists are good sources of information and have links to city, county, and state resource agencies. State experiment stations publish information sheets on specific topics available through the extension services. Results from research projects at land-grant institutions are often published in professional journals and symposium proceedings.

8-4 Foundations, Societies, and State Agencies

There is considerable information available from private organizations that publish journals, conference proceedings, and material presented at national conventions that are available and can be obtained by directly contacting their headquarters or a state chapter. Professional organizations usually require a membership to receive copies of their published documents but authors of articles in these publications will often have copies for distribution on request. A good source for names of conservation agencies in the U.S. is the National Wildlife Federation's Conservation Directory 2005, published by Island Press. The following list provides names of several organizations that resource managers might contact for information:

- The Nature Conservancy,
- The National Wildlife Federation,
- The Wildlife Management Institute,
- The Society of American Foresters,
- The Society for Range Management,
- The Audubon Society,
- The Wildlife Society,
- The Society of Mammalogy,
- The American Ornithological Society, and
- The National Association of State Foresters.

All 50 states have agencies dedicated to the conservation and protection of forests and wildlife within a state border. Names vary by state when referencing a wildlife or forestry entity. For example in wildlife there is the California Department of Fish and Wildlife, Oklahoma Department of Wildlife Conservation, and Florida Fish and Wildlife Commission. Forestry departments follow the same naming trend; for example, the Oregon Department of Forestry, Iowa Department of Natural Resources, and South Carolina Forestry Commission all fulfill similar functions. State agencies generally have their own series of publications to document research results and have websites to provide management information to the public.

8-5 A Working Library

Considerable information exists for managers and planners to absorb and understand that it is worthwhile to have a working library as a quick reference for some of the major topics. The need for information is not limited to employees of federal or state agencies. Private landowners also may want a reference library for background information they need in making plans for their property or applying for cooperative assistance from state or federal programs.

A working library does not have to include a large collection of books; a modest selection that increases your knowledge base and enables you to stay current in your profession is practical and beneficial. Each passing year improves the number of references and volume of information available electronically. But it also is useful to have a few classics to remind us of how much the work of earlier field biologists and ecologists is still relevant today. These classics were influential in setting the direction for forestry and wildlife education, research, administration, and management.

Suggested references include books in the general field of ecology, forestry, and wildlife. Other topics of interest can be found in the references for each chapter. Having several references covering the same topic will provide you with a set of ideas to compare and a history from literature that can facilitate developing your own perspective. References with an asterisk (*) are books that private land owners might want to have as a desk copy.

Some Classics

*Allen, D.L. 1974. *Our wildlife legacy.* Funk and Wagnalls, New York.

Clements, F.E. 1916. *Plant succession: An analysis of the development of vegetation.* Carnegie Inst. Pub. 242, Washington, DC.

Dasmann, R.F. 1981. *Wildlife biology.* John Wiley and Sons, New York.

Elton, C. 1927. *Animal ecology.* Sidgwich and Jackson, Ltd., London.

Leopold, A. 1933. *Game management.* Charles Scribner's Sons, New York.

*Leopold, A. 1939. A biotic view of land. *J. For.* 37:727–730.

Pinchot, G. 1947. *Breaking new grounds.* Harcourt, Brace, and Co., New York.

Trippensee, R.E. 1948. *Wildlife management: Upland game and general principles,* Vol. 1. McGraw-Hill, New York.

Suggested References

Bailey, J.A. 1984. *Principles of wildlife management.* John Wiley and Sons, New York.

Barbour, M.G., and W.D. Billings. 2000. *North American terrestrial vegetation,* Second edition. Cambridge University Press, Cambridge, UK.

*Beattie, M., C. Thompson, and L. Levine. 1993. *Working with your woodland.* University Press of New England, Lebanon, NH.

Boyce, M.D., and Alan Haney (eds.). 1997. *Ecosystem management: Applications for sustainable forest and wildlife resources.* Yale University Press, New Haven and London.

Burkhart, H.E., T.E. Avery, and B.P. Bullock. 2019. *Forest measurement,* Sixth edition. Waveland Press, Inc., Long Grove, IL.

Cooperrider, A.Y., R.J. Boyd, and H.R. Stuart (eds.). 1986. *Inventory and monitoring of wildlife habitat.* USDI, Bureau of Land Management, Service Center, Denver, CO.

Daubenmire, R.F. 1968. *Plant communities.* Harper and Row, New York.

Davis, L.S., and K.N. Johnson. 1987. *Forest management,* Third edition. McGraw-Hill, New York.

Deal, R. (ed.). 2019. *Dictionary of forestry,* Second edition. Society of American Foresters, Bethesda, MD.

DeSteffano, S., and R.C. Haight (eds.). 2002. *Forest wildlife-habitat relationships: Population and community responses to forest management.* Society of American Foresters, Bethesda, MD.

Franklin, J.F., and R.T.T. Forman. 1987. Creating landscape patterns by forest cutting: Ecological consequences and principles. *Landscape Ecology* 1:5–18.
Gaston, K.J., and J.I. Spicer. 2004. *Biodiversity: An introduction.* Blackwell Scientific Publishing, Oxford, UK.
*Hanson, A.L., M. Severson, and D.L. Waterman. 2011. *A landowner's guide to managing your woodlands.* Story Publishing, North Adams, MA.
Harris, L.D. 1984. *The fragmented forest: Island biogeography theory and the preservation of biotic diversity.* The University of Chicago Press, IL.
*Hilts, S., and P. Mitchell. 2009. *The woodlot management handbook.* A Firefly Book, Ontario, Canada.
Hunter, M.L., Jr., and F.K.A. Schmiegelow. 2011. *Wildlife, forests, and forestry: Principles of managing forests for biological diversity.* Prentice-Hall, Upper Saddle River, NJ.
Kendeigh, S.C. 1961. *Animal ecology.* Prentice-Hall, Englewood Cliffs, NJ.
Kimmins, J.P. 1987. *Forest ecology: A foundation for sustainable management,* Second edition. Prentice-Hall, Upper Saddle River, NJ.
Krausman, P.R. 2002. *Wildlife management: The basics.* Prentice-Hall, Upper Saddle River, NJ.
Lindenmayer, D.B., and J.F. Franklin. 2002. *Conserving forest biodiversity: A comprehensive multi-scaled approach.* Island Press, Washington, DC.
McComb, B.C. 2016. *Wildlife habitat management: Concept and applications in forestry.* CRC Press, Taylor and Francis Group, Boca Raton, London, New York.
Morrison, M.L. 2002. *Wildlife restoration: Techniques for habitat analysis and animal monitoring.* Soc. for Ecol. Restoration, Island Press, Washington, DC.
Morrison, M.L., B.G. Marcot, and R.W. Mannan. 1998. *Wildlife-habitat relationships, concepts and applications.* The University of Wisconsin Press, Madison.
Nyland, R.D. 2016. *Silviculture: Concepts and applications,* Third edition. Waveland Press, Inc., Long Grove, IL.
Odum, E.P. 1971. *Fundamentals of ecology.* W. B. Sanders Co., Philadelphia, PA.
Patton, D.R. 1992. *Wildlife habitat relationships in forested ecosystems.* Timber Press, Portland, OR.
*Payne, N.F., and F.C. Bryant. 1994. *Techniques for wildlife habitat management of uplands.* McGraw-Hill, Inc., New York.
Peek, J.M. 1986. *A review of wildlife management.* Prentice-Hall, Englewood Cliffs, NJ.
Perry, D.A., R.O. Oren, and S.C. Hart. 2008. *Forest ecosystems,* Second edition. The John Hopkins University Press, Baltimore, MD.
Primack, R.B. 2006. *Essentials of conservation biology.* Sinauer Associates, Inc., Sunderland, MA.
*Robinson, G. 1988. *The forest and the trees: A guide to excellent forestry.* Island Press, Washington, DC.
Shaw, J.H. 1985. *Introduction to wildlife management.* McGraw-Hill Book Company, New York.
Shelford, V.E. 1963. *The ecology of North America.* University of Illinois Press, Urbana.
Silvy, N.J. (ed.). 2012. *The wildlife techniques manual.* Research, Vol 1, Seventh edition. The John Hopkins University Press, Baltimore, MD.
Silvy, N.J. (ed). 2012. *The wildlife techniques manual.* Management, Vol 2, Seventh edition. The John Hopkins University Press, Baltimore, MD.
Smith, R.L., and T.M. Smith. 2009. *Elements of ecology.* Benjamin Cummings, New York.
Tappeiner, J.C., D.A. Maguire, T.B. Harrington, and J.D. Bailey. 2015. *Silviculture and ecology of western U.S. forests,* Second edition. Oregon State University Press, Corvallis.
Walters, C. 1986. *Adaptive management of renewable resources.* MacMillan Publishing Co., New York.
Wilson, E.O. (ed.). 1976. *Biodiversity.* The National Academy of Science, National Academy Press, Washington, DC.
Young, R.A., and R.L. Giese (eds.). 2003. *Introduction to forest ecosystem science and management,* Third edition. John Wiley and Sons, Inc., New York.

8-6 From Literature to Data Models

In the previous section, we recommend a selection of books for a working library. These books contain concepts and ideas that collectively form a knowledge base that leads to a better understanding of how the phenomenon of life in the natural world of trees, forests, environment, and habitat affects the ability of wildlife to **survive and reproduce**. One problem with such a large body of information is how it can be organized and presented in a way that positions it to be interpreted and used for planning and management purposes (Figure 8-1). This can be accomplished by creating a data model containing information in a scheme of connecting tables with columns, rows, and data cells. An intuitive process is to create lists in different categories, such as animals in a given forest type or forest types in a state or region. A well-designed data model can elicit a mental image from information in data cells of what might occur in a field situation.

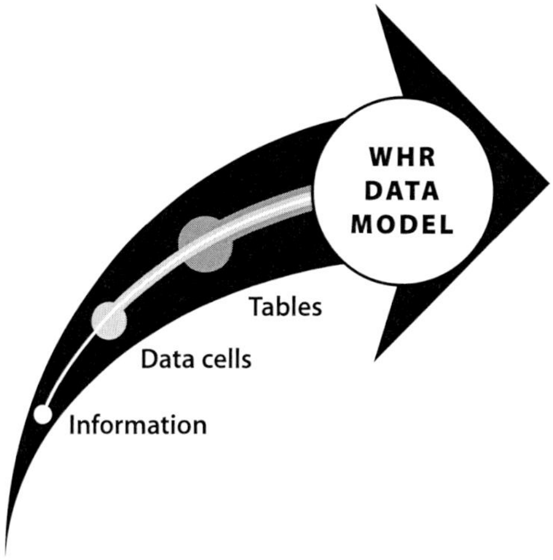

Figure 8-1 From information to data cells and tables that define attributes of the subjects, a Wildlife Habitat Data Model can be created.

Items in a list create a column in a table with each item in one cell. Other columns can be added to create a table with different items related to other columns. Creating a table is the first step in developing a database management system containing many tables and subject headings. Most word processing software will have a menu system for creating a table that:

- organizes information for ease of presentation,
- can group items into different categories,
- provides a lot of information in a small space,
- are easy to copy to transfer into other documents, and
- can easily be exported to database management systems.

Tables become even more important and useful when a database is continually updated as part of a management planning process. One of the management strategies discussed in Chapter 6 (Section 6-6) was Wildlife Habitat Relationships (WHR). This approach depends on research results contained in many different sources of literature from professional journals, government publications, and educational institutions. In a database scheme tables that are connected can identify areas containing physical components of wildlife habitat used for food, cover, water, and space. While the body of data is stored in tables they are linked in a variety of ways similar to ecological food chains and webs. Data file format and links are largely defined by the many ways in which the relational software packages have built upon Codd's (1990) initial formulations.

The term **relation** refers to an accumulation of facts (attributes) in a table describing the same subject. An example of a table subject might be students. The question is: What do students have in common? They all have a first and last name, weight, height, date of birth, social security number, and so forth. All the characteristics pertaining only to students form a set of relations. The different characteristics just mentioned are attributes of each student. Therefore, an **attribute** is a fact about an inherent characteristic, property, or measurement of a particular subject. In a database scheme, a table is a relation and an attribute is a column in that table.

Data cells exist in a table at the intersection of a vertical column with a horizontal row. Information in a data cell can be numerical or text. One of the challenges in developing a usable relational database is the use of coded texts that relate to items in the data cells. Most of the text to describe wildlife habitat relationships is too long for a data cell and it is common practice to use an abbreviation that has some recognition, such as **ma** for mammals or **bi** for birds.

A relational data model can combine the single, multispecies, and functional groups into a detailed but easy-to-use format. Information can be obtained from the level needed to meet management objectives. The different levels of content allow **nested questions** to be asked about the relationships. A nested question contains several questions in one statement. Such as (1) what mammals, (2) live in the ponderosa pine forest type, (3) in the state of Colorado, (4) that nests in trees? The power of a relational data model is how data are connected from one table to another. Manipulating data is accomplished by selecting, combining, and filtering through a menu system developed for relational databases, such as Microsoft Access (Groh et al. 2007). The desired outcome of the data model is to provide the basic structure that can account for forest components and how they can be associated with different animals or groups of animals. It is easy to filter and sort information in the tables and columns that can be developed for a local or project database for analysis and application.

Structure and Organization

Data for storage in a relational database must be structured and organized in a logical and consistent way. The fundamental structure is a record devoted to an identifiable entity (e.g., species). Within this record a varying number of fields are located, each dedicated to holding information or a specific characteristic associated with the entity. Some advantages of relational databases are:

- they retain the simplicity and intellectually comfortable style associated with data in tabular form,
- two relations can be used to develop a third relation for either permanent or temporary use,
- they have the ability to link and associate different relations, and
- new relations can be added or deleted at any time and no information is lost.

The use of relational databases implicitly imparts a very particular way of thinking and conceptualizing information. The procedures followed to link information forces the user to take into account the way that the natural world is organized not simply as hierarchical chains, but also as webs in a complex of relationships (e.g., as in an ecosystem).

Design Criteria

The categories of information identified for inclusion in a relational database, along with individual items in each category, should be developed in consultation with working professionals to determine what subjects generally are used at the local level for decision making. Broad categories such as Animals, Occurs, and Life-Needs can be used as subjects of a table to define the contents of columns containing more detailed information. Discussions with foresters and wildlife biologists over many years have produced design criteria that a data model should meet:

- information in the model must be suitable for use in a table format using rows and columns;
- the model must be adaptable for storing in a microcomputer for use with relational database software;
- the model must have the capability to answer nested questions;
- the model must have the capability of identifying animal species using the same or overlapping physical habitat factors; and
- data in the model must be displayed without redundancy.

The above criteria can all be met if restrictions are enforced to describe habitat by relations and attributes conforming to relational theory. The problem is one of storing summary data in tables, columns, and rows for ease in understanding and for reporting purposes. Each data cell has a logical and physical domain. The logical domain corresponds to the subject of the data in a column. For example, data cells containing common names should not contain scientific names. The physical domain refers to the types of data: real, integer, date, time, text, etc.

Data cells can have one item or several items in a group that can be repeated with a delimiter such as a comma or period. Ordinarily data cells do not contain repeating groups, but this is a trade-off in design and how the information is to be visually presented. Several characteristics are common to tables in relational databases:

- every row has the same number of columns;
- each column contains the same kind of fact (logical or physical domain) in each row;
- the order of the rows is not important;
- the order of the columns is not important; and
- no two rows are the same.

How Relationship Systems Work

Relations have key attributes for retrieving data from more than one table simultaneously. Data in one table is linked to other tables by direct and indirect routes as in Figure 8-2. Direct links are provided by a unique identification code (key) linking every row in the one table with every row in another table where the codes are identical. Indirect links occur when one relation contains links to other relations not directly linked by the unique identification code.

One-to-One Relationships: In this case, one row in one table is linked only to one row in another table. The link between two tables is indicated by a solid line.

One-to-Many Relationships: One row in one table is linked to many rows in another table. To indicate the relationship in a diagram, no arrow is the one side of the relation and two arrows (>>) identifies the many side.

Many-to-Many Relationships: This relationship can be identified by many rows in one table linking many rows in another table. Many-to-many relationships are not as efficient as one-to-many relationships and need to be redesigned by including a linking table.

In one-to-one relationships the link should be a stable attribute or it should be taken from the table that by definition contains a key column. In one-to-many relationships the linking column should be whatever column is defined as the key column. Information from two tables can be queried if there is a link between tables. When making a query for information from more than one table, the relationships between tables are important because they determine how the information will be displayed when retrieved.

The power and utility of a data model are in the combinations of attributes within tables associating animals with their use of forest components. Information can be projected into new tables that can be deleted after they have served their purpose. All types of habitat information can be entered into relational databases for developing useful wildlife habitat relationships. As more information is added to the wildlife literature and as management decisions become more complex, the use of increasingly powerful microcomputers using relational database software will provide wildlife biologists and foresters the ability to compare, select, and make decisions based on the essential facts contained in massive databanks.

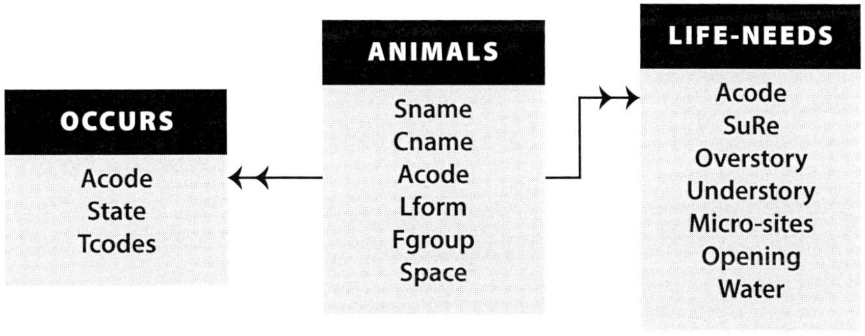

Figure 8-2 WHR-BASIC, a relational data model structure.

Determining the proper links and relationships between tables is as much an art as a science. Intuition plays an important role in determining the links between tables and the dependencies created. Small test databases well "worked over" will in time identify the anomalies that exist and once known, a solution to problems can be found.

8-7 A Basic Forest Wildlife Habitat Relationships Model

The Wildlife Habitat Relationships (WHR) model presented below resulted from the work of many wildlife biologists and foresters in federal and state agencies to meet the needs of the Forest and Rangeland Renewable Resources Planning Act (1973). The Act required a national assessment of natural resources in the United States and the information used is still available today in many publications and databases. New relational software, which uses structured query language (SQL) in a menu system, now make it easier to design an efficient relational scheme to manage large amounts of data for field applications. SQL is used to store and retrieve data from tables in a relational database. Manipulating data occurs by initiating queries using a menu embedded in the software. For example, extracting information from the Animals table in WHR-BASIC might be:

SELECT Cname (common name) from Animals **WHERE** Lform (life-form) equals ma (mammals).

The result would be a list of all the common names of mammals in the Animals table. The **WHR-BASIC** data model incorporates information from published literature usable in forest planning and managing (Table 8-2). When defined in a relational data model, wildlife habitat relationships are a generalization of habitats and environmental characteristics that describe the phenomenon of life that exists in forest ecosystems. However, when including environmental characteristics in a WHR model, it is tempting to describe items in finer detail than necessary for decision-making but a simple and practical data model contains information in three basic relations: Animals, Life-Needs, and Occurs. Table 8-2 contains all the acronyms in the data cell example and in the data model structure in Figure 8-2.

ANIMALS RELATION. Animals is a core table containing wildlife species in the United States. Every table in the data model is ultimately connected to Animals through a series of linked tables that contain basic life history information. Animals has a one-to-many relation with **Occurs** and a one-to-many link with **Life-Needs**. The default column width in WHR when a table is created is 255 spaces. In addition to text, other format choices exist such as: computed, number, memo, currency, etc.

When creating a column it is best to accept the default value. Later it is more efficient to change width to a specific value depending on the potential length of a data string because leaving a default value wastes space when text string characters are low, such as in Animals. A good practice is to increment five spaces at a

Table 8-2 Directory and menu for the WHR-BASIC relational data model.

ANIMALS

Sname: Scientific name

Cname: Common name

Acode: Derived from scientific name

Lform: Life-form
am = amphibian
bi = bird
fi = fish
ma = mammal
re = reptile

Fgroup: Functional group
bp = birds of prey
cu = cavity using
cc = cavity creator
gn = ground nesting
pe = birds-perching
cl = carnivore—large
cs = carnivore—small
In = insectivore
hl = herbivore-large
rl = rodent—large
rs = rodent—small
ts = tree squirrel

Space: Home range
S1 = <1 ac
S2 = 1–10 ac
S3 = 11–100 ac
S4 = 101–1,000 ac
S5 = 1,001–10,000 ac
S6 = 10,1001–100,000 ac

LIFE-NEEDS

Acode: Animal code

SuRe: Survive and reproduce
Fa = Feeding area
Ca = Cover area

Overstory
sa = stand attributes
tcr = tree crown
tt = tree trunk
tb = tree bark
wt = wolf tree
sn = dead tree/snag

Understory
gf = grass/forb
sh = shrub
ls = log/stump
dh = duff/humus
bo = boulders
rf = rockfield/talus/crevices
wd = woody debris
st = small trees

Micro-sites (shelter)
tn = tree/shrub nest
tc = tree cavity
tp = tree perch/roost
pt = platform nest
ug = underground cavity/tunnel
gn = ground nest
gb = ground bed
cl = elevated cliff/ledge nest

Opening
gf = grass/forb
sh = shrub
wm = wet meadow

Water
ri = riparian
nw = near water
wa = water medium
aq = aquatic vegetation
wi = within home range

(continued)

OCCURS

Acode: Animal code

State: State abbreviations from postal codes.

Tcode:
10—White-red-jack pine
11—Spruce fir
12—Longleaf-slash pine
13—Loblolly-shortleaf pine
14—Oak-pine
15—Oak-hickory
16—Oak, gum-cypress
17—Elm-ash-cottonwood
18—Maple-beech-birch
19—Aspen-birch
20—Douglas-fir
21—Hemlock-Sitka spruce
22—Ponderosa pine
23—Western white pine
24—Lodgepole pine
25—Larch
26—Fir-spruce
27—Redwood
28—Pinyon-juniper
29—Western hardwoods

MENU ADDITIONS TO CONSIDER

A stand structure attribute in Overstory in Life-Needs can replace the "sa" code when the information is available.

Stand
Vertical layers
1s = single story
2s = two stories
ms = multi-story

Tree crown
up = upper crown
gc = general crown
lc = lower crown

Canopy closure
op = opening (<10%)
th = thin (10–40%)
tk = thick (41–70%)
cl = closed (>70%)

Structure
1sa = seedling sapling (<4-in dbh)
2yo = young (5–10 in)
3in = intermediate (11–17 in)
4ma = mature (>17 in)
5ol = old forest (ma + conditions)

time to set the needed value. Additional space can always be added later. In Animals there are six columns of information (Sname, Cname, Acode, Lform, Fgroup, Space) relating to an individual species. The Animals table can be expanded by adding columns when other information is needed, such as legal status (endangered or threatened species), sub-species designations, or a common name alias. A list of common and scientific names and life-forms is available on the book's web page at waveland.com.

Life-Needs Relation. The Life-Needs relation brings together an animal's association with use of forest components. The structural components of a forest are the first level of abstraction when identifying food and cover areas. Food and cover are columns (attributes) with descriptors. Each data cell must have a food or a cover designation in the Life-Needs table. Data cells can have more than one entry, such as in Understory, when separated by a period or comma (see the following section: Examples of Data Cells).

Common name (Cname) and Life-form (Lform) are entered into the Animals core table but it is convenient to continue to use Cname and Lform in other tables for ease in identification although this can be done with the connecting Acode link. While this violates the one-time entry protocol, convenience to be able to recognize a common name without looking up the name by an Acode reference, overrides protocol. A link between tables establishes relationships and how a change in one can affect another.

Life-Needs is where environmental characteristics come together to indicate how much defined habitat overlap exists from one animal to another. While considerable information exists about some species, such information is mostly from those species traditionally studied as game animals. Information on non-game and threatened and endangered species is still developing so the data model has many vacant spaces in its data cells. Data for categorizing how wildlife is associated with forest components was patterned after a Species Habitat Profile (SHP) system developed for a mainframe computer (Patton 1975b), the Life-Form system for the Blue Mountains of Oregon and Washington (Thomas 1979), and the work by Robert Hoover and Dale Wills (1984).

The terms food and cover are the survive and reproduce (SuRe) components. Cover is well established in the wildlife literature and there exists a general understanding that cover refers to security or protection. In many cases cover is associated with vegetation. However, wildlife also use non-vegetation components such as rocks, down woody material, caves, and water for protection. Cover can be separated into roosting, resting, thermal, breeding, etc. when information is available and there is a need for a finer resolution. More than one vertical forest layer can be associated with a single species.

The horizontal components have descriptive characteristics across the forest floor associated with the vertical strata in horizontal space. Conveniently for managers and researchers, forest ecosystems with stands and open areas described by vertical and horizontal components define a matrix with columns and data cells in a table. Stand attributes (sa) is used in the data model to indicate a lack of specifics that can be added later by users when these attributes are available for tree size, density, and layers.

OCCURS RELATION. The definition of habitat includes a place where an animal lives. A place provides survival components of food and cover as listed in the Life-Needs relation. In a forest, the place can be hierarchically identified as a forest type in a state for a low level of resolution. Type names are long and use a lot of space; therefore, the information is entered only once in the type name column. All forest types have a numerical code (Tcode) associated with the type name. The type name follows the naming convention on the Atlas Map of 1967 and the Society of American Foresters forest cover type designation. However, the Tcode used in the data model does not follow the codes shown on the 1967 map because of the need for format continuity in using a relational system. Codes beginning with a "0" or having only one digit such as a "1" can sometimes cause problems when using structured query language and therefore, the Tcodes are all two-digit numbers from 10 to 29.

EXAMPLES OF DATA CELLS

Animals Relation

Sname	Cname	Acode	Lform	Fgroup	Space
Odocoileus virginianus	white-tailed deer	odovir	ma	hl	S4
Sciurus carolinensis	eastern gray squirrel	scicar	ma	ts	S3

Life-Needs Relation

Acode	SuRe	Overstory	Understory	Micro-sites	Opening	Water
odovir	Fa	sa.	gf. sh.		gf. sh.	wi.
odovir	Ca	sa.	sh. st.	gb.		
scicar	Fa	sa. tcr.				
scicar	Ca	sa. tcr.		tn. tc.		

Occurs Relation

Acode	State	Tcodes
odovir	VA	10. 13. 14. 15. 16

Depending on need other relations can be added to the WHR-BASIC data model. The **occurs** relation identifies forest types where animals occur but does not contain a listing of **plants** that occur in each type. This relation would contain a listing of several hundred plants with their scientific and common name along with life-form and might be needed for regional or local projects. Another relation is the **food** needs of animals to include scientific and common names of plants, animals, and life-forms that constitute the food chains and webs in a particular geographical area or forest type.

Readers are encouraged to invest in one of the many books that provide the basics for developing an Access database, but free tutorials can be found online. One worth considering is by the Goodwill Community Foundation, Inc. Sign up is required to use their tutorial at **GCFLearnFree.org**. In addition, readers might want to consider public domain and commercial software programs as alternatives to Access. The references listed below contain useful information in developing wildlife habitat relationships.

Sources for Basic Information

Benyus, J.M. 1989. *The field guide to wildlife habitats of the eastern United States*. Simon and Schuster, Inc., New York.

Benyus, J.M. 1989. *The field guide to wildlife habitats of the western United States*. Simon and Schuster, Inc., New York.

Brown, E.R. (tech. ed.). 1985. Management of wildlife and fish habitats in forests of western Oregon and Washington. USDA Forest Service R6F&WL 192, Parts 1 and 2. PNW Region, Portland, OR.

Burns, R.M., and B.H. Honkala. 1990. Silvics of trees of North America. Vol. 1, Conifers, Vol. 2, Hardwoods. Agric. Hdbk. 654. USDA Forest Service, Washington, DC.

Conant, R. 1998. *Reptiles and amphibians: Eastern/central North America*. Houghton Mifflin, New York.

DeGraff, R. M. (tech. coord.). 1978. *Management of southern forests for nongame birds*. Proceedings of the workshop. Gen. Tech. Report SE-14. USDA Forest Service, Asheville, NC.

DeGraff, R.M., and K.E. Evans. 1979. *Management of north central and northeastern forests for nongame birds*. Proceedings of the workshop. Gen. Tech. Report NC-51, USDA Forest Service, Minneapolis, MN.

DeGraff, R.M., V.E. Scott, R.H. Hamre, et al. 1991. Forest and rangeland birds of the United States: Natural history and habitat use. Agri. Hdbk. 688. USDA Forest Service, Washington, DC.

DeGraff, R.M., and N.G. Tilghman. 1980. *Management of western forests and grasslands for nongame birds*. Proceedings of the workshop. USDA Forest Service Gen. Tech. Report INT-86. Intermountain Region, Ogden, UT.

DeGraff, R.M., M. Yamasaki, W.B. Leak, et al. 1989. *New England wildlife: Management of forested habitats*. Gen. Tech. Report NE-144, USDA Forest Service, Northeastern Forest Experimental Station, Radnor, PA.

Ehrlich, P.R., D.S. Dobkin, and D.W. Naeem. 1988. *The birder's handbook*. Simon and Schuster, New York.

Eyre, F.H. (ed.). 1980. *Forest cover types of the United States and Canada*. Society of American Foresters, Washington, DC.

Feldhamer, G.A., B.C. Thompson, and J.A. Chapman. 2003. *Wild mammals of North America: Biology, management and conservation*. Johns Hopkins University Press, Baltimore, MD.

Garrison, G.A., A.J. Bjustad, D.A. Duncan, et al. 1977. Vegetation and environmental features of forest and range ecosystems. Agri. Hdbk 475. USDA Forest Service, Washington, DC.

Hall, E.R. 1981. *The mammals of North America*, Vol 1 and 2. John Wiley and Sons, New York. Reprint by Blackburn Press, Caldwell, NJ.

Hoover, R. L., and D. L. Wills (eds.). 1984. *Managing forested lands for wildlife*. USDA Forest Service, Rocky Mountain Region and Colorado Division of Wildlife, Denver, CO.

Kircher, J.C. 1993. *Ecology of western forests*. Houghton Mifflin, New York.

Kircher, J.C. 1998. *California and Pacific Northwest forests*. Houghton Mifflin, New York.

Kircher, J.C. 1998. *Eastern forests*. Houghton Mifflin, New York.

Kircher, J.C. 1998. *Rocky Mountain and southwest forests*. Houghton Mifflin, New York.

National Geographic Society. 1983. *Birds of North America*. National Geographic Society, Washington, DC.

Patton, D.R. 1978. *Run Wild: A storage and retrieval system for wildlife habitat information*. USDA Forest Service Gen. Tech. Report RM-51. USDA Forest Service, Fort Collins, CO.

Patton, D.R. 1992. *Wildlife habitat relationships in forested ecosystems*. Timber Press, Portland, OR.

Payne, N.F., and F.C. Bryant. 1994. *Techniques for wildlife habitat management of uplands*. McGraw-Hill, Inc. New York.

Peterson, R.T. 1980. *A field guide to eastern birds*. Houghton Mifflin, Boston.

Peterson, R.T. 1990. *A field guide to western birds*. Houghton Mifflin, Boston.

Petrides, G.A. 1988. *A field guide to eastern trees*. Houghton Mifflin, Boston.

Petrides, G.A. 1992. *A field guide to western trees*. Houghton Mifflin, Boston.

Reid, F.A. 2006. *Mammals of North America*. Houghton Mifflin, Boston.

Scott, V.E., K.E. Evans, D.R. Patton, et al. 1977. Cavity nesting birds of North American forests. Agri. Hdbk 511. USDA Forest Service, Washington, DC.

Smith, D.R. (tech. coord.). 1975. *Proceedings of symposium on management of forest and range habitats for nongame birds*. Gen. Tech. Report WO-1. USDA Forest Service, Washington, DC.

Stebbins, R.C. 2003. *Western reptiles and amphibians*. Houghton Mifflin, Boston.

Sutton, A., and M. Sutton. 1985. *Eastern forests.* Alfred A. Knopf, New York.

Szaro, R.C., K.S. Severson, and D.R. Patton (tech. coords.). 1988. *Management of amphibians, reptiles, and small mammals in North America.* Gen. Tech. Report RM-166. USDA Forest Service, Ft. Collins, CO.

Thomas, J.W. (ed.). 1979. Wildlife habitats in managed forests, the Blue Mountains of Oregon and Washington. Agri. Hdbk 553. USDA Forest Service, Washington, DC.

Udvardy, M.D.F. 1977. *Field guide to North American birds, western region.* Alfred A. Knopf, New York.

Verner, J., and A.S. Boss (tech. coord.). 1980. *California wildlife and their habitats, Western Sierra Nevada.* Gen. Tech. Report PSW-37. USDA Forest Service, Berkeley, CA.

Whitaker, J.O., Jr., and W. Hamilton Jr. 1998. *Mammals of the eastern United States.* Comstock. Cornell University Press, Ithaca, NY.

Whitney, S. 1985. *Western forests.* Alfred A. Knopf, New York.

Zabel, C.J., and R.G. Anthony (eds.). 2003. *Mammal community dynamics, management and conservation in the coniferous forests of western North America.* Cambridge University Press, Cambridge, UK.

8-8 Where Are the Alligators?

One of the challenges facing resource managers is the large amount of information available in regional and local databases such as for a National Forest. This information is stored either in a main frame or desktop computer for use in routine activities or in response to federal legislation requiring a management plan or scheduled assessment. Whatever the need, managers should understand the types of stored data that may be used to select information for a report. Three issues exist to keep in mind when accessing information in a relational database.

1. A computer can only process information programmed in a specific way to obtain information stored in tables, rows, columns, and data cells with strict entry criteria.
2. Any error that violates entry criteria will result in an output error that may not always be easily recognized. The most common error is in a spelling or unidentified code.
3. A computer lacks the ability to identify results that may contain errors that can only be detected by managers who have knowledge of the subject matter.

Subtle errors can occur, such as the code for an animal's life-form associated with a scientific name. An example of an obvious error occurred during the first Resource Planning Act assessment for the Forest Service in Arizona and New Mexico (Patton 1975a). In examining several thousand lines of data matching species with forest types in regions and states, a Florida alligator showed up in ponderosa pine in Arizona. As it turns out, an employee wanted to test the ability to find errors in a large data file. After the entry error was discovered and corrected the assessment team often used the expression "Where are the alligators?" when testing the accuracy of a database.

A data model is a powerful tool for organizing and sorting information to use in planning and decision-making. WHR-BASIC is a place to start for creating a data model with three tables (relations) that ultimately can become more complex as new relations are added to define attributes of habitats that affect the ability of animals to survive and reproduce in a specific place. And finally, while a computer can sort and organize thousands of pieces of information in just seconds of time the most powerful computer continues to be the human brain with sensory systems and an intuition used in decision-making not possessed by computers.

8-9 Summary

The information contained in the literature is essential for working professionals to stay current and to increase their knowledge by learning from the experience of others. Knowing where and how to find such information is a necessary part of the learning experience. One place to start obtaining information on resource management are the field and regional offices of the state and federal organizations that have authority to manage forest lands. In addition, state land-grant colleges and universities have extension services available to land owners and in some cases can provide financial support under a cooperative management agreement. The experiment stations in the Departments of Agriculture and Interior have publications available on forestry and wildlife subjects on request or for downloading from websites.

The magnitude of the basic information on all aspects for forestry and wildlife necessitates that this information has to be made available in a condensed form for understanding and practical use. The way to do this is through a relational database system that can be easily adapted for use in all fields of natural resource management. A basic wildlife habitat relationship data model, WHR-BASIC, is presented in this chapter to provide readers with a basic design and a place to start for their use at a local, landscape, or regional levels for management purposes.

References

Codd, E.F. 1990. *The relational model for database management: Version 2.* Addison-Wesley, New York.

Groh, R.R., J.C. Stockman, G. Powell, et al. 2007. *Microsoft access.* Wiley, Indianapolis, IN.

Hoover, R.L., and D.L. Wills (eds.). 1984. *Managing forested lands for wildlife.* USDA Forest Service, Rocky Mountain Region and Colorado Division of Wildlife, Denver, CO.

Patton, D.R. 1975a. *Run Wild: A storage and retrieval system for wildlife habitat information.* Gen. Tech. Report RM-GTR-51. USDA Forest Service, Fort Collins, CO.

Patton, D.R. 1975b. Run Wild System 2000 storage and retrieval system. Computer printouts. USDA Forest Service, Rocky Mt. For. and Range Exp. Stn., Wildlife Research Unit RM-1710, Arizona State University Campus, Tempe.

Thomas, J.W. (ed.). 1979. Wildlife habitats in managed forests, the Blue Mountains of Oregon and Washington. Agric. Hdbk 553. USDA Forest Service, Washington, DC.

Epilogue
We Are Not at the End—Yet

We started this book with the words "in the beginning" and we have reached the end of our project to provide undergraduate students with the basic concepts of wildlife and forest management, to provide working professionals with information for a continuing education program, and to provide private land owners resource information and a framework for developing management plans for their property. So, what does the future hold?

In 1933 Aldo Leopold suggested that our ability to change the land was advancing at a rate where the land, through succession, could no longer heal itself. If true, does any hope exist for future generations to have the quality of life that most people in developed countries enjoy today? Two areas hold promise for maintaining a healthy planet for future generations: reducing deforestation and slowing human population growth. However, these areas cross religious and political boundaries that now and will likely continue to provide major challenges because they affect individual and collective human values.

The first area involves slowing down deforestation by curtailing some of the economic pressures and human activities responsible for the rapid land-use changes taking place in the world ecosystem. The practices responsible for the changes include the large-scale modification of forest lands to meet demands for forest products of all kinds and the clearing of forest land worldwide for agriculture and infrastructure expansion. To slow down these activities, humans will have to reduce their consumption of natural resources—an event not likely to happen very soon, alas!

The second area, slowing down or keeping stable the world human population, is rooted in the same philosophy as controlling wildlife populations: maintaining a balance of resources needed for reproduction and survival with the sustainable productive capacity of the Earth itself. Without this balance, populations of both humans and animals face extinction much like the polar bear in the arctic is experiencing today. The early and mid-1970s gave rise to major concerns about the world's population. This concern still exists today as mega-wildfires become more common and increasingly attributed (at least in part) to human-caused global climate change; however, countervailing economic and political forces have become stronger. So where do we go from here?

First, biologists have experience with and knowledge of what goes on in nature at the local level with the introduction of animals into a new environment (see Chapter 3), and they know and understand the consequences of the effects on habitat of overpopulations of animals (Chapter 6). By analogy, these same effects extend to human populations. Natural resource professionals such as field biologists, researchers, and academicians can broaden their horizons to look beyond local projects and become more involved at the national and international levels. If the choice is not to get involved, then the opportunities to work with and preserve forests and wildlife at the local level may disappear as more harvesting is done to provide more forest products of all kinds and more clearings are used to produce more food for more people.

Evidence exists that wildlife populations will crash when animal density exceeds the ability of the habitat to produce food and cover. Whether eventually this same ecological regulatory mechanism will contribute to reducing human populations is certainly a matter of concern—or should be. As humans mature from the individualistic concept of survival to a more thoughtful view that we all share this planet together, then a more cooperative attitude of sharing and caring may develop. But how long will it take for this new attitude to come about? Probably longer than the lifetime of anyone reading this text—but the question remains.

Prior to several thousand years ago, major changes brought about on earth were through the forces of nature: hurricanes, tornados, fire, volcanoes, earthquakes, floods, glaciers, and so forth. Humans noticed how these forces affected the landscape that provided the plants and animals contributing food and clothing for their survival. At that time, humans lacked the ability, using only the axe, cow, and plough, to drastically alter their environment to an extent that prevented recovery, and the forces of nature by themselves largely did not change the world landscapes at the current rate.

Slowing down resource consumption, including nonrenewable resources such as fossil fuels, and slowing human population increases to reach a degree of equilibrium at the local, regional, and national level—and ultimately in the world ecosystem—is not unrealistic. However, this requires the involvement of the sciences in the decision-making process by providing sound ecological, social, political, and economic data to the decision makers. Getting involved is relatively easily done in the planning process discussed in previous chapters (e.g., Chapter 7). The planning process is the place to start because it offers opportunities to influence people who make decisions. Other ways to get involved include working outside one's own "habitat" by going on field trips sponsored by professional organizations, attending conferences and workshops, discussing issues with scientists in one's own organization, and being enrolled in a continuing education program to obtain a certificate or license.

Within the planning and management process, the value of science in making decisions has increased in importance. As human populations rise along with increasing per capita consumption of natural resources, the margin for error is critical in considering options for management. In the past, mistakes could be ignored by moving to new lands (as described in Chapter 3), but this is no longer a viable option. The Earth has a finite amount of natural resources and although technology can increase the ability to effectively use resources and increase productivity, a limit still exists. Forestry and wildlife professionals understand these limits—even if they may not know their exact nature and magnitude.

Although decision-making at the landscape level may ultimately be a political process, resource management professionals have an obligation to provide the best information available to the decision makers. Decision makers may also ask for recommendations for a course of action. Managers may face the situation where their recommendations are rejected. Although tempting to disengage when professional opinions are ignored, this is exactly the time when resource managers, researchers, and the academic community need to be involved in the process. Natural resource professionals have considerable experience, information, and facts about resource management and they need to be involved to ensure that their expertise is included in the planning process.

Regardless of the engagement of many people in the management of natural resources, the Earth is likely in store for rapid and often chaotic changes to our physical climate—general warming and all of the associated changes with that warming (e.g., drought) and the supplemental energy in the atmosphere and oceans. The earth is projected to have periods of unprecedented climatic shifts, which will drive ecological, social, political, and economic changes. At this point, natural resource management must again consider the concepts of sustainable and adaptable management, conservation of our natural resources and associated biological diversity, and the maintenance of options for future generations of children and grandchildren (Chapter 5).

The way forward is to continually be reminded of why we became foresters and wildlife professionals—a sense of wonder, awe, respect, and deep concern about and love of forests and wildlife. Let us not forget the passion that led many natural resource managers to their chosen professions. They may never become iconic leaders of environmental stewardship, such as John Muir, Aldo Leopold, Gifford Pinchot, and Rachel Carson, but collectively they can contribute to a better world of forests and wildlife through their ethical professionalism, active engagement, and how they live. Our shared world deserves nothing less.

Appendix A
Common and Scientific Name, Scientific Code, Life-Form, and Kingdom (A and P), Alphabetized by Common Name

Bi-Ma-Tr-US

Cname	Sname	Scode	Lform	King
Abert's squirrel	Sciurus aberti	sciabe	ma	A
Abert's towhee	Pipilo aberti	pipabe	bi	A
Acadian flycatcher	Empidonax virescens	empvir	bi	A
Acorn woodpecker	Melanerpes formicivorus	melfor	bi	A
Agile kangaroo rat	Dipodomys agilis	dipagi	ma	A
Alaskan hare	Lepus othus	lepoth	ma	A
Alder flycatcher	Empidonax alnorum	empaln	bi	A
Allegheny woodrat	Neotoma magister	neomag	ma	A
Allen's big-eared bat	Idionycteris phyllotis	idiphy	ma	A
Allen's chipmunk	Tamias senex	tamsen	ma	A
Allen's hummingbird	Selasphorus sasin	selsas	bi	A
Alligator juniper	Juniperus deppeana	jundep	tr	P
Alpine chipmunk	Tamias alpinu	tamalp	ma	A
American avocet	Recurvirostra americana	recame	bi	A
American beaver	Castor canadensis	cascan	ma	A
American bison	Bison bison	bisbis	ma	A
American bittern	Botaurus lentiginosus	botlen	bi	A
American black duck	Anas rubripes	anarub	bi	A
American coot	Fulica americana	fulame	bi	A
American crow	Corvus brachyrhynchos	corbra	bi	A
American dipper	Cinclus mexicanus	cinmex	bi	A
American elm	Ulmus americana	ulmame	tr	P
American goldfinch	Carduelis tristis	cartri	bi	A
American kestrel	Falco sparverius	falspa	bi	A
American pika	Ochotona princeps	ochpri	ma	A
American pipit	Anthus rubescens	antrub	bi	A
American plum	Prunus americana	pruame	tr	P
American redstart	Setophaga ruticilla	setrut	bi	A
American robin	Turdus migratorius	turmig	bi	A
American shrew mole	Neurotrichus gibbsii	neugib	ma	A
American swallow-tailed kite	Elanoides forficatus	elafor	bi	A
American tree sparrow	Spizella arborea	spiarb	bi	A
American water shrew	Sorex palustris	sorpal	ma	A
American wigeon	Anas americana	anaame	bi	A
American woodcock	Scolopax minor	scomin	bi	A
Anhinga	Anhinga anhinga	anhanh	bi	A
Anna's hummingbird	Calypte anna	calann	bi	A
Antelope jack rabbit	Lepus alleni	lepall	ma	A
Apache pine	Pinus engelmannii	pineng	tr	P

(continued)

Cname	Sname	Scode	Lform	King
Apache pocket mouse	Perognathus apacher	perapa	ma	A
Aplomado falcon	Falco femoralis	falfem	bi	A
Appalachian cottontail	Sylvilagus obscurus	sylobs	ma	A
Arctic fox	Alopex lagopus	alolag	ma	A
Arctic ground squirrel	Spermophilus parryii	spepar	ma	A
Arctic hare	Lepus arcticus	leparc	ma	A
Arctic loon	Gavia arctica	gavarc	bi	A
Arctic shrew	Sorex arcticus	sorarc	ma	A
Arctic willow	Salix arctica	salarc	tr	P
Arizona cotton rat	Sigmodon arizonae	sigari	ma	A
Arizona alder	Alnus oblongifolia	alnobl	tr	P
Arizona cypress	Cupressus arizonica	cupari	tr	P
Arizona gray squirrel	Sciurus arizonensis	sciari	ma	A
Arizona honeysuckle	Lonicera arizonica	lonari	vi	P
Arizona madrone	Arbutus arizonica	arbari	tr	P
Arizona mountain-ash	Sorbus dumosa	sordum	tr	P
Arizona myotis	Myotis occultus	myoocc	ma	A
Arizona shrew	Sorex arizonae	sorari	ma	A
Arizona sycamore	Platanus wrightii	plawri	tr	P
Arizona walnut	Juglans major	jugmaj	tr	P
Arizona white oak	Quercus arizonica	queari	tr	P
Arizona woodrat	Neotoma devia	neodev	ma	A
Arroyo willow	Salix lasiolepis	salla2	tr	P
Ash-throated flycatcher	Myiarchus cinerascens	myicin	bi	A
Attwater's pocket gopher	Geomy's attwateri	geoatt	ma	A
Aztec thrush	Ridgwayia pinicola	ridpin	bi	A
Bachman's sparrow	Aimophila aestivalis	aimaes	bi	A
Bachman's warbler	Vermivora bachmanii	verbac	bi	A
Badger	Taxidea taxus	taxtax	ma	A
Bailey's pocket mouse	Perognathus baileyi	perbai	ma	A
Baird's pocket gopher	Geomys breviceps	geobre	ma	A
Baird's sandpiper	Calidris bairdii	calbai	bi	A
Baird's shrew	Sorex bairdi	sorbai	ma	A
Baird's sparrow	Ammodramus bairdii	ammbai	bi	A
Bald eagle	Haliaeetus leucocephalus	halleu	bi	A
Band-tailed pigeon	Columba fasciata	colfas	bi	A
Bank swallow	Riparia riparia	riprip	bi	A
Banner-tailed kangaroo rat	Dipodomys spectabilis	dipspe	ma	A
Barbary sheep	Ammotragus lervia	ammler	ma	A
Baren ground shrew	Sorex ugyunak	sorugy	ma	A
Barn owl	Strix alba	stralb	bi	A
Barn swallow	Hirundo rustica	hirrus	bi	A
Barred owl	Strix varia	strvar	bi	A
Barrenground willow	Salix brachycarpa	salbra	tr	P
Barrow's goldeneye	Bucephala islandica	bucisl	bi	A
Bay-breasted warbler	Dendroica castanea	dencas	bi	A
Beaked hazel	Corylus cornuta	corcor	tr	P
Bearberry honeysuckle	Lonicera involucrata	loninv	vi	P
Beardless flycatcher	Camptostoma imberbe	camimb	bi	A

Cname	Sname	Scode	Lform	King
Bebb willow	Salix bebbiana	salbeb	tr	P
Belding's ground squirrel	Spermophilus beldingi	spebel	ma	A
Bell's vireo	Vireo bellii	virbel	bi	A
Belted kingfisher	Ceryle alcyon	ceralc	bi	A
Bendire's thrasher	Toxostoma bendirei	toxben	bi	A
Berylline hummingbird	Amazilia beryllina	amaber	bi	A
Bessey plum	Prunus besseyi	prubes	tr	P
Bewick's wren	Thryomanes bewickii	thrbew	bi	A
Big birch	Betula glandulosa	betgla	tr	P
Big brown bat	Eptesicus fuscus	eptfus	ma	A
Big free-tailed bat	Nyctinomop macrotis	nycmac	ma	A
Bigeared woodrat	Neotoma macrotis	neomac	ma	A
Bigtooth maple	Acer grandidentatum	acegra	tr	P
Bitter cherry	Prunus emarginata	pruema	tr	P
Black bear	Ursus americanus	ursame	ma	A
Black cherry	Prunus serotina	pruser	tr	P
Black cottonwood	Populus trichocarpa	poptri	tr	P
Black hawk	Buteogallus anthracinus	butant	bi	A
Black locust	Robinia pseudo-acacia	robpse	tr	P
Black phoebe	Sayornis nigricans	saynig	bi	A
Black rail	Laterallus jamaicensis	latjam	bi	A
Black scoter	Melanitta nigra	melnig	bi	A
Black skimmer	Rynchops nigra	rynnig	bi	A
Black swift	Cypseloides niger	cypnig	bi	A
Black tern	Chlidonias niger	chlnig	bi	A
Black vulture	Coragyps atratus	coratr	bi	A
Black whiskered vireo	Vireo altiloquus	viralt	bi	A
Black-and-white warbler	Mniotilta varia	mnivar	bi	A
Black-backed woodpecker	Picoides arcticus	picarc	bi	A
Black-bellied plover	Pluvialis squatarola	plusqu	bi	A
Black-bellied whistling duck	Dendrocygna autumnalis	denaut	bi	A
Black-billed cuckoo	Coccyzus erythropthalmus	cocery	bi	A
Black-billed magpie	Pica pica	picpic	bi	A
Blackburnian warbler	Dendroica fusca	denfus	bi	A
Black-capped chickadee	Parus atricapillus	paratr	bi	A
Black-capped gnatcatcher	Polioptila nigriceps	polnig	bi	A
Black-capped vireo	Vireo atricapillus	viratr	bi	A
Black-chinned hummingbird	Archilochus alexandri	arcale	bi	A
Black-chinned sparrow	Spizella atrogularis	apiatr	bi	A
Black-crowned night-heron	Nycticorax nycticorax	nycnyc	bi	A
Black-footed ferret	Mustela nigripes	musnig	ma	A
Black-headed grosbeak	Pheucticus melanocephalus	phemel	bi	A
Black-necked stilt	Himantopus mexicanus	himmex	bi	A
Blackpoll warbler	Dendroica striata	denstr	bi	A
Black-shouldered kite	Elanus caeruleus	elacae	bi	A
Black-tailed gnatcatcher	Polioptila melanura	polmel	bi	A
Black-tailed jack rabbit	Lepus californicus	lepca1	ma	A
Black-tailed prairie dog	Cynomys ludovicianus	cynlud	ma	A
Black-throated blue warbler	Dendroica caerulescens	dencae	bi	A

(continued)

Cname	Sname	Scode	Lform	King
Black-throated gray warbler	Dendroica nigrescens	dennig	bi	A
Black-throated green warbler	Dendroica virens	denvir	bi	A
Black-throated sparrow	Amphispiza bilineata	ampbil	bi	A
Blue grosbeak	Guiraca caerulea	gurcae	bi	A
Blue grouse	Dendragapus obscurus	denobs	bi	A
Blue jay	Cyanocitta cristata	cyacri	bi	A
Blue paloverde	Cercidium floridum	cerflo	tr	P
Blue spruce	Picea pungens	picpun	tr	P
Blue-footed booby	Sula nebouxii	sulneb	bi	A
Blue-gray gnatcatcher	Polioptila caerulea	polcae	bi	A
Blue-throated hummingbird	Lampornis clemenciae	lamcle	bi	A
Blue-winged teal	Anas discors	anadis	bi	A
Blue-winged warbler	Vermivora pinus	verpin	bi	A
Bobcat	Lynx rufus	lynruf	ma	A
Bobolink	Dolichonyx oryzivorus	dolory	bi	A
Bohemian waxwing	Bombycilla garrulus	bomgar	bi	A
Bonaparte's gull	Larus philadelphia	alrphi	bi	A
Bonpland willow	Salix bonplandiana	salbon	tr	P
Boreal owl	Aegolius funereus	aegfun	bi	A
Borel chickadee	Parus hudsonicus	parhud	bi	A
Botta's pocket gopher	Thomomys bottae	thobot	ma	A
Botteri's sparrow	Aimophila botterii	aimbot	bi	A
Boxelder	Acer negundo	aceneg	tr	P
Brant	Branta bernicla	braber	bi	A
Brasilian free-tailed bat	Tadarida brasiliensis	tadbra	ma	A
Brewer's blackbird	Euphagus cyanocephalus	eupcya	bi	A
Brewer's sparrow	Spizella breweri	spibre	bi	A
Bridled titmouse	Parus wollweberi	parwol	bi	A
Bristlecone pine	Pinus aristata	pinari	tr	P
Broad-billed hummingbird	Cynanthus latirostris	cynlat	bi	A
Broad-footed mole	Scapanus latimanus	scalat	ma	A
Broad-tailed hummingbird	Selasphorus platycercus	selpla	bi	A
Broad-winged hawk	Buteo platypterus	butpla	bi	A
Bronzed cowbird	Molothrus aeneus	molaen	bi	A
Brown bear	Ursus arctos	ursarc	ma	A
Brown booby	Sula leucogaster	sulleu	bi	A
Brown creeper	Certhia americana	cerame	bi	A
Brown lemming	Lemmus trimucronatus	lemtri	ma	A
Brown pelican	Pelecanus occidentalis	pelocc	bi	A
Brown thrasher	Toxostoma rufum	toxruf	bi	A
Brown-crested flycatcher	Myiarchus tyrannulus	myityr	bi	A
Brown-headed cowbird	Molothrus ater	molate	bi	A
Brown-headed nuthatch	Sitta pusilla	sitpus	bi	A
Brush mouse	Peromyscus boylii	perboy	ma	A
Brush rabbit	Sylvilagus bachmani	sylbac	ma	A
Buff-breasted flycatcher	Empidonax fulvifrons	empful	bi	A
Buff-collared nightjar	Caprimulgus ridgwayi	caprid	bi	A
Bufflehead	Bucephala albeola	bucalb	bi	A
Bumblebee hummingbird	Atthis heloisa	atthel	bi	A

Cname	Sname	Scode	Lform	King
Bur oak	Quercus macrocarpa	quemac	tr	P
Burro	Equus asinus	equasi	ma	A
Burrowing owl	Athene cunicularia	athcun	bi	A
Bushtit	Psaltriparus minimus	psamin	bi	A
Bushy-tailed woodrat	Neotoma cinerea	neocin	ma	A
Cackling goose	Branta hutchinsii	brahut	bi	A
Cactus mouse	Peromyscus eremicus	perere	ma	A
Cactus wren	Campylorynchus brunneicapillus	cambru	bi	A
California chipmunk	Tamias obscurus	tamobs	ma	A
California condor	Gymnogyps californianus	gymcal	bi	A
California gnatcatcher	Polioptila californica	popcal	bi	A
California ground squirrel	Spermophilus beecheyi	spebee	ma	A
California gull	Larus californicus	larcal	bi	A
California kangaroo rat	Dipodomys californicus	dipcal	ma	A
California leaf-nosed bat	Macrotus californicus	maccal	ma	A
California mouse	Peromyscus californicus	percal	ma	A
California myotis	Myotis californicus	myocal	ma	A
California pocket mouse	Chaetodipus californicus	chacal	ma	A
California quail	Callipepla calfornica	calcal	bi	A
California redbud	Cercis occidentalis	cerocc	tr	P
California thrasher	Toxostoma dedivivum	toxded	bi	A
California towhee	Pipilo crissalis	pipcri	bi	A
California vole	Microtus californicus	miccal	ma	A
California washingtonia	Washingtonia filifera	wasfil	tr	P
Calliope hummingbird	Stellula calliope	stecal	bi	A
Camas pocket gopher	Thomlys bulbivorus	thobul	ma	A
Canada goose	Branta canadensis	bracan	bi	A
Canada lynx	Lynx canadensis	lyncan	ma	A
Canada warbler	Wilsonia canadensis	wilcan	bi	A
Canvasback	Aythya valisineria	aytval	bi	A
Canyon grape	Vitis arizonica	vitari	vi	P
Canyon live oak	Quercus chrysolepis	quechr	tr	P
Canyon mouse	Peromyscus crinitus	percri	ma	A
Canyon towhee	Pipilo fuscus	pipfus	bi	A
Canyon wren	Catherpes mexicanus	catmex	bi	A
Cape May warbler	Dendroica tigrina	dentri	bi	A
Caracara	Cacara cheriway	cacher	bi	A
Cardinal	Cardinalis cardinalis	carca3	bi	A
Caribou	Rangifer tarandus	ranta1	ma	A
Carolina chickadee	Parus carolinensis	parcar	bi	A
Carolina wren	Thryothorus ludovicianus	thrlud	bi	A
Cascade ground squirrel	Spermophilus saturatus	spesat	ma	A
Cascade willow	Salix cascadensis	salcas	tr	P
Caspian tern	Sterna caspia	stecas	bi	A
Cassin's finch	Carpodacus cassinii	carcas	bi	A
Cassin's kingbird	Tyrannus vociferans	tyrvoc	bi	A
Cassin's sparrow	Aimophila cassinii	aimcas	bi	A
Cattle egret	Bubulcus ibis	bubibi	bi	A
Cave myotis	Myotis velifer	myovel	ma	A

(continued)

Cname	Sname	Scode	Lform	King
Cave swallow	Hirundo fulva	hirful	bi	A
Cedar waxwing	Bombycilla cedrorum	bomced	bi	A
Cerulean warbler	Dendroica cerulea	dencer	bi	A
Chaparral honeysuckle	Lonicera interrupta	lonint	vi	P
Chestnut-backed chickadee	Parus rufescens	parruf	bi	A
Chestnut-collared longspur	Calcarius ornatus	calorn	bi	A
Chestnut-sided warbler	Dendroica pensylvanica	denpen	bi	A
Chihuahua ash	Fraxinus papillosa	frapap	tr	P
Chihuahua pine	Pinus leiophylla	pinlei	tr	P
Chihuahuan pocket mouse	Chaetodipus eremicus	chaere	ma	A
Chihuahuan raven	Corvus cryptoleucus	corcry	bi	A
Chimney swift	Chaetura pelagica	chapel	bi	A
Chinkapin oak	Quercus muhlenbergii	quemuh	tr	P
Chipping sparrow	Spizella passerina	spipas	bi	A
Chisel-toothed kangaroo rat	Dipodomys microps	dipmic	ma	A
Chuck-will's-widow	Caprimulgus carolinensis	capcar	bi	A
Chukar	Alectoris chukar	alechu	bi	A
Cinnamon teal	Anas cyanoptera	anacya	bi	A
Clapper rail	Rallus longirostris	rallon	bi	A
Clark's nutcracker	Nucifraga columbiana	nuccol	bi	A
Clay-colored sparrow	Spizella pallida	spipal	bi	A
Clematis columbiana	Clematis columbiana	clecol	vi	P
Clematis hirsutissima	Clematis hirsutissima	clehir	vi	P
Cliff chipmunk	Eutamias dorsalis	eutdor	ma	A
Cliff swallow	Hirundo pyrrhonota	hirpyr	bi	A
Coast mole	Scapanus orarius	scaora	ma	A
Collared peccary	Tayassu tajacu	taytaj	ma	A
Collared pika	Ochotona collaris	ochcol	ma	A
Colorado chipmunk	Tamias quadrivittatus	tamqu1	ma	A
Columbia Plateau ground squirrel	Spermophilus canus	specan	ma	A
Columbian ground squirrel	Spermophilus columbianus	specol	ma	A
Common barn-owl	Tyto alba	tytalb	bi	A
Common chokecherry	Prunus virginiana	pruvir	tr	P
Common gallinule	Gallinule chloropus	galcho	bi	A
Common goldeneye	Bucephala clangula	buccla	bi	A
Common grackle	Quiscalus quiscula	quiqui	bi	A
Common ground-dove	Columbina passerina	colpas	bi	A
Common loon	Gavia immer	gavimm	bi	A
Common merganser	Mergus merganser	mermer	bi	A
Common moorhen	Gallinula chloropus	galchl	bi	A
Common muskrat	Ondatra zibethicus	ondzib	ma	A
Common nighthawk	Chordeiles minor	chomin	bi	A
Common poorwill	Phalaenoptilus nuttallii	phanut	bi	A
Common raven	Corvus corax	corcor	bi	A
Common redpoll	Carduelis flammea	carfla	bi	A
Common snipe	Gallinago gallinago	galgal	bi	A
Common tern	Sterna hirundo	stehir	bi	A
Common yellowthroat	Geothlypis trichas	geotri	bi	A
Connecticut warbler	Oporornis agilis	opoagi	bi	A

Cname	Sname	Scode	Lform	King
Cooper's hawk	Accipiter cooperii	acccoo	bi	A
Costa's hummingbird	Calypte costae	calcos	bi	A
Cotton mouse	Peromyscus gosspinus	pergos	ma	A
Coue's rice rat	Oryzomys couesi	orycou	ma	A
Coyote	Canis latrans	canlat	ma	A
Coyote willow	Salix exigua	salexi	tr	P
Coypu	Myocastor coypus	myocoy	ma	A
Creeping vole	Microtus oregoni	micore	ma	A
Crested caracara	Polyborus plancus	polpla	bi	A
Crissal thrasher	Toxostoma crissale	toxcri	bi	A
Curve-billed thrasher	Toxostoma curvirostre	toxcur	bi	A
Dall's sheep	Ovia dalli	ovidal	ma	A
Dark kangaroo mouse	Microdipodops megacephalus	micmeg	ma	A
Dark-eyed junco	Junco hyemalis	junhye	bi	A
Deer mouse	Peromyscus maniculatus	perman	ma	A
Desert cottontail	Sylvilagus audubonii	sylaud	ma	A
Desert kangaroo rat	Dipodomys deserti	dipdes	ma	A
Desert pocket gopher	Geomys arenarius	geoare	ma	A
Desert pocket mouse	Perognathus penicillatus	perpen	ma	A
Desert shrew	Notiosorex crawfordi	notcra	ma	A
Desert woodrat	Neotoma lepida	neolep	ma	A
Dickcissel	Spiza americana	spiame	bi	A
Double-crested cormorant	Phalacrocorax auritus	phaaur	bi	A
Douglas' squirrel	Tamiasciurus douglasii	tamdou	ma	A
Douglas-fir	Pseudotsuga menziesii	psemen	tr	P
Downy woodpecker	Picoides pubescens	picpub	bi	A
Drummond willow	Salix drummondiana	saldru	tr	P
Dulzure kangaroo rat	Dipodomys simulana	dipsim	ma	A
Dunlin	Calidris alpina	calalp	bi	A
Dusky flycatcher	Empidonax oberholseri	empobe	bi	A
Dusky shrew	Sorex monticolus	sormon	ma	A
Dusky-capped flycatcher	Myiarchus tuberculifer	myitub	bi	A
Dusky-footed woodrat	Netoma fuscipes	neofus	ma	A
Dwarf shrew	Sorex nanus	sornan	ma	A
Eared grebe	Podiceps nigricollis	podnig	bi	A
Eared trogon	Euptilotis neoxenus	eupneo	bi	A
Eastern bluebird	Sialia sialis	siasia	bi	A
Eastern chipmunk	Tamias striatus	tamstr	ma	A
Eastern cottontail	Sylvilagus floridanus	sylflo	ma	A
Eastern cottonwood	Populus deltoides	popdel	tr	P
Eastern fox squirrel	Sciurus niger	scinig	ma	A
Eastern gray squirrel	Sciurus carolinensis	scicar	ma	A
Eastern harvest mouse	Reithrodontomys hummulis	reihum	ma	A
Eastern heather vole	Phenacomys ungava	pheung	ma	A
Eastern hophornbeam	Ostrya virginiana	ostvir	tr	P
Eastern kingbird	Tyrannus tyrannus	tyrtyr	bi	A
Eastern meadowlark	Sturnella magna	stumag	bi	A
Eastern mole	Scalopus aquaticus	scaaqu	ma	A
Eastern phoebe	Sayornis phoebe	saypho	bi	A

(continued)

Cname	Sname	Scode	Lform	King
Eastern pipistrelle	Pipistrellus subflavus	pipsub	ma	A
Eastern red bat	Lasiurus borealis	lasbor	ma	A
Eastern screech-owl	Otus asio	otuasi	bi	A
Eastern small-footed myotis	Myotis leibii	myolei	ma	A
Eastern spotted skunk	Spilogale gracilis	spigra	ma	A
Eastern timber wolf	Canis lycaon	canlyc	ma	A
Eastern white-throated woodrat	Neotoma leucodon	neoleu	ma	A
Eastern wood-pewee	Contopus virens	convir	bi	A
Eastern woodrat	Neotoma floridana	neoflo	ma	A
Elegant trogon	Trogon elegans	troele	bi	A
Elf owl	Micrathene whitneyi	micwhi	bi	A
Elk	Cervus elaphus	cerela	ma	A
Elliot's short-tailed shrew	Blarina hylophaga	blahyl	ma	A
Emory oak	Quercus emoryi	queemo	tr	P
Engelmann spruce	Picea engelmannii	piceng	tr	P
Ermine	Mustela erminea	muserm	ma	A
European birdcherry	Prunus padus	prupad	tr	P
European hare	Lepus capensis	lepcap	ma	A
European mountain-ash	Sorbus aucuparia	sorauc	tr	P
European rabbit	Oryctolagus cunniculus	orycun	ma	A
European starling	Sturnus vulgaris	stuvul	bi	A
Evening bat	Nycticeius humeralis	nychum	ma	A
Evening grosbeak	Coccothraustes vespertinus	cocves	bi	A
Fan-tailed warbler	Euthlypis lachrymose	eutlac	bi	A
Ferruginous hawk	Buteo regalis	butreg	bi	A
Ferruginous pygmy-owl	Glaucidium brasilianum	glabra	bi	A
Field sparrow	Spizella pusilla	spipus	bi	A
Fish crow	Corvus ossifragus	coross	bi	A
Fisher	Martes pennanti	marpen	ma	A
Five-striped sparrow	Amphispiza quinquestriata	ampqui	bi	A
Flammulated owl	Otus flammeolus	otufla	bi	A
Florida bonneted bat	Eumops floridanus	eumflo	ma	A
Florida mouse	Podomys floridanus	podflo	ma	A
Fog shrew	Sorex sonomae	sorson	ma	A
Forster's tern	Sterna forsteri	stefor	bi	A
Fox sparrow	Passerella iliaca	pasili	bi	A
Fragrant ash	Fraxinus cuspidata	fracus	tr	P
Franklin's ground squirrel	Spermophilus franklinii	spefra	ma	A
Franklin's gull	Larus pipixcan	larpip	bi	A
Fremont cottonwood	Populus fremontii	popfre	tr	P
Fringed myotis	Myotis thysanodes	myothy	ma	A
Frost grape	Vitis vulpina	vitvul	vi	P
Fulvous harvest mouse	Reithrodontomys fulvescens	reiful	ma	A
Fulvous whistling-duck	Dendrocygna bicolor	denbic	bi	A
Gadwall	Anas strepera	anastr	bi	A
Gambel oak	Quercus gambelii	quegam	tr	P
Gambel's quail	Callipepla gambelii	calgam	bi	A
Gaspe shrew	Sorex gaspensis	sorgas	ma	A
Geyer willow	Salix geyeriana	salgey	tr	P

Cname	Sname	Scode	Lform	King
Giant kangaroo rat	Dipodomys ingens	diping	ma	A
Gila woodpecker	Melanerpes uropygialis	meluro	bi	A
Glaucous-winged gull	Larus glaucescens	largla	bi	A
Glossy ibis	Plegadis falcinellus	plefal	bi	A
Golden eagle	Aquila chrysaetos	aquchr	bi	A
Golden mouse	Ochrotomys nuttalli	ochnut	ma	A
Golden-crowned kinglet	Regulus satrapa	regsat	bi	A
Golden-crowned sparrow	Zonotrichia atricapilla	zonatr	bi	A
Golden-mantled ground squirrel	Spermophilus lateralis	spelat	ma	A
Golden-winged warbler	Vermivora chrysoptera	verchr	bi	A
Goodding ash	Fraxinus gooddingii	fragoo	tr	P
Goodding willow	Salix gooddingii	salgoo	tr	P
Grace's warbler	Dendroica graciae	dengra	bi	A
Grasshopper sparrow	Ammodramus savannarum	ammsav	bi	A
Gray catbird	Dumetella carolinensis	dumcar	bi	A
Gray flycatcher	Empidonax wrightii	empwri	bi	A
Gray fox	Urocyon cinereoargenteus	urocin	ma	A
Gray hawk	Buteo nitidus	butnit	bi	A
Gray jay	Perisoreus canadensis	percan	bi	A
Gray kingbird	Tyrannus dominiicensis	tyrdom	bi	A
Gray myotis	Myotis grisescens	myogri	ma	A
Gray oak	Quercus grisea	quegri	tr	P
Gray partridge	Perdix perdix	perper	bi	A
Gray vireo	Vireo vicinior	virvic	bi	A
Gray wolf	Canus lupus	canlup	ma	A
Gray-breasted jay	Aphelocoma ultramarina	aphult	bi	A
Gray-cheeked thrush	Catharus minimus	catmin	bi	A
Gray-collared chipmunk	Eutamias cinereicollis	eutcin	ma	A
Gray-footed chipmunk	Tamias canipes	tamcan	ma	A
Grayleaf willow	Salix glauca	salgla	tr	P
Gray-tailed vole	Microtus canicaudus	miccan	ma	A
Great Basin ground squirrel	Spermophilus mollis	spemol	ma	A
Great Basin pocket mouse	Perognathus parvus	perpar	ma	A
Great blue heron	Ardea herodias	ardher	bi	A
Great crested flycatcher	Myiarchus crinitus	myicri	bi	A
Great egret	Casmerodius albus	casalb	bi	A
Great gray owl	Strix nebulosa	strneb	bi	A
Great horned owl	Bubo virginianus	butvir	bi	A
Great kiskadee	Pitangus sulphuratus	pitsul	bi	A
Greater pewee	Contopus pertinax	conper	bi	A
Greater prairie-chicken	Tympanuchus cupido	tymcup	bi	A
Greater roadrunner	Geococcys californianus	geocal	bi	A
Greater scaup	Aythya marila	aytmar	bi	A
Greater yellowlegs	Tringa melanoleuca	trimel	bi	A
Great-tailed grackle	Quiscalus mexicanus	quimex	bi	A
Green ash	Fraxinus pennsylvanica	frapen	tr	P
Green kingfisher	Chloroceryle americana	chlame	bi	A
Green-backed heron	Butorides striatus	butstr	bi	A
Greenes mountain-ash	Sorbus scopulina	sorsco	tr	P

(continued)

Cname	Sname	Scode	Lform	King
Green-tailed towhee	Pipilo chlorurus	pipchl	bi	A
Green-winged teal	Anas crecca	anacre	bi	A
Groove-billed ani	Crotophaga sulcirostris	crosul	bi	A
Gulf Coast kangaroo rat	Dipodomys compactus	dipcom	ma	A
Gull-billed tern	Sterna nilotica	stenil	bi	A
Gunnison's prairie dog	Cynomys gunnisoni	cyngun	ma	A
Hairy woodpecker	Picoides villosus	picvil	bi	A
Hairy-legged vampire bat	Diphylla escudata	dipesc	ma	A
Hairy-tailed mole	Parascalops breweri	parbre	ma	A
Hammond's flycatcher	Empidonax hammondii	empham	bi	A
Harlequin duck	Histrionicus histrionicus	hishis	bi	A
Harris's antelope squirrel	Ammospermophilus harrisii	ammhar	ma	A
Harris's hawk	Parabuteo unicinctus	paruni	bi	A
Harris's sparrow	Zonotrichia querula	zonque	bi	A
Havard oak	Quercus havardii	quehav	tr	P
Hayden's shrew	Sorex haydeni	sorhay	ma	A
Heermann's gull	Larus heermanni	larhee	bi	A
Heermann's kangaroo rat	Dipodomys heermanni	dipher	ma	A
Henslow's sparrow	Ammodramus henslowii	ammhen	bi	A
Hepatic tanager	Piranga flava	pirfla	bi	A
Hermit thrush	Catharus guttatus	catgut	bi	A
Hermit warbler	Dendroica occidentalis	denocc	bi	A
Herring gull	Larus argentatus	lararg	bi	A
Hispid cotton rat	Sigmodon hispidus	sighis	ma	A
Hispid pocket mouse	Perognathus hispidus	perhis	ma	A
Hoary bat	Lasiurus cinereus	lascin	ma	A
Hoary marmot	Marmota caligata	marcal	ma	A
Hog-nosed skunk	Conepatus mesoleucus	conmes	ma	A
Hooded merganser	Lophodytes cucullatus	lopcuc	bi	A
Hooded oriole	Icterus cucullatus	ictcuc	bi	A
Hooded skunk	Mephitis macroura	mepmac	ma	A
Hooded warbler	Wilsonia citrina	wilcit	bi	A
Hopi chipmunk	Tamias rufus	tamru1	ma	A
Hoptree	Ptelea trifoliata	ptetri	tr	P
Horned grebe	Podiceps auritus	podaur	bi	A
Horned lark	Eremophila alpestris	erealp	bi	A
House finch	Carpodacus mexicanus	carmex	bi	A
House mouse	Mus musculus	musmus	ma	A
House sparrow	Passer domesticus	assdom	bi	A
House wren	Troglodytes aedon	troaed	bi	A
Hudsonian godwit	Limosa haemastica	limhae	bi	A
Hutton's vireo	Vireo huttoni	virhut	bi	A
Ibex	Capra ibex	capibe	ma	A
Idaho ground squirrel	Spermophilus brunneus	spebru	ma	A
Idaho pocket gopher	Geomys bursarius	feobur	ma	A
Inca dove	Columbina inca	colinc	bi	A
Indiana myotis	Myotis sodalis	myosod	ma	A
Indigo bunting	Passerina cyanea	pascya	bi	A
Insular vole	Microtus abbreviatus	micabb	ma	A

Cname	Sname	Scode	Lform	King
Inyo shrew	Sorex tenellus	sorten	ma	A
Jaguar	Felis onca	felonc	ma	A
Jaguarundi	Felis yagouaroundi	felyag	ma	A
Jamaican fruit-eating bat	Artibeus jamaicensis	artjam	ma	A
Japanese honeysuckle	Lonicera japonica	lonjap	vi	P
Jerusalem-thorn	Parkinsonia aculeata	paracu	tr	P
Jones's pocket gopher	Geomy's knoxjonesi	geokno	ma	A
Jumping-bean sapium	Sapium biloculare	sapbil	tr	P
Keen's mouse	Peromyscus keeni	perkee	ma	A
Keen's myotis	Myotis keenii	myokee	ma	A
Kentucky warbler	Oporornis formosus	opofor	bi	A
Killdeer	Charadrius vociferus	chavoc	bi	A
King rail	Rallus elegans	ralele	bi	A
Kirkland's warbler	Dendroica kirtlandii	denkir	bi	A
Kit fox	Vulpes macrotis	vulmac	ma	A
Knowlton hophornbeam	Ostrya knowltonii	ostkno	tr	P
Ladder-backed woodpecker	Picoides scalaris	picsca	bi	A
Lanceleaf cottonwood	Populus acuminata	popacu	tr	P
Lapland longspur	Calcarius lapponicus	callap	bi	A
Lark bunting	Calamospiza melanocorys	calme1	bi	A
Lark sparrow	Chondestes grammacus	chogra	bi	A
Laughing gull	Larus atricilla	laratr	bi	A
Laurel willow	Salix pentandra	salpen	tr	P
Lawrence's goldfinch	Carduelis lawrencei	carlaw	bi	A
Lazuli bunting	Passerina amoena	pasamo	bi	A
Le Conte's sparrow	Ammodramus leconteii	ammlec	bi	A
Le Conte's thrasher	Toxostoma lecontei	toslec	bi	A
Least bittern	Ixobrychus exilis	ixoexi	bi	A
Least chipmunk	Eutamias minimus	eutmin	ma	A
Least flycatcher	Empidonax minimus	empmin	bi	A
Least grebe	Tachybaptus dominicus	tacdom	bi	A
Least sandpiper	Calidris minutilla	calmin	bi	A
Least shrew	Cryptotis parva	crypar	ma	A
Least tern	Sterna antillarum	steant	bi	A
Least weasel	Mustela nivalis	musniv	ma	A
Lesser golden-plover	Pluvialis dominica	pludom	bi	A
Lesser goldfinch	Carduelis psaltria	carpsa	bi	A
Lesser long-nosed bat	Leptonycteris yerbabuenae	lepyer	ma	A
Lesser nighthawk	Chordeiles acutipennis	choacu	bi	A
Lesser prairie-chicken	Tympanuchus pallidicinctus	tympal	bi	A
Lesser scaup	Aythya affinis	aytaff	bi	A
Lesser yellowlegs	Tringa flavipes	trifla	bi	A
Lewis's woodpecker	Melanerpes lewis	mellew	bi	A
Limber pine	Pinus flexilis	pinfle	tr	P
Limkin	Aramus guarauna	aragua	bi	A
Lincoln's sparrow	Melospiza lincolnii	mellin	bi	A
Little blue heron	Egretta caerulea	egrcae	bi	A
Little brown myotis	Myotis lucifugus	myoluc	ma	A
Little mastiff bat	Molossus molossus	molmol	ma	A

(continued)

Cname	Sname	Scode	Lform	King
Little pocket mouse	Perognathus longimembris	perlon	ma	A
Little walnut	Juglans microcarpa	jugmic	tr	P
Llano pocket gopher	Geomy's taxensis	geotax	ma	A
Lodgepole chipmunk	Tamias speciosus	tamspe	ma	A
Lodgepole pine	Pinus contorta	pincon	tr	P
Loggerhead shrike	Lanius ludovicianus	lanlud	bi	A
Long-billed curlew	Numenius americanus	numame	bi	A
Long-billed dowitcher	Limnodromus scolopaceus	limsco	bi	A
Long-billed marsh wren	Cistothorus palustris	cispal	bi	A
Long-eared chipmunk	Tamias quadrimaculatus	tamqu2	ma	A
Long-eared myotis	Myotis evotis	myoevo	ma	A
Long-eared owl	Asio otus	asiotu	bi	A
Long-legged myotis	Myotis volans	myovol	ma	A
Long-tailed pocket mouse	Perognathus formosus	perfor	ma	A
Long-tailed vole	Microtus longicaudus	miclon	ma	A
Long-tailed weasel	Mustela frenata	musfre	ma	A
Louisiana waterthrush	Seiurus motacilla	seimot	bi	A
Lucifer hummingbird	Calothorax lucifer	calluc	bi	A
Lucy's warbler	Vermivora lucia	verluc	bi	A
MacGillivray's warbler	Oporornis tolmiei	opotol	bi	A
Magnificent frigatebird	Fregata magnificens	freman	bi	A
Magnificent hummingbird	Eugenes fulgens	eugmag	bi	A
Magnolia warbler	Dendroica magnolia	denmag	bi	A
Mallard	Anas platyrhynchos	anapla	bi	A
Marbled godwit	Limosa fedoa	limfed	bi	A
Marbled murrelet	Brachyramphus marmoratus	bramar	bi	A
Marsh hawk	Circus cyaneus	ciccya	bi	A
Marsh rabbit	Sylvilagus palustris	sylpal	ma	A
Marsh rice rat	Oryzomys palustris	orypal	ma	A
Marsh shrew	Sorex bendirii	sorben	ma	A
Marten	Martes americana	marame	ma	A
Masked shrew	Sorex cinereus	sorcin	ma	A
McCown's longspur	Calcarius mccownii	calmcc	bi	A
Meadow jumping mouse	Aapus hudsonicus	aaphud	ma	A
Meadow vole	Microtus pennsylvanicus	micpen	ma	A
Meadow willow	Salix petiolaris	salpet	tr	P
Mearn's grasshopper mouse	Onychomys arenicola	onyare	ma	A
Merlin	Falco columbarius	falcol	bi	A
Merriam's chipmunk	Tamias merriami	tammer	ma	A
Merriam's kangaroo rat	Dipodomys merriami	dipmer	ma	A
Merriam's pocket mouse	Perognathus merriami	perme1	ma	A
Merriam's shrew	Sorex merriami	sormer	ma	A
Mesquite	Prosopis juliflora	projul	tr	P
Mesquite mouse	Peromyscus merriami	perme2	ma	A
Mew gull	Larus canus	larcan	bi	A
Mexican blue oak	Quercus oblongifolia	queobl	tr	P
Mexican chickadee	Parus sclateri	parscl	bi	A
Mexican fox squirrel	Sciurus nayaritensis	scinay	ma	A
Mexican ground squirrel	Spermophilus mexicanus	spemex	ma	A

Cname	Sname	Scode	Lform	King
Mexican Long-nosed bat	Leptonycteris nivalis	lepniv	ma	A
Mexican Long-tongued bat	Choeronycteris mexicana	chomex	ma	A
Mexican pinyon	Pinus cembroides	pincem	tr	P
Mexican spiney pocket mouse	Liomys irroratus	lioirr	ma	A
Mexican vole	Microtus mexicanus	micmex	ma	A
Mexican woodrat	Neotoma mexicana	neomex	ma	A
Mexican buckeye	Ungnadia speciosa	ungspe	tr	P
Mink	Mustela vison	musvis	ma	A
Mississippi kite	Ictinia mississippiensis	ictmis	bi	A
Missouri River willow	Salix eriocephala	saleri	tr	P
Mohave ground squirrel	Spermophilus mohavensis	spemoh	ma	A
Montane vole	Microtus montanus	micmon	ma	A
Montezuma quail	Cyrtonyx montezumae	crymon	bi	A
Moose	Alces alces	alcalc	ma	A
Mountain bluebird	Sialia currucoides	siacur	bi	A
Mountain chickadee	Parus gambeli	pargam	bi	A
Mountain cottontail	Sylvilagus nuttallii	sylnut	ma	A
Mountain goat	Oreamnos americanus	oreame	ma	A
Mountain lion	Puma concolor	pumcon	ma	A
Mountain plover	Charadrius montanus	chamon	bi	A
Mountain pocket gopher	Thomomys monticola	thomon	ma	A
Mountain quail	Oreortyx pictus	orepic	bi	A
Mountain willow	Salix monticola	salmon	tr	P
Mourning dove	Zenaida macroura	zenmac	bi	A
Mourning warbler	Oporornis philadelphia	opophi	bi	A
Mt. Lyell shrew	Sorex lyelli	sorly	ma	A
Mule deer	Odocoileus hemionus	odohem	ma	A
Narrow-faced kangaroo rat	Dipodomys venustus	dipven	ma	A
Narrowleaf cottonwood	Populus angustifolia	popang	tr	P
Nashville warbler	Vermivora ruficapilla	verfur	bi	A
Nelson's antelope squirrel	Ammospermophilus nelsoni	ammnel	ma	A
Nelson's pocket mouse	Perognathus nelsoni	pernel	ma	A
Netleaf hackberry	Celtis reticulata	celret	tr	P
Netleaf oak	Quercus reticulata	queret	tr	P
New England cottontail	Sylvilagus transitionals	syltra	ma	A
New Mexican locust	Robinia neomexicana	robneo	tr	P
New Mexico shrew	Sorex neomexicanus	sorneo	ma	A
Nine-banded armadillo	Dasypus novemcinctus	dasnov	ma	A
North American Porcupine	Erethizon dorsatum	eredor	ma	A
Northern bobwhite	Colinus virginianus	colvir	bi	A
Northern bog lemming	Synaptomys borealis	synbor	ma	A
Northern cardinal	Cardinalis cardinalis	carca2	bi	A
Northern collared lemming	Dicrostonyx groenlandicus	dicgro	ma	A
Northern flicker	Colaptes auratus	colaur	bi	A
Northern flying squirrel	Glaucomys sabrinus	glasab	ma	A
Northern goshawk	Accipiter gentilis	accgen	bi	A
Northern grasshopper mouse	Onychomys leucogaster	onyleu	ma	A
Northern harrier	Circus cyaneus	circya	bi	A
Northern hawk-owl	Surnia ulula	surula	bi	A

(continued)

Cname	Sname	Scode	Lform	King
Northern mockingbird	Mimus polyglottos	mimpol	bi	A
Northern myotis	Myotis septentrionalis	myosep	ma	A
Northern oriole	Icterus galbula	ictgal	bi	A
Northern parula	Parula americana	parame	bi	A
Northern phalarope	Lobipes lobatus	loblob	bi	A
Northern pocket gopher	Thomomys talpoides	thotal	ma	A
Northern pygmy mouse	Baiomys taylori	baitay	ma	A
Northern pygmy-owl	Glaucidium gnoma	glagno	bi	A
Northern red-backed vole	Clethrionomys rutilus	clerut	ma	A
Northern rock mouse	Peromyscus nasutus	pernas	ma	A
Northern rough-winged swallow	Stelgidopteryx serripennis	steser	bi	A
Northern saw-whet owl	Aegolius acadicus	aegaca	bi	A
Northern short-tailed shrew	Blarina brevicauda	blabre	ma	A
Northern shoveler	Anas clypeata	anacly	bi	A
Northern shrike	Lanius excubitor	lanexc	bi	A
Northern waterthrush	Seiurus noveboracensis	seinov	bi	A
Northern yellow bat	Lasiurus intermedius	lasint	ma	A
Northern river otter	Lontra canadensis	loncan	ma	A
Northwestern crow	Corvus caurinus	corcau	bi	A
Norway rat	Rattus norvegicus	ratnor	ma	A
Nuttall's woodpecker	Picoides nuttallii	picnu1	bi	A
Nutting's flycatcher	Myiarchus nuttingi	myinut	bi	A
Ocelot	Felis pardalis	felpar	ma	A
Oldfield mouse	Peromyscus polinonotus	perpol	ma	A
Oldsquaw	Clangula hyemalis	clahye	bi	A
Olivaceous cormorant	Phalacrocorax olivaceus	phaoli	bi	A
Olive warbler	Peucedramus taeniatus	peutae	bi	A
Olive-backed pocket mouse	Perognathus fasciatus	perfas	ma	A
Olive-sided flycatcher	Contopus borealis	conbor	bi	A
Olympic marmot	Marmota broweri	marbro	ma	A
One-seed juniper	Juniperus monosperma	junmon	tr	P
Orange-crowned warbler	Vermivora celata	vercel	bi	A
Orchard oriole	Icterus spurius	ictspu	bi	A
Ord's kangaroo rat	Dipodomys ordii	dipord	ma	A
Ornate shrew	Sorex ornatus	sororn	ma	A
Oryx	Oryx gazella	orygaz	ma	A
Osgood's mouse	Peromyscus gratus	pergra	ma	A
Osprey	Pandion haliaetus	panhal	bi	A
Ovenbird	Seiurus aurocapillus	seicur	bi	A
Pacific jumping mouse	Zapus trinotatus	zaptri	ma	A
Pacific shrew	Sorex pacifieus	sorpac	ma	A
Pacific willow	Salix lasiandra	salla1	tr	P
Painted bunting	Passerina ciris	pascir	bi	A
Painted redstart	Myioborus pictus	myipic	bi	A
Pale kangaroo mouse	Microdipodops pallidus	micpal	ma	A
Pallid bat	Antrozous pallidus	antpal	ma	A
Palm warbler	Dendroica palmarum	denpal	bi	A
Palmer's chipmunk	Tamias palmeri	tampal	ma	A
Panamint chipmunk	Tamias panamintinus	tampan	ma	A

Cname	Sname	Scode	Lform	King
Panamint kangaroo rat	Dipodomys panamintinus	dippan	ma	A
Paper birch	Betula papyrifera	betpap	tr	P
Parasitic jaeger	Stercorarius parasiticus	stepar	bi	A
Peachleaf willow	Salix amygdaloides	salamy	tr	P
Pectoral sandpiper	Calidris melanotos	calme2	bi	A
Peregrine falcon	Falco peregrinus	falper	bi	A
Persian ibex	Capra aegagrus	capaeg	ma	A
Peters ghost-faced bat	Mormoops megalophylla	mormeg	ma	A
Phainopepla	Phainopepla nitens	phanit	bi	A
Philadelphia vireo	Vireo philadelphicus	virphi	bi	A
Pied-billed grebe	Podilymbus podiceps	podpod	bi	A
Piliated woodpecker	Dryocopus pileatus	drypil	bi	A
Pin cherry	Prunus pensylvanica	prupen	tr	P
Pine grosbeak	Pinicola enucleator	pinenu	bi	A
Pine siskin	Carduelis pinus	carpin	bi	A
Pine warbler	Dendroica pinus	denpin	bi	A
Pinon mouse	Peromyscus truei	pertru	ma	A
Pintail	Anas acuta	anaacu	bi	A
Pinyon	Pinus edulis	pinedu	tr	P
Pinyon jay	Gymnorhinus cyanocephalus	gymcyn	bi	A
Piping plover	Charadrius melodus	chamel	bi	A
Plain titmouse	Parus inornatus	parino	bi	A
Plain-capped starthroat	Heliomaster constantii	helcon	bi	A
Plains cottonwood	Populus sargentii	popsar	tr	P
Plains harvest mouse	Reithrodontomys montanus	reimon	ma	A
Plains pocket gopher	Geomys bursarius	geobur	ma	A
Plains pocket mouse	Perognathus flavescens	perfl2	ma	A
Planeleaf willow	Salix planifolia	salpla	tr	P
Pocket mouse	Perognathus amplus	peramp	ma	A
Pocketed free-tailed bat	Tadarida femorosacca	tadfem	ma	A
Polar bear	Ursus maritimus	ursmar	ma	A
Pomarine jaeger	Stercorarius pomarinus	stepom	bi	A
Ponderosa pine	Pinus ponderosa	pinpon	tr	P
Prairie falcon	Falco mexicanus	falmex	bi	A
Prairie vole	Microtus ochrogaster	micoch	ma	A
Prairie warbler	Dendroica discolor	dendis	bi	A
Preble's shrew	Sorex prebei	sorpre	ma	A
Prickly oxytenia	Oxytenia acerosa	oxyace	tr	P
Pronghorn	Antilocapra americana	antame	ma	A
Prothonotary warbler	Protonotaria citrea	procit	bi	A
Purple finch	Carpodacus purpureus	carpur	bi	A
Purple gallinule	Porphyrula martinica	pormar	bi	A
Purple martin	Progne subis	prosub	bi	A
Pussy willow	Salix discolor	saldis	tr	P
Pygmy nuthatch	Sitta pygmaea	sitpyg	bi	A
Pygmy rabbit	Brachylagus idahoensis	braida	ma	A
Pygmy shrew	Sorex hoyi	sorhoy	ma	A
Pyrrhuloxia	Cardinalis sinuatus	arnsin	bi	A
Quaking aspen	Populus tremuloides	poptre	tr	P

(continued)

Cname	Sname	Scode	Lform	King
Raccoon	Procyon lotor	prolot	ma	A
Rafinesque big-eared bat	Corynorhinus rafinesquii	corraf	ma	A
Read-headed woodpecker	Melanerpes erythrocephalus	melery	bi	A
Red crossbill	Loxia curvirostra	loxcur	bi	A
Red fox	Vulpes vulpes	vulvul	ma	A
Red knot	Calidris canutus	calcan	bi	A
Red phalarope	Phalaropus fulicaria	phaful	bi	A
Red squirrel	Tamiasciurus hudsonicus	tamhud	ma	A
Red tree vole	Arborimus longicaudus	arblon	ma	A
Red willow	Salix laevigata	sallae	tr	P
Red wolf	Canis rufus	canruf	ma	A
Red-bellied woodpecker	Melanerpes carolinus	melcar	bi	A
Red-billed tropicbird	Phaethon aethereus	phaaet	bi	A
Red-breasted sapsucker	Sphyrapicus ruber	sphrub	bi	A
Red-breasted merganser	Mergus serrator	merser	bi	A
Red-breasted nuthatch	Sitta canadensis	sitcan	bi	A
Red-cockaded woodpecker	Picoides borealis	picbor	bi	A
Reddish egret	Egretta rufescens	egrruf	bi	A
Red-eyed vireo	Vireo olivaceus	viroli	bi	A
Red-faced warbler	Cardellina rubrifrons	carrub	bi	A
Redhead	Aythya americana	aytame	bi	A
Red-headed woodpecker	Melanerpes erythroceohalus	melnuc	bi	A
Red-naped sapsucker	Sphyrapicus nuchalis	sphnuc	bi	A
Red-shouldered hawk	Buteo lineatus	butlin	bi	A
Red-tailed chipmunk	Tamias ruficaudus	tamru2	ma	A
Red-tailed hawk	Buteo jamaicensis	butjam	bi	A
Red-winged blackbird	Agelaius phoeniceus	agepho	bi	A
Richardson's collared lemming	Dicrostonyx richardsoni	dicric	ma	A
Richardson's ground squirrel	Spermophilus richardsonii	speric	ma	A
Ring-billed gull	Larus delawarensis	lardel	bi	A
Ring-necked duck	Aythya collaris	aytcol	bi	A
Ring-necked pheasant	Phasianus colchicus	phacol	bi	A
Ringtail	Bassariscus astutus	basast	ma	A
River otter	Lutra canadensis	lutcan	ma	A
Rock dove	Columba livia	colliv	bi	A
Rock mouse	Peromyscus difficilis	perdif	ma	A
Rock pocket mouse	Perognathus intermedius	perint	ma	A
Rock shew	Sorex dispar	sordis	ma	A
Rock squirrel	Spermophilus variegatus	spevar	ma	A
Rock vole	Microtus chrotorrhinus	micchr	ma	A
Rock wren	Salpinctes obsoletus	salobs	bi	A
Rocky Mountain bighorn sheep	Ovis canadensis	ovican	ma	A
Rocky Mountain clematis	Clematis pseudoalpina	clepse	vi	P
Rocky Mountain juniper	Juniperus scopulorum	junsco	tr	P
Rocky Mountain maple	Acer glabrum	acegla	tr	P
Roof rat	Rattus rattus	ratrat	ma	A
Roseate spoonbill	Ajaia ajaja	ajaaja	bi	A
Rose-breasted grosbeak	Pheucticus ludovicianus	phelud	bi	A
Rose-throated becard	Pachyramphus aglaiae	pacagl	bi	A

Cname	Sname	Scode	Lform	King
Ross's goose	Chen rossii	cheros	bi	A
Rosy finch	Leucosticte arctoa	leuarc	bi	A
Rough-legged hawk	Buteo lagopus	butlag	bi	A
Rough-winged swallow	Riparia ruficollis	ripruf	bi	A
Round-tailed ground squirrel	Spermophilus tereticaudus	speter	ma	A
Round-tailed muskrat	Neofiber alleni	neoall	ma	A
Ruby-crowned kinglet	Regulus calendula	regcal	bi	A
Ruby-throated hummingbird	Archilochus colubris	arccol	bi	A
Ruddy duck	Oxyura jamaicensis	oxyjam	bi	A
Ruffed grouse	Bonasa umbellus	boaumb	bi	A
Rufous hummingbird	Selasphorus rufus	selruf	bi	A
Rufous-backed robin	Turdus rufopalliatus	turruf	bi	A
Rufous-capped warbler	Basileuterus rufifrons	basruf	bi	A
Rufous-crowned sparrow	Aimophila ruficeps	aimruf	bi	A
Rufous-sided towhee	Pipilo erythrophthalmus	pipery	bi	A
Rufous-winged sparrow	Aimophila carpalis	aimcar	bi	A
Russian olive	Elaeagnus angustifolia	elaang	tr	P
Rusty blackbird	Euphagus carolinus	eupcar	bi	A
Sabine's gull	Xema sabini	xemsab	bi	A
Sage grouse	Centrocercus urophasianus	cenuro	bi	A
Sage sparrow	Amphispiza belli	ampbel	bi	A
Sage thrasher	Oreoscoptes montanus	oremon	bi	A
Sagebrush vole	Lemmiscus curtatus	lemcut	ma	A
Salt-marsh harvest mouse	Reithrodontomys raviventris	reirav	ma	A
San Diego pocket mouse	Chaetodipus fallax	chafal	ma	A
San Joaquin kangaroo rat	Dipodomys nitratoides	dipnit	ma	A
San Joaquin pocket mouse	Perognathus inornatus	perino	ma	A
Sanborn's long-nosed bat	Leptonycteris sanborni	lepsan	ma	A
Sandbar willow	Salix interior	salint	tr	P
Sanderling	Calidris alba	calalb	bi	A
Sandhill crane	Grus canadensis	grucan	bi	A
Sandpaper oak	Quercus pungens	quepun	tr	P
Savannah sparrow	Passerculus sandwichensis	passan	bi	A
Say's phoebe	Sayornis saya	saysay	bi	A
Scaled quail	Callipepla squamata	calsqu	bi	A
Scarlet tanager	Piranga olivacea	piroli	bi	A
Scissor-tailed flycatcher	Tyrannus forficatus	tyrfor	bi	A
Scott's oriole	Icterus parisorum	ictpar	bi	A
Scouler willow	Salix scouleriana	salsco	tr	P
Screwbean mesquite	Prosopis pubescens	propub	tr	P
Scrub jay	Aphelocoma coerulescens	aphcoe	bi	A
Sedge wren	Cistothorus platensis	cispla	bi	A
Seminole bat	Lasiurus seminolus	lassem	ma	A
Semipalmated plover	Charadrius semipalmatus	chasem	bi	A
Semipalmated sandpiper	Calidris pusilla	calpus	bi	A
Sewell	Aplodontia rufa	aplruf	ma	A
Sharp-shinned hawk	Accipiter striatus	accstr	bi	A
Sharp-tailed grouse	Tympanuchus phasianellus	tympha	bi	A
Sharp-tailed sparrow	Ammospiza caudacuta	ammcau	bi	A

(continued)

Cname	Sname	Scode	Lform	King
Short-billed dowitcher	Limnodromus griseus	limgri	bi	A
Short-eared owl	Asio flammeus	asifla	bi	A
Short-tailed hawk	Buteo brachyurus	butbra	bi	A
Shrub live oak	Quercus turbinella	quetur	tr	P
Silky pocket mouse	Perognathus flavus	perfl1	ma	A
Silverberry	Elaeagnus commutata	elacom	tr	P
Silver-haired bat	Lasionycteris noctivagans	lasnoc	ma	A
Silverleaf oak	Quercus hypoleucoides	quehyp	tr	P
Singing vole	Microtus miurus	micmiu	ma	A
Singleleaf ash	Fraxinus anomala	fraano	tr	P
Singleleaf pinyon	Pinus monophylla	pinmon	tr	P
Siskiyou chipmunk	Tamias siskiyou	tamsis	ma	A
Slate-throated redstart	Myioborus miniatus	myimin	bi	A
Smith's longspur	Calcarius pictus	calpic	bi	A
Smokethorn	Dalea spinosa	dalspi	tr	P
Smoky shrew	Sorex fumeus	sorfum	ma	A
Snow bunting	Plectrophenax nivalis	pleniv	bi	A
Snow goose	Chen caerulescens	checae	bi	A
Snow willow	Salix nivalis	salniv	tr	P
Snowshoe hare	Lepus americanus	lepame	ma	A
Snowy egret	Egretta thula	egrthu	bi	A
Snowy plover	Charadrius alexandrinus	chaale	bi	A
Solitary sandpiper	Tringa solitaria	trisol	bi	A
Solitary vireo	Vireo solitarius	virsol	bi	A
Song sparrow	Melospiza melodia	melmel	bi	A
Sonoma chipmunk	Tamias sonomae	tamson	ma	A
Sonoma tree vole	Arborimus pomo	arbpom	ma	A
Sora	Porzana carolina	porcar	bi	A
Southeastern myotis	Myotis austroriparius	myoaus	ma	A
Southeastern pocket gopher	Geomys pinetis	geopin	ma	A
Southeastern shrew	Sores longirostris	sorlon	ma	A
Southern bog lemming	Synaptomys cooperi	syncoo	ma	A
Southern flying squirrel	Glaucomys volans	glavol	ma	A
Southern grasshopper mouse	Onychomys torridus	onytor	ma	A
Southern plains woodrat	Neotoma micropus	neomic	ma	A
Southern pocket gopher	Thomomys umbrinus	thoumb	ma	A
Southern red-backed vole	Clethrionomys gapperi	clegap	ma	A
Southern short-tailed shrew	Blarina carolinensis	blacar	ma	A
Southern yellow bat	Lasiurus ega	lasega	ma	A
Southwestern myotis	Myotis auriculus	myoaur	ma	A
Spiney pocket mouse	Chaetodipus spinatus	chaspin	ma	A
Spiny hackberry	Celtis pallida	celpal	tr	P
Spotted bat	Euderma maculatum	eudmac	ma	A
Spotted ground squirrel	Spermophilus spilosoma	spespi	ma	A
Spotted owl	Strix occidentalis	strocc	bi	A
Spotted sandpiper	Actitis macularia	actmac	bi	A
Sprague's pipit	Anthus spragueii	antspr	bi	A
Spruce grouse	Dendragapu canadensis	dencan	bi	A
St. Lawrence Island shrew	Sorex jacksoni	sorjac	ma	A

Cname	Sname	Scode	Lform	King
Star-nosed mole	Condylura cristata	concri	ma	A
Steller's jay	Cyanocitta stelleri	cyaste	bi	A
Stephen's kangaroo rat	Dipodomys stephensi	dipste	ma	A
Stephen's wood rat	Neotoma stephensi	neoste	ma	A
Stilt sandpiper	Calidris himantopus	calhim	bi	A
Streak-backed oriole	Icterus pustulatus	ictpus	bi	A
Strickland's woodpecker	Picoides stricklandi	picstr	bi	A
Striped skunk	Mephitis mephitis	mepmep	ma	A
Subalpine fir	Abies lasiocarpa	abilas	tr	P
Sulphur-bellied flycatcher	Myiodynastes luteiventris	myilut	bi	A
Summer tanager	Piranga rubra	pirrub	bi	A
Surf scoter	Melanitta perspicillata	melper	bi	A
Swainson's hawk	Buteo swainsoni	butswa	bi	A
Swainson's thrush	Catharus ustulatus	catust	bi	A
Swainson's warbler	Limnothlypis swainsonii	limswa	bi	A
Swamp rabbit	Sylvilagus aquaticus	sylaqu	ma	A
Swamp sparrow	Melospiza georgiana	melgeo	bi	A
Swift fox	Vulpes velox	vulvel	ma	A
Taiga vole	Microtus xanthognathus	micxan	ma	A
Tall blueberry willow	Salix novae-angliae	salnov	tr	P
Tawney-bellied cotton rat	Sigmodon fulviventer	sigful	ma	A
Tennessee warbler	Vermivora peregrina	verper	bi	A
Tesota	Olneya tesota	olntes	tr	P
Texas antelope squirrel	Ammospermophilus interpres	ammint	ma	A
Texas kangaroo rat	Dipodomys elator	dipela	ma	A
Texas madrone	Arbutus texana	arbtex	tr	P
Texas mouse	Peromyscus attwateri	peratt	ma	A
Texas mulberry	Morus microphylla	mormic	tr	P
Texas pocket gopher	Geomys personatus	geoper	ma	A
Thayer's gull	Larus thayeri	lartha	bi	A
Thick-billed kingbird	Tyrannus crassirostris	tyrcra	bi	A
Thicket creeper	Parthenocissus inserta	parins	vi	P
Thinleaf alder	Alnus tenuifolia	alnten	tr	P
Thirteen-lined ground squirrel	Spermophilus tridecemlineatus	spetri	ma	A
Three-toed woodpecker	Picoides tridactylus	pictri	bi	A
Toumey oak	Quercus toumeyi	quetou	tr	P
Townsend's big-eared bat	Plecotus townsendii	cortow	ma	A
Townsend's chipmunk	Tamias townsendii	tantow	ma	A
Townsend's ground squirrel	Spermophilus townsendii	spetow	ma	A
Townsend's mole	Scapanus townsendii	scatow	ma	A
Townsend's solitaire	Myadestes townsendi	myatow	bi	A
Townsend's vole	Microus townsendii	mictow	ma	A
Townsend's warbler	Dendroica townsendi	dentow	bi	A
Towsend's pocket gopher	Thomomys townsendii	thotow	ma	A
Tree swallow	Tachycineta bicolor	tacbic	bi	A
Tricolored blackbird	Agelaius tricolor	agetri	bi	A
Tricolored heron	Egretta tricolor	gretri	bi	A
Tropical kingbird	Tyrannus melancholicus	tyrmel	bi	A
Trowbridge's shrew	Sorex trowbridgii	sortro	ma	A

(continued)

Cname	Sname	Scode	Lform	King
Tufted titmouse	Parus bicolor	parbic	bi	A
Tundra shrew	Sorex tundrensis	sortun	ma	A
Tundra swan	Cygnus columbianus	cygcol	bi	A
Tundra vole	Microtus oconomus	micoco	ma	A
Turkey vulture	Cathartes aura	cataur	bi	A
Tweedy willow	Salix tweedyi	saltwe	tr	P
Uinta chipmunk	Eutamias umbrinus	eutumb	ma	A
Uinta ground squirrel	Spermophilus armatus	spearm	ma	A
Underwood's bonneted bat	Eumops underwoodi	eumund	ma	A
Ungava collared lemming	Dicrostonys hudsonius	dichud	ma	A
Upland sandpiper	Bartramia longicauda	barlon	bi	A
Utah honeysuckle	Lonicera utahensis	lonuta	vi	P
Utah juniper	Juniperus osteosperma	junost	tr	P
Utah prairie dog	Cynomys parvidens	cynpav	ma	A
Vagrant shrew	Sorex vagrans	sorvag	ma	A
Varied bunting	Passerina versicolor	pasver	bi	A
Varied thrush	Ixoreus neaevius	ixonae	bi	A
Vaux's swift	Chaetura vauxi	chavau	bi	A
Veery	Catharus fuscescens	catfus	bi	A
Velvet ash	Fraxinus velutina	fravel	tr	P
Verdin	Auriparus flaviceps	aurfla	bi	A
Vermilion flycatcher	Pyrocephalus rubinus	pyrrub	bi	A
Vesper sparrow	Pooecetes gramineus	poogra	bi	A
Violet-crowned hummingbird	Amazilia violiceps	amavio	bi	A
Violet-green swallow	Tachycineta thalassina	tactha	bi	A
Virginia opossum	Didelphis virginiana	didvir	ma	A
Virginia rail	Rallus limicola	rallim	bi	A
Virginia's warbler	Vermivora virginiae	vervir	bi	A
Wandering tattler	Heteroscelus incanus	hetinc	bi	A
Warbling vireo	Vireo gilvus	virgil	bi	A
Washington ground squirrel	Spermophilus washingtoni	spewas	ma	A
Water birch	Betula occidentalis	betocc	tr	P
Water pipit	Anthus spinoletta	antspi	bi	A
Water vole	Microtus richardsoni	micric	ma	A
Wavyleaf oak	Quercus undulata	queund	tr	P
Western bluebird	Sialia mexicana	siamex	bi	A
Western bonneted bat	Eumops perotis	eumper	ma	A
Western flycatcher	Empidonax difficilis	empdif	bi	A
Western gray squirrel	Sciurus griseus	scigri	ma	A
Western grebe	Aechmophorus occidentalis	aecocc	bi	A
Western gull	Larus occidentalis	larocc	bi	A
Western harvest mouse	Reithrodontomys megalotis	reimeg	ma	A
Western heather vole	Phenacomys intermedius	pheint	ma	A
Western jumping mouse	Zapus princeps	zappri	ma	A
Western kingbird	Tyrannus verticalis	tyrver	bi	A
Western meadowlark	Sturnella neglecta	stuneg	bi	A
Western pipistrel	Pipistrellus hesperus	piphes	ma	A
Western pocket gopher	Thomomys mazama	thomaz	ma	A
Western red bat	Lasiurus blosservillii	lasblo	ma	A

Cname	Sname	Scode	Lform	King
Western red-backed vole	Clethrionomys californicus	clecal	ma	A
Western sandpiper	Calidris mauri	calpro	bi	A
Western screech-owl	Otus kennicottii	calmau	bi	A
Western small-footed myotis	Myotis ciliolabrum	myocil	ma	A
Western spotted skunk	Spilogale putorius	spiput	ma	A
Western tanager	Piranga ludoviciana	pirlud	bi	A
Western virginsbower	Clematis ligusticifolia	clelig	vi	P
Western wood-pewee	Contopus sordidulus	consor	bi	A
Whimbrel	Numenius phaeopus	numpha	bi	A
Whip-poor-will	Caprimulgus vociferus	capvoc	bi	A
Whiskered screech-owl	Otus trichopsis	otutri	bi	A
Whistling swan	Olor columbianus	olocol	bi	A
White ankled mouse	Ochtrotomys pectoralis	perpec	ma	A
White crowned sparrow	Zonotrichia leucophrys	zonleu	bi	A
White fir	Abies concolor	abicon	tr	P
White ibis	Eudocimis albus	eudalb	bi	A
White nosed coati	Nasua nasua	nasnas	ma	A
White pelican	Pelecanus erythrorhynchos	pelery	bi	A
White spruce	Picea glauca	picgla	tr	P
White willow	Salix alba	salalb	tr	P
White-backed hog-nosed skunk	Conepatus leuconotus	conleu	ma	A
Whitebark pine	Pinus albicaulis	pinalb	tr	P
White-breasted nuthatch	Sitta carolinensis	sitcar	bi	A
White-crowned sparrow	Zonotrichia leucophrys	zonlew	bi	A
White-eared hummingbird	Hylocharis leucotis	hylleu	bi	A
White-eared pocket mouse	Perognathus alticola	peralt	ma	A
White-eyed vireo	Vireo griseus	virgri	bi	A
White-faced ibis	Plegadis chihi	plechi	bi	A
White-footed mouse	Peromyscus leucopus	perleu	ma	A
White-footed vole	Arborimus albipes	arbalb	ma	A
White-fronted goose	Anser albifrons	ansalb	bi	A
White-headed woodpecker	Picoides albolarvatus	picalb	bi	A
White-sided jack rabbit	Lepus callotis	lepca2	ma	A
White-tailed antelope squirrel	Ammospermophilus leucurus	ammleu	ma	A
White-tailed deer	Odocoileus virginianus	odovir	ma	A
White-tailed hawk	Buteo albicaudatus	butalb1	bi	A
White-tailed jack rabbit	Lepus townsendii	leptow	ma	A
White-tailed prairie dog	Cynomys leucurus	cynleu	ma	A
White-tailed ptarmigan	Lagopus leucurus	lagleu	bi	A
White-throated sparrow	Zonotrichia albicollis	zonalb	bi	A
White-throated swift	Aeronautes saxatalis	aersax	bi	A
White-throated woodrat	Neotoma albigula	neoalb	ma	A
White-winged crossbill	Loxia leucoptera	loxleu	bi	A
White-winged dove	Zenaida asiatica	zenasi	bi	A
White-winged scoter	Melanitta fusca	melfus	bi	A
Whooping crane	Grus americana	gruame	bi	A
Wild boar	Sus scrofa	susscr	ma	A
Wild horse	Equus caballus	equcab	ma	A
Wild turkey	Meleagris gallopavo	melgal	bi	A

(continued)

Cname	Sname	Scode	Lform	King
Willet	Catoptrophorus semipalmatus	catsem	bi	A
Williamson's sapsucker	Sphyrapicus thyroideus	sphthy	bi	A
Willow flycatcher	Empidonax traillii	emptra	bi	A
Wilson's phalarope	Phalaropus tricolor	phatri	bi	A
Wilson's warbler	Wilsonia pusilla	wilpus	bi	A
Winter wren	Troglodytes troglodytes	trotro	bi	A
Wolfs willow	Salix wolfii	salwol	tr	P
Wolverine	Gulo gulo	gulgul	ma	A
Wood duck	Aix sponsa	aixspo	bi	A
Wood ibis	Mycteria americana	mycame	bi	A
Wood thrush	Hylocichla mustelina	hylmus	bi	A
Woodchuck	Marmota monax	marmon	ma	A
Woodland jumping mouse	Napaeozapus insignis	napins	ma	A
Woodland vole	Microtus pinetorium	micpin	ma	A
Worm-eating warbler	Helmitheros vermivorus	helver	bi	A
Wrentit	Chamaea fasciata	chafas	bi	A
Wright silktassel	Garrya wrightii	garwri	tr	P
Wyoming ground squirrel	Spermophilus elegans	speele	ma	A
Wyoming pocket gopher	Thomomys clusius	thoclu	ma	A
Yellow grosbeak	Pheucticus chrysopeplus	phechr	bi	A
Yellow paloverde	Cercidium microphyllum	cermic	tr	P
Yellow rail	Coturnicops noveboracensis	cotnov	bi	A
Yellow warbler	Dendroica petechia	denpet	bi	A
Yellow willow	Salix lutea	sallut	tr	P
Yellow-bellied flycatcher	Empidonax flaviventris	empfla	bi	A
Yellow-bellied marmot	Marmota flaviventris	marfla	ma	A
Yellow-bellied sapsucker	Sphyrapicus varius	sphvar	bi	A
Yellow-billed cuckoo	Coccyzus americanus	cocame	bi	A
Yellow-billed magpie	Pica nuttalli	picnu2	bi	A
Yellow-breasted chat	Icteria virens	ictvir	bi	A
Yellow-checked chipmunk	Tamias ochrogenys	tamoch	ma	A
Yellow-crowned night-heron	Nyctanassa violacea	nycvio	bi	A
Yellow-eyed junco	Junco phaeonotus	junpha	bi	A
Yellow-faced pocket gopher	Pappogeomys castanops	papcas	ma	A
Yellow-headed blackbird	Xanthocephalus xanthocephalus	xanxan	bi	A
Yellowleaf silktassel	Garrya flavescens	garfla	tr	P
Yellow-nosed cotton rat	Sigmodon ochrognathus	sigoch	ma	A
Yellow-pine chipmunk	Tamias amoenus	tamamo	ma	A
Yellow-rumped warbler	Dendroica coronata	dencor	bi	A
Yellow-throated vireo	Vireo flavifrons	virfla	bi	A
Yellow-throated warbler	Dendroica dominica	dendom	bi	A
Yewleaf willow	Salix taxifolia	saltax	tr	P
Yuma myotis	Myotis yumanensis	myoyum	ma	A
Zone-tailed hawk	Buteo albonotatus	butalb2	bi	A

Appendix B
Tree Species and Their Tolerance to Shade

Common Name	Group*	Tolerance**	Common Name	Group*	Tolerance**
Ailanthus	H	2	Cedro hembra	H	2
Alaska-cedar	C	4	Cherrybark oak	H	2
Alpine larch	C	1	Chestnut oak	H	3
American basswood	H	1	Chinkapin oak	H	2
American beech	H	5	Common persimmon	H	5
American elm	H	3	Cucumbertree	H	3
American holly	H	5	Digger pine	C	1
American hornbeam	H	5	Douglas-fir	C	3
Atlantic white-cedar	C	3	Eastern cottonwood	H	1
Baldcypress	C	3	Eastern hemlock	C	5
Balsam fir	C	5	Eastern hophornbeam	H	4
Balsam poplar	H	1	Eastern redbud	H	4
Bigcone Douglas-fir	C	2	Eastern redcedar	C	2
Bigleaf maple	H	1	Eastern white pine	C	3
Bigtooth aspen	H	1	Engelmann spruce	C	4
Bitternut hickory	H	2	Eucalyptus grandis	H	1
Black ash	H	2	European alder	H	4
Black cherry	H	2	European black pine	H	2
Black cottonwood	H	1	Florida maple	H	4
Black locust	H	1	Florida torreya	C	4
Black maple	H	1	Flowering dog wood	H	5
Black oak	H	3	Fraser fir	C	5
Black spruce	C	4	Fraser magnolia	H	3
Black tupelo	H	4	Giant chinkapin	H	3
Black walnut	H	2	Giant sequoia	C	2
Black willow	H	1	Grand fir	C	4
Blue oak	H	2	Green ash	H	4
Blue spruce	C	3	Hackberry	H	3
Bluegum eucalyptus	H	2	Honeylocust	H	2
Boxelder	H	4	Incense-cedar	C	3
Brewer spruce	C	4	Jack pine	C	2
Bur oak	H	3	Jeffrey pine	C	2
Butternut	H	2	Kiawe	H	2
Cabbage palmetto	H	4	Koa	H	2
California black oak	H	2	Laurel oak	H	4
California laurel	H	4	Limber pine	C	2
California red fir	C	4	Live oak	H	3
Canyon live oak	H	4	Loblolly-bay	H	4
Carolina silverbell	H	4	Loblolly pine	C	2
Casuarina	H	2	Lodgepole pine	C	1
Cedar elm	H	3	Longleaf pine	C	2

(continued)

*H = Hardwood, C = Conifer

**1 = Very intolerant, 2 = Intolerant, 3 = Intermediate, 4 = Tolerant, 5 = Very tolerant

Source: Burns, R.M., and B.H. Honkala. 1990. *Silvics of trees of North America*. Vol. 1, conifers, Vol. 2, hardwoods. Agric. Hdbk. 654. USDA Forest Service, Washington, DC.

Common Name	Group*	Tolerance**	Common Name	Group*	Tolerance**
Maria	H	3	Shellbark hickory	H	5
Melaleuca	H	2	Shortleaf pine	C	2
Mockernut hickory	H	2	Shumard oak	H	2
Monkey-pod	H	2	Silk oak	H	1
Monterey pine	C	3	Silver maple	H	4
Mountain hemlock	C	4	Singleleaf pinyon	C	2
Noble fir	C	3	Sitka spruce	C	4
Northern red oak	H	3	Slash pine	C	2
Northern white-cedar	C	4	Slippery elm	H	4
Nutmeg hickory	H	2	Sourwood	H	4
Nuttall oak	H	2	Southern magnolia	H	4
Ogeechee tupelo	H	2	Southern red oak	H	3
Ohi'a lehua	H	2	Southern redcedar	C	2
Ohio buckeye	H	4	Spruce pine	C	5
Oregon ash	H	3	Striped maple	H	5
Oregon white oak	H	2	Subalpine fir	C	4
Osage-orange	H	2	Sugar maple	H	5
Overcup oak	H	3	Sugar pine	C	3
Pacific madrone	H	3	Sugarberry	H	4
Pacific silver fir	C	5	Swamp chestnut oak	H	2
Pacific yew	C	5	Swamp cottonwood	H	2
Paper birch	H	2	Swamp white oak	H	3
Pecan	H	2	Sweet birch	H	2
Pignut hickory	H	3	Sweetbay	H	3
Pin cherry	H	1	Sweetgum	H	2
Pin oak	H	2	Sycamore	H	3
Pinyon pine	C	2	Table mountain pine	C	2
Pitch pine	C	2	Tamarack	C	1
Pond pine	C	2	Tanoak	H	4
Ponderosa pine	C	2	Turkey oak	H	2
Port-Orford-cedar	C	4	Virginia pine	C	2
Post oak	H	2	Water hickory	H	3
Pumpkin ash	H	3	Water oak	H	2
Quaking aspen	H	1	Water tupelo	H	2
Red alder	H	2	Western hemlock	C	5
Red maple	H	4	Western juniper	C	2
Red mulberry	H	4	Western larch	C	1
Red pine	C	2	Western red cedar	C	5
Red spruce	C	4	Western white pine	C	3
Redwood	C	5	White ash	H	2
River birch	H	2	White basswood	H	4
Rock elm	H	3	White fir	C	4
Rocky Mountain juniper	C	1	White oak	H	3
Royal paulowinia	H	2	White spruce	C	3
Saligna eucalyptus	H	1	Whitebark pine	C	3
Sand pine	C	3	Willow oak	H	2
Sassafras	H	2	Winged elm	H	4
Scarlet oak	H	1	Yellow birch	H	3
Scotch pine	C	2	Yellow buckeye	H	4
September elm	H	4	Yellow poplar	H	2
Shagbark hickory	H	3			

Appendix C
Natural Range of Hardwoods in the Continental U.S. and Alaska

Common name	No. of States	State
American basswood	29	AR, CT, IA, IL, IN, KS, KY, MA, MD, ME, MI, MN, MO, NC, ND, NE, NH, NJ, NY, OH, OK, PA, RI, SD, TN, VA, VT, WI, WV
American beech	26	AL, AR, DE, FL, LA, MA, MD, ME, MI, MO, MS, NC, NH, NJ, NY, OH, OK
American elm	37	AL, AR, CT, DE, FL, GA, IA, IL, IN, KS, KY, LA, MA, MD, ME, MI, MN, MO, MS, NC, ND, NE, NH, NJ, NY, OH, OK, PA, RI, SC, SD, TN, TX, VA, VT, WI, WV
American holly	18	AL, AR, DE, FL, GA, KY, LA, MD, MO, MS, NC, NJ, OK, SC, TN, TX, VA, WV
American hornbeam	33	AL, AR, CT, DE, FL, GA, IA, IL, IN, KY, LA, MA, MD, ME, MI, MN, MO, MS, NC, NH, NJ, NY, OH, OK, PA, RI, SC, TN, TX, VA, VT, WI, WV
Balsam poplar	9	ME, MI, MN, NH, NY, UT, VT, WI, WV
Bigleaf maple	3	CA, OR, WA
Bigtooth aspen	22	CT, DE, IA, IL, IN, KY, MA, MD, ME, MI, MN, NH, NJ, NY, OH, PA, RI, TN, VA, VT, WI, WV
Bitternut hickory	35	AL, AR, CT, DE, FL, GA, IA, IL, IN, KS, KY, LA, MA, MD, MI, MN, MO, MS, NC, ND, NE, NJ, NY, OH, OK, PA, RI, SC, SD, TN, TX, VA, VT, WI, WV
Black ash	15	CT, IA, IL, IN, MA, ME, MI, MN, NH, NY, OH, PA, VT, WI, WV
Black cherry	38	AL, AR, AZ, CT, DE, FL, GA, IA, IL, IN, KS, KY, LA, MA, MD, ME, MI, MN, MO, MS, NC, ND, NH, NJ, NM, NY, OH, OK, PA, RI, SC, SD, TN, TX, VA, VT, WI, WV
Black cottonwood	5	CA, ID, NV, OR, WA
Black locust	12	AL, AR, GA, KY, NC, OH, OK, PA, SC, TN, VA, WV
Black maple	12	IL, IN, MI, MN, MO, NY, OH, PA, SC, TN, WI, WV
Black oak	34	AL, AR, CT, DE, FL, GA, IA, IL, IN, KS, KY, LA, MA, MD, ME, MI, MN, MO, MS, NC, NE, NH, NJ, NY, OH, OK, PA, RI, SC, TN, TX, VA, WI, WV
Black tupelo	29	AL, AR, CT, DE, FL, GA, IL, IN, KY, LA, MA, MS, ME, MI, MO, MS, NC, NH, NJ, NY, OH, OK, PA, RI, SC, TN, TX, VA, WV
Black walnut	29	AL, AR, DE, FL, GA, IA, IL, IN, KS, KY, MD, MI, MN, MO, MS, NC, NE, NJ, NY, OH, OK, PA, SC, SD, TN, TX, VA, WI, WV
Black willow	35	AL, AR, CT, DE, FL, GA, IA, IL, IN, KS, KY, LA, MA, MD, ME, MI, MN, MO, MS, NC, NE, NH, NJ, NY, OH, OK, PA, RI, SC, TN, TX, VT, WI, WV
Blue oak	1	CA
Boxelder	39	AL, AR, AZ, CO, DE, FL, GA, IA, IL, IN, KS, KY, LA, MA, MD, MI, MN, MO, MT, NC, ND, NE, NH, NJ, NM, NY, OH, OK, PA, SC, SD, TN, TX, UT, VA, VT, WI, WV, WY
Bur oak	22	AR, IA, IL, IN KS, KY, MD, ME, MI, MN, MO, ND, NE, NY, OH, OK, PA, SD, TN, TX, WI, WV
Butternut	26	AL, AR, CT, GA, IA, IL, IN, KY, MA, ME, MI, MN, MO, MS, NC, NH, NJ, NY, OH, PA, RI, TN, VA, VT WI, WV
Cabbage palmetto	3	FL, GA, SC
California black oak	2	CA, OR

(continued)

Does not include minimally present, disjunct populations, or introduced species. Data was interpreted from maps in Burns, R.M., and B.H. Honkala. 1990. *Silvics of trees of North America*. Vol. 2, hardwoods. Agric. Hdbk. 654. USDA Forest Service, Washington, DC.

Common name	No. of States	State
California laurel	2	CA, OR
Canyon live oak	3	AZ, CA, OR
Carolina silverbell	10	AL, AR, FL, GA, NC, OK, SC, TN, VA, WV
Cedar elm	5	AR, LA, MS, OK, TX
Chestnut oak	20	AL, CT, DE, IL, IN, KY, MA, MD, ME, MS, NH, NJ, NY, OH PA, RI, SC, TN, VA, WV
Chinkapin oak	21	AL, GA, IA, KS, KY, LA, MD, MI, MO, MS, NC, NY, OH, OK, PA, SC, TN, TX
Cucumbertree	16	AL, AR, GA, IN, KY, LA, MO, MS, NC, NY, OH, PA, SC, TN, VA, WV
Eastern cottonwood	27	AL, AR, CT, FL, GA, IA, IL, IN, KS, LA, MI, MO, MS, NC, ND, NE, NY, OH, OK, PA, SC, SD, TN, TX, VA, WI, WV
Eastern hophornbeam	32	AL, AR, CT, GA, IL, IN, KS, KY, LA, MA, ME, MI, MN, MO, MS, NC, ND, NE, NH, NJ, NY, OK, PA, RI, SC, SD, TN, TX VA, VT, WI, WV
Eastern redbud	24	AL, AR, FL, GA, IA, IL, IN, KS, KY, LA, MI, MO, MS, NC, NE, NY, OH, OK, PA, SC, TN, TX, VA, WV
Florida maple	11	AL, AR, FL, GA, LA, MS, NC, OK, SC, TX, VA
Flowering dogwood	26	AL, AR, DE, FL, GA, IL, IN, KY, LA, MA, MD, MI, MO, MS, NC, NJ, NY, OH, OK, PA, RI, SC, TN, TX, VA, WV
Fraser magnolia	6	GA, KY, NC, TN, VA, WV
Gambel oak	1	AZ
Giant chinkapin	2	CA, OR
Green ash	37	AL, AR, CT, DE, FL, GA, IA, IL, IN, KS, KY, LA, MA, MD, ME, MI, MN, MO, MS, MT, NC, ND, NE, NH, NJ, OH, OK, PA, RI, SC, SD, TN, TX, VA, WI, WV, WY
Hackberry	27	AR, CT, DE, IA, IL, IN, KS, MA, MD, MI, MN, MO, ND, NE, NH, NJ, NY, OH, OK PA, RI, SD, TN, VA, VT, WI, WV
Honeylocust	18	AL, AR, IL, IN, KS, KY, LA, MI, MO, MS, NE, OH, OK, PA, TN, TX, WI, WV
Laurel oak	4	AL, NC, SC, TX
Live oak	9	AL, FL, GA, LA, MS, NC, SC, TX, VA
Loblolly-bay	6	AL, FL, GA, MS, NC, SC
Mockernut hickory	27	AL, AR, CT, DE, FL, GA, IA, IL, IN, MA, MD, MO, MS, NC, ND, NJ, NY, OH, OK, PA, SC, SD, TN, TX, VA, WI, WV
Northern red oak	33	AL, AR, CT, DE, GA, IA, IL, IN, KS, KY, MA, MD, ME, MI, MN, MO, MS, NC, NE, NH, NJ, NY, OH, OK, PA, RI, SC, TN, TX, VA, VT, WI, WV
Nutmeg hickory	4	AR, LA, OK, TX
Nuttall oak	6	AL, AR, LA, MO, MS, TX
Ogeechee tupelo	3	FL, GA, SC
Ohio buckeye	13	AL, AR, IL, IN, KS, KY, MO, OH, OK, PA, TN, TX, WV
Oregon ash	3	CA, OR, WA
Oregon white oak	3	CA, OR, WA
Osage-orange	2	OK, TX
Overcup oak	16	AL, AR, DE, FL, GA, IL, IN, KY, LA, MO, MS, NC, SC, TN, TX, VA
Pacific madrone	3	CA, OR, WA
Paper birch	18	CT, IA, ID, MA, ME, MI, MN, MO, MT, NH, NY, OR, PA, RI, SD, VT, WA, WI
Pecan	9	AR, IL, IN, KS, LA, MO, OK, TN, TX
Persimmon	23	AL, AR, DE, FL, GA, IL, IN, KS, KY, LA, MD, MO, MS, NC, NJ, OH, OK PA, SC, TN, TX, VA, WV
Pignut hickory	21	AL, AR, CT, IL, IN, KY, MA, MD, MI, MO, MS, NC, NJ, NY, OH, PA, RI, TN, VA, WV

Common name	No. of States	State
Pin cherry	23	CO, CT, IL, IN, MA, MD, ME, MI, MN, NC, ND, NH, NJ, NY, PA, RI, SD, TN, VA, VT, WI, WV
Pin oak	20	AR, CT, DE, IA, IL, IN, KS, KY, MA, MD, MI, NC, NJ, NY, OH, OK, SC, TN, VA, WV
Post oak	24	AL, AR, CT, DE, FL, GA, IA, IL, IN, KY, LA, MD, MO, MS, NC, NJ, OH, OK, PA, SC, TN
Pumpkin ash	9	AR, FL, IL, IN, MO, NC, SC, TN, VA
Quaking aspen	30	AZ, CA, CO, CT, ID, IL, IN, MA, ME, MI, MN, MT, ND, NE, NH, NJ, NM, NV, NY, OH, OR, PA, RI, SD, UT, VT, WA, WI, WV, WY
Red alder	4	AK, CA, OR, WA
Red maple	32	AL, CT, DE, FL, GA, IL, IN, KY, LA, MA, MD, ME, MI, MN, MO, MS, NC, NH, NJ, NY, OH, OK, PA, RI, SC, TN, TX, VA, VT, WI, WV
Red mulberry	28	AL, AR, DE, FL, GA, IA, IL, IN, KS, KY, LA, MA, MD, MI, MN, MO, MS, NC, NE, NJ, OH, OK, SC, TN, TX, VA, WI, WV
Redbay	9	AL, FL, GA, LA, MS, NC, SC, TX, VA
River birch	25	AL, AR, DE, FL, GA, IA, IL, IN, KY, LA, MD, MO, MS, NC, NJ, NY, OH, OK, PA, RI, SC, TN, TX, VA, WV
Rock elm	13	IL, IN, KY, MI, MN, MO, NY, OH, TN, VA, VT, WI, WV
Sassafras	29	AL, AR, CT, DE, FL, GA, IL, IN, KY, MA, MD, MI, MO, MS, NC, ND, NH, NJ, NY, OH, OK, PA, RI, SC, SD, TN, TX, VA, WV
Scarlet oak	23	AL, CT, DE, IL, IN, LA, MA, MD, ME, MI, MO, MS, NC, NH, NJ, NY, OH, PA, RI, SC, TN, VA, WV
September elm	4	AL, AR, OK, TN
Shagbark hickory	34	AL, AR, CT, DE, IL, IN, KS, KY, LA, MA, MD, ME, MI, MN, MO, MS, NC, ND, NE, NH, NJ, NY, OH, OK, PA, RI, SC, SD, TN, TX, VA, VT, WI, WV
Shellbark hickory	11	AR, IL, IN, KS, KY, MI, MO, NY, OH, PA, WV
Shumard oak	17	AL, AR, FL, GA, IL, KS, KY, LA, MO, MS, NC, OH, OK, SC, TN, TX, VA
Silver maple	31	AL, AR, CT, GA, IA, IL, IN, KS, KY, MA, MD, ME, MI, MO, MS, NC, NE, NH, NJ, NY, OH, OK, PA, SC, SD, TN, TX, VA, VT, WI, WV
Slippery elm	33	AL, AR, CT, DE, FL, GA, IA, IL, IN, KS, KY, LA, MA, MD, MI, MN, MO, MS, NC, ND, NJ, NY, OH, OK, PA, RI, SC, SD, TN, TX, VA, WI, WV
Sourwood	12	AL, FL, GA, LA, MS, NC, OH, PA, SC, TN, VA, WV
Southern magnolia	7	AL, FL, GA, LA, MS, SC, TX
Southern red oak	20	AL, AR, DE, FL, GA, IL, IN, LA, MD, MO, MS, NC, NJ, OH, OK, SC, TN, TX, VA, WV
Striped maple	15	CT, KY, MA, ME, MI, NC, NH, NJ, NY, OH, PA, SC, VA, VT, WV
Sugar maple	23	CT, DE, IA, IL, IN, MA, MD, ME, MI, MN, MO, NC, NH, NJ, NY, OH, PA, RI, TN, VA, VT, WI, WV
Sugarberry	15	AL, AR, FL, GA, IL, KY, LA, MO, MS, NC, OK, SC, TN, TX, VA
Swamp chestnut oak	15	AL, DE, FL, GA, IL, IN, MD, MO, MS, NC, NJ, SC, TN, TX, VA
Swamp cottonwood	16	AR, CT, DE, FL, IL, IN, KY, LA, MD, MO, MS, NC, OH, SC, TN, VA
Swamp white oak	22	CT, DE, IL, IN, MA, MD, ME, MI, MN, MO, NC, NH, NJ, NY, OH, PA, RI, TN, VA, VT, WI, WV
Sweet birch	17	CT, GA, KY, MA, MD, ME, NC, NH, NJ, NY, OH, PA, RI, TN, VA, VT, WV
Sweetbay	13	AL, AR, DE, FL, GA, LA, MD, MS, NC, NJ, SC, TN, VA
Sweetgum	20	AL, AR, DE, FL, GA, IL, IN, LA, MD, MO, MS, NC, NJ, OH, OK, SC, TN, TX, VA, WV

(continued)

Common name	No. of States	State
Sycamore	31	AL, AR, CT, FL, GA, IA, IL, IN, MA, MD, ME, MI, MO, MS, NC, NE, NH, NJ, NY, OH, OK, PA, RI, SC, TN, TX VA, VT, WI, WV
Tanoak	2	CA, OR
Turkey oak	8	AL, FL, GA, LA, MS, NC, SC, VA
Water hickory	11	AL, AR, FL, GA, LA, MO, MS, NC, SC, TN, TX
Water oak	13	AL, DE, FL, GA, LA, MD, MS, NC, OK, SC, TN, TX, VA
Water tupelo	13	AL, AR, FL, GA, IL, KY, LA, MO, MS, NC, SC, TN, TX
White ash	27	AL, AR, CT, DE, GA, KS, LA, MA, MD, ME, MI, MO, MS, NH, NJ, NY, OH, OK, PA, RI, SC, TN, TX, VA, VT, WI, WV
White basswood	14	AL, AR, FL, GA, IN, KY, MO, NC, OH, PA, SC, TN, VA, WV
White oak	34	AL, AR, CT, DE, FL, GA, IA, IL, IN, KS, KY, LA, MA, MD, ME, MI, MN, MO, MS, NC, NH, NJ, NY, OH, OK, PA, RI, SC, TN, TX, VA, VT, WI, WV
Willow oak	16	AL, AR, DE, FL, GA, LA, MD, MO, MS, NC, NJ, OK, SC, TN, TX, VA
Winged elm	15	AL, AR, FL, GA, IL, IN, KY, LA, MO, MS, NC, SC, TN, TX, VA
Yellow birch	17	CT, IN, MA, ME, MI, MN, NC, NH, NY, OH, PA, RI, TN, VA, VT, WI, WV
Yellow buckeye	6	NC, OH, PA, TN, VA, WV
Yellow-poplar	22	AL, AR, CT, DE, FL, GA, IL, IN, MD, MI, MO, MS, NC, NJ, NY, OH, PA, RI, SC, TN, VA, WV

Appendix D
Natural Range of Conifers in the Continental U.S. and Alaska

Common Name	No. of States	State
Alaska-cedar	4	AK, ID, OR, WA
Atlantic white-cedar	10	AL, CT, DE, FL, MA, MD, NC, NJ, NY, RI
Baldcypress	14	AL, AR, FL, GA, IL, KY, LA, MO, MS, NC, SC, TN, TX, VA
Balsam fir	7	ME, MI, MN, NH, NY, VT, WI
Bigcone Douglas-fir	1	CA
Black spruce	9	AK, MA, ME, MI, MN, NH, NY, VT, WI
Blue spruce	4	CO, NM, UT, WY
California red fir	1	CA
Digger pine	1	CA
Douglas-fir	10	AZ, CA, CO, ID, MT, NM, OR, UT, WA, WY
Eastern hemlock	21	AL, CT, GA, IN, KY, MA, ME, MI, MN, NC, NH, NJ, NY, OH, PA, RI, TN, VA, VT, WI, WV
Eastern redcedar	35	AL, AR, CT, DE, GA, IA, IL, IN, KS, KY, LA, MA, MD, ME, MI, MN, MO, MS, NC, NE, NH, NJ, NY, OH, OK, PA, RI, SC, SD, TN, TX, VA, VT, WI, WV
Eastern white pine	20	CT, DE, GA, IA, KY, MA, MD, MI, MN, NC, NJ, NY, OH, PA, RI, TN, VA, VT, WI, WV
Engelmann spruce	7	CO, ID, MT, OR, UT, WA, WY
Giant sequoia	1	CA
Grand fir	3	ID, OR, WA
Incense-cedar	2	CA, OR
Jack pine	4	ME, MI, MN, WI
Jeffrey pine	2	CA, OR
Limber pine	7	CO, ID, MT, NM, NV, UT, WY
Loblolly pine	12	AL, DE, FL, GA, LA, MD, MS, NC, SC, TN, TX, VA
Lodgepole pine	8	AK, CA, CO, ID, MT, OR, WA, WY
Longleaf pine	8	AL, FL, GA, LA, MS, NC, SC, TX
Mountain hemlock	4	AK, ID, OR, WA
Noble fir	2	OR, WA
Northern white-cedar	10	ME, MI, MN, NC, NH, NY, OH, TN, VT, WI
Pacific silver fir	2	OR, WA
Pacific yew	3	CA, ID, OR, WA
Pinyon pine	4	AZ, CO, NM, UT
Pitch pine	16	CT, DE, KY, MA, MD, ME, NC, NH, NJ, NY, OH, PA, RI, TN, VA, WV
Pond pine	7	DE, FL, GA, MD, NC, SC, VA
Ponderosa pine	12	AZ, CA, CO, ID, MT, NE, NM, OR, SD, UT, WA, WY
Port-Orford-cedar	1	OR
Red pine	8	ME, MI, MN, NH, NY, PA, VT, WI
Red spruce	7	MA, ME, NC, NH, NY, TN, VT

(continued)

Does not include minimally present or introduced species. Data was interpreted from maps in Burns, R.M., and B.H. Honkala. 1990. *Silvics of trees of North America*. Vol. 1, conifers. Agric. Hdbk. 654. USDA Forest Service, Washington, DC.

Common Name	No. of States	State
Redwood	2	CA, OR
Rocky Mountain juniper	8	AZ, CO, ID, MT, NM, SD, UT, WY
Sand pine	1	FL
Shortleaf pine	19	AL, AR, DE, FL, GA, KY, LA, MD, MO, MS, NC, NJ, OH, OK, SC, TN, TX, VA, WV
Singleleaf pinyon	3	CA, NV, UT
Sitka spruce	3	AK, OR, WA
Slash pine	6	AL, FL, GA, LA, MS, SC
Southern redcedar	2	FL, TX
Spruce pine	6	AL, FL, GA, LA, MS, SC
Subalpine fir	7	CA, OR
Sugar pine	2	CA, OR
Table mountain pine	4	NC, PA, VA, WV
Tamarack	11	CT, IN, MA, ME, MI, MN, NH, NY, OH, VT, WI
Virginia pine	13	AL, DE, GA, KY, MD, NC, NJ, OH, PA, SC, TN, VA, WV
Western hemlock	5	AK, CA, ID, OR, WA
Western juniper	2	CA, OR
Western larch	4	ID, MT, OR, WA
Western red cedar	5	AK, ID, MT, OR, WA
Western white pine	5	CA, ID, MT, OR, WA
White fir	5	AZ, CA, CO, NM, UT
White spruce	7	ME, MI, MN, NH, NY, VT, WI
Whitebark pine	6	CA, ID, MT, OR, WA, WY

Appendix E
Bird and Mammal Checklist of Species in Forest Regions

Species	Regions
Birds	
Acadian flycatcher	6-7
Acorn woodpecker	2-4
American crow	2-3-4-5-6-7
American dipper	1-2-3-4
American redstart	1-2-3-4-5-6-7
American robin	1-2-3-4-5-6-7
American swallow-tailed kite	7
Anhinga	7
Anna's hummingbird	2-3
Ash-throated flycatcher	2-3-4
Bald eagle	1-2-3-4-5-6-7
Band-tailed pigeon	2-4
Barred owl	5-6-7
Bay-breasted warbler	5
Bewick's wren	2-4-5-6-7
Black-and-white warbler	1-2-3-5
Black-backed woodpecker	1-2-3-5
Black-billed cuckoo	3-5-6
Black-capped chickadee	1-2-3-4-5-6
Black-chinned hummingbird	2-3-4-6
Black-headed grosbeak	3-4-5
Black-throated blue warbler	6-7
Black-throated gray warbler	2-3-4
Black-throated green warbler	5-6
Blackburnian warbler	5-6
Blackpoll warbler	1-5
Blue-gray gnatcatcher	2-3-4-5-6-7
Blue grouse	1-2-3-4
Blue jay	5-6-7
Boreal chickadee	1-5
Bridled titmouse	4
Broad-tailed hummingbird	2-4
Broad-winged hawk	5-6-7
Brown creeper	1-2-3-4-5-6-7
Brown-crested flycatcher	4-6
Brown-headed nuthatch	6-7
Brown thrasher	5-6-7
Bushtit	2-4
California quail	2
Calliope hummingbird	2-3-4
Canada warbler	5-6
Cape May warbler	5
Carolina chickadee	6-7
Cassin's finch	2-3-4
Cedar waxwing	1-2-3-4-5-6-7
Chestnut-backed chickadee	1-2-3
Chestnut-sided warbler	5-6
Chipping sparrow	1-2-3-4-5-6-7
Chuck-will's-widow	6-7
Clark's nutcracker	2-3-4
Common loon	1-2-3-5-6-7
Common raven	1-2-3-4-5-6
Cooper's hawk	2-3-4-5-6-7
Dark-eyed junco	1-2-3-4-5-6-7
Downy woodpecker	2-3-4-5-6-7
Dusky-capped flycatcher	1
Eastern screech-owl	2-3-4-5-6-7
Eastern wood-pewee	5-6-7
Elegant trogon	1
Evening grosbeak	2-3-4-5-6-7
Flammulated owl	2-3-4
Fox sparrow	1-2-3-4-5-6-7
Golden-crowned kinglet	1-2-3-4-5-6-7
Grace's warbler	4
Gray jay	1-2-3-4-5
Great crested flycatcher	5-6-7
Great gray owl	1-2-3-5
Great horned owl	1-2-3-4-5-6-7
Green-tailed towhee	2-3-4
Hairy woodpecker	1-2-3-4-5-6-7
Hermit thrush	1-2-3-4-5-6-7
Hermit warbler	2
Hooded warbler	6-7
Kentucky warbler	6-7
Kirkland's warbler	5
Lazuli bunting	2-3-4
Lewis's woodpecker	2-3-4
Lincoln's sparrow	1-2-3-4-5-7
Long-eared owl	2-3-4-5-6-7
MacGillivray's warbler	1-2-3-4
Magnolia warbler	5-6
Merlin	1-3-4-7
Mississippi kite	7
Mountain bluebird	1-2-3-4
Mountain chickadee	2-3-4
Mountain quail	2-3
Mourning dove	2-3-4-5-6-7

(continued)

Species	Regions	Species	Regions
Nashville warbler	2-3-5-6	Swainson's thrush	1-2-3-5-6
Northern cardinal	4-5-6-7	Tennessee warbler	5
Northern flicker	1-2-3-4-5-6-7	Townsend's solitaire	1-2-3-4
Northern goshawk	1-2-3-4-5-6	Townsend's warbler	1-2-3
Northern oriole	1-2-3-4-5-6-7	Tree swallow	1-2-3-4-5-6-7
Northern parula	5-6-7	Tufted titmouse	5-6-7
Northern pygmy-owl	1-2-3-4	Varied thrush	1-2
Northern saw-whet owl	1-2-3-4-5-6-7	Veery	3-4-5-6
Northern shrike	1-2-3-4-5-6	Violet-green swallow	1-2-3-4
Nuttall's woodpecker	2	Virginia's warbler	3
Olive-sided flycatcher	1-2-3-4-5-6	Warbling vireo	2-3-4-5-6-7
Orange-crowned warbler	1-2-3-4-7	Western bluebird	2-3-4
Ovenbird	5-6	Western flycatcher	1-2-3-4
Painted bunting	6-7	Western screech-owl	2-3-4
Pileated woodpecker	2-3-5-6-7	Western tanager	2-3-4
Pine grosbeak	1-3-4-5	Western wood-pewee	1-2-3-4
Pine siskin	1-2-3-4-5-6-7	Whip-poor-will	2-4-5-6-7
Pine warbler	5-6-7	White-breasted nuthatch	2-3-4-5-6-7
Pinyon jay	2-3-4	White-crowned sparrow	1-2-3-4-5-6-7
Plain titmouse	2-4	White-eyed vireo	6-7
Prairie warbler	5-6-7	White-headed woodpecker	2-3
Prothonotary warbler	5	White-throated sparrow	2-6-7
Purple finch	2-3-5-6-7	White-winged crossbill	1-2-3-5-6
Pygmy nuthatch	2-3-4	Wild turkey	2-4-6-7
Red-bellied woodpecker	6-7	Williamson's sapsucker	2-4
Red-breasted nuthatch	1-2-3-4-5-7	Wilson's warbler	1-2-3-4-5
Red-cockaded woodpecker	6-7	Winter wren	1-2-3-5-6-7
Red crossbill	1-2-3-4-5-6	Wood duck	2-3-5-6-7
Red-eyed vireo	3-4-5-6-7	Wood thrush	5-6-7
Red-faced warbler	4	Worm-eating warbler	6-7
Red-headed woodpecker	4-5-6-7	Yellow-bellied sapsucker	2-3-4-5-6-7
Red-shouldered hawk	2-5-6-7	Yellow-breasted chat	2-3-4-5-6-7
Red-tailed hawk	1-2-3-4-5-7	Yellow-rumped warbler	1-2-3-4-5-6-7
Rose-breasted grosbeak	5-7	Yellow-throated vireo	5-6-7
Ruby-crowned kinglet	1-2-4-5-6-7	Yellow-throated warbler	6-7
Ruby-throated hummingbird	6-7	***Mammals***	
Ruffed grouse	1-2-3-4-5-6	Abert's squirrel	4
Rufous hummingbird	1-2-3	American beaver	1-2-3-4-5-6-7
Rufous-sided towhee	2-3-4-5-6-7	Bighorn sheep	2-3-4
Scarlet tanager	5-6	Black bear	1-2-3-4-5-6-7
Scrub jay	2-3-7	Bobcat	2-3-4-5-6-7
Sharp-shinned hawk	1-2-3-4-5-6-7	Brush rabbit	2
Solitary vireo	2-3-4-5-6-7	Bushy-tailed woodrat	2-3-4
Song sparrow	1-2-3-4-5-6-7	White-nosed coati	4
Spotted owl	2-4	Collared peccary	4
Spruce grouse	1-2-3-5	Cotton mouse	6-7
Steller's jay	1-2-3-4	Coyote	1-2-3-4-5-6-7
Strickland's woodpecker	4	Deer mouse	1-2-4-5-6
Summer tanager	2-4-6-7	Douglas's squirrel	2

Species	Regions	Species	Regions
Eastern chipmunk	5-6	Northern flying squirrel	1-2-3-4-6
Eastern cottontail	4-5-6-7	Virginia opossum	2-4-5-6-7
Eastern spotted skunk	5-6-7	American pika	2-3-4
Eastern woodrat	6-7	North American porcupine	1-2-3-4-5
Elk	2-3-4	Pygmy shrew	1-3-5-6
Ermine	1-2-3-4-5-6	Raccoon	2-3-4-5-6-7
Fisher	2-3-5	Red fox	1-2-3-4-5-6-7
Eastern fox squirrel	5-6-7	Red squirrel	1-2-3-4-5
Gray-collared chipmunk	4	Red tree vole	2
Gray fox	2-4-5-6-7	Ringtail	2-4-7
Eastern gray squirrel	5-6-7	River otter	1-2-3-4-5-6-7
Gray wolf	1-2-5	Southern short-tailed shrew	5-6-7
Brown bear	1-3	Snowshoe hare	1-2-3-4-5-6
Hoary bat	2-3-4-5-6-7	Southern flying squirrel	5-6-7
Hoary marmot	2-3	Striped skunk	2-3-4-5-6-7
Least chipmunk	2-3-4	Townsend's chipmunk	2
Long-tailed vole	1-2-3-4	Uinta chipmunk	2-3-4
Long-tailed weasel	2-3-4-5-6-7	Vagrant shrew	2-3-4
Canada lynx	1-3-4	Western gray squirrel	2
Golden-mantled ground squirrel	2-3-4	Western jumping mouse	2-3-4
Marten	1-2-3-4-5	White-footed mouse	3-4-5-6-7
Mexican woodrat	4	White-tailed deer	2-3-4-5-6-7
Mink	1-2-3-4-5-6-7	Wild boar	6
Moose	1-3-4-5	Wolverine	2-3
Sewell	2	Woodchuck	1-5-6-7
Mountain lion	2-3-4-7	Woodland jumping mouse	5-6
Mule deer	1-2-3-4	Woodland vole	5-6-7
New England cottontail	5-6	Yellow-bellied marmot	2-3-4
Nine-banded armadillo	6-7	Yellow-pine chipmunk	2-3

Appendix F
Synopsis of Bird and Mammal Life History

Birds

American dipper
- General habitat: Mountain streams with fast flowing water
- Food habits: Insects and aquatic invertebrates
- Comments: Bobbing motion, walks on stream bottom

American robin
- General habitat: Meadows, cut areas, open woods
- Food habits: Insects and plant material
- Comments: Best known of all U.S. birds

Bald eagle
- Breeding season: December–January
- Mating behavior: Mates for life
- Young per season: 1–2
- General habitat: Near large lakes and rivers
- Food habits: Primarily fish, small rodents, carrion

Bewick's wren
- General habitat: On the ground in brush areas
- Food habits: Insects and spiders
- Comments: Cavity nester, uses a singing perch

Black-capped chickadee
- General habitat: Mixed coniferous and deciduous forests
- Food habits: Insect and coniferous plant material
- Comments: Cavity nester

Black-chinned hummingbird
- General habitat: Mountain meadows and woodlands
- Food habits: Insects
- Comments: Male has a mating and feeding territory

Blue grouse
- Breeding season: April–May
- Mating behavior: Polygamous
- Young per season: 7–10
- General habitat: Coniferous forest
- Food habits: Insects, buds, fruits
- Comments: Nests on ground

Blue-gray gnatcatcher
- General habitat: Pinyon-juniper and deciduous woodland
- Food habits: Insects
- Comments: Flicks tail while feeding

Brown creeper
- General habitat: Coniferous and mixed forests
- Food habits: Bark insects, moves upward when feeding

Chipping sparrow
- General habitat: Edges of coniferous and deciduous forests
- Food habits: Insects

Common raven
- General habitat: All types of forest land
- Food habits: Generalist
- Comments: Intelligence matches that of a dog

Cooper's hawk
- General habitat: Primarily deciduous forest types with meadows and clearings
- Food habits: Small birds and mammals

Dark-eyed junco
- General habitat: Coniferous and mixed forests
- Food habits: Ground feeder on seeds and fruits

Downey woodpecker
- General habitat: Deciduous woodlands
- Food habits: Insects and mast
- Comments: Cavity nester

(Patton 1997).

Eastern screech-owl
General habitat: Deciduous and mixed forests
Food habits: Rodents and insects
Comments: Defends a nesting and feeding territory

Evening grosbeak
General habitat: Coniferous and mixed forests
Food habits: Seeds and fruits

Fox sparrow
General habitat: Undergrowth in coniferous forests
Food habits: Insects and seeds
Comments: Make considerable noise when scratching for food under brush

Golden-crowned kinglet
General habitat: Old coniferous and deciduous forests
Food habits: Primarily insects

Great horned owl
General habitat: All forest types
Food habits: Birds and small mammals
Comments: Largest of the eared owls

Hairy woodpecker
General habitat: Deciduous forests
Food habits: Insects and mast
Comments: Cavity nester

Hermit thrust
General habitat: Coniferous and mixed forests
Food habits: Ground insect feeder

Lincoln's sparrow
General habitat: Deciduous forest with meadows and bogs
Food habits: Seed

Long-eared owl
General habitat: Open deciduous and coniferous forests
Food habits: Small mammals

Mountain bluebird
General habitat: Open areas near forest edge
Food habits: Insects
Comments: Cavity nester

Mourning dove
Breeding season: April–September
Mating behavior: Monogamous
Young per season: 1–3
General habitat: Open forest areas
Food habits: Seeds, fruits, mast

Northern cardinal
General habitat: Mixed woodland
Food habits: Primarily seeds but some insects

Northern flicker
General habitat: Open forested areas
Food habits: Insects, especially fond of ants
Comments: Cavity nester

Northern goshawk
General habitat: Coniferous forests
Food habits: Birds and small mammals

Northern oriole
General habitat: Deciduous forests
Food habits: Insects, seeds, fruits

Northern saw-whet owl
General habitat: Coniferous and deciduous forests
Food habits: Insects and small mammals
Comments: Cavity nester

Northern shrike
General habitat: Coniferous woodlands
Food habits: Insects, small rodents
Comments: Kills prey with bill

Olive-sided flycatcher
General habitat: Spruce-fir forests
Food habits: Flying insects
Comments: Perches on tops of tall trees

Pileated woodpecker
General habitat: Deciduous and mixed coniferous trees
Food habits: Insects and mast

Pine grosbeak
General habitat: Coniferous forest and mixed woodlands
Food habits: Seed and fruits
Comments: Slow moving and tame

Pine siskin
General habitat: Coniferous forests and mixed woodlands
Food habits: Seeds of hemlock, alder, birch
Comments: Likes salt

Purple finch
General habitat: Coniferous forests and woodlands
Food habits: Seeds

Red crossbill
General habitat: Coniferous forests
Food habits: Conifer seeds

Red-breasted nuthatch
General habitat: Coniferous forests
Food habits: Insects
Comments: Cavity nester

Red-eyed vireo
General habitat: Deciduous forests
Food habits: Insects

Red-headed woodpecker
General habitat: Open park-like conditions
Food habits: Insects and mast

Ruby-crowned kinglet
General habitat: Coniferous and deciduous forests
Food habits: Insects
Comments: Small bird, able to feed on tips of twigs

Ruffed grouse
Breeding season: April–May
Mating behavior: Promiscuous
Young per season: 10–12
General habitat: Coniferous and deciduous forests near edge
Food habits: Buds, fruits, seeds

Rufous-sided towhee
General habitat: Brush around forest edges
Food habits: Insects and seeds
Comments: Ground feeder

Sharp-shinned hawk
General habitat: Dense coniferous forests
Food habits: Small birds, rodents, and insects

Solitary vireo
General habitat: Coniferous and deciduous forests
Food habits: Insects

Song sparrow
General habitat: Forest edges and streamside vegetation
Food habits: Insects and seeds

Spruce grouse
Breeding season: April–May
Mating behavior: Monogamous
Young per season: 10–12
General habitat: Coniferous forests
Food habits: Needles, buds, fruit

Steller's jay
General habitat: Coniferous and deciduous mixed forests
Food habits: Insects and seeds

Three-toed woodpecker
General habitat: Coniferous forests
Food habits: Insects
Comments: Important species in controlling insects; cavity nester

Violet-green swallow
General habitat: Coniferous and deciduous forests
Food habits: Flying insects
Comments: Ponderosa pine is major habitat; cavity nester

Western flycatcher
General habitat: Deciduous or coniferous forests near water
Food habits: Flying insects
Comments: Cavity nester

Whip-poor-will
General habitat: Deciduous forests
Food habits: Insects
Comments: Ground nester

White-crowned sparrow
General habitat: Forest edges, meadows, clearings
Food habits: Insects
Comments: Nests on or near ground

Wild turkey
Breeding season: April–May
Mating behavior: Polygamous
Young per season: 10–12
General habitat: Deciduous or coniferous forests
Food habits: Insects, mast, fruits, seeds

Wilson's warbler
General habitat: Deciduous forests with shrubs
Food habits: Insects
Comments: Nests on or near ground

Winter wren
General habitat: Coniferous and deciduous forests
Food habits: Insects
Comments: Cavity nester

Wood duck
General habitat: Deciduous trees near water
Food habits: Acorns, fleshy fruit, vegetable matter
Comments: Cavity nester

Yellow-bellied sapsucker
- General habitat: Deciduous and coniferous forests
- Food habits: Tree sap and insects
- Comments: Cavity nester

Mammals

Abert's squirrel
- Mating behavior: Promiscuous, mating bouts
- Young per season: 2–4
- General habitat: Primarily ponderosa pine
- Food habits: Inner bark of ponderosa pine, buds
- Comments: Several subspecies including the Kaibab squirrel

Beaver
- Breeding season: January–February
- Mating behavior: Monogamous
- Breeding age: 2
- Young per season: 4
- General habitat: Lakes or slow moving water
- Food habits: Deciduous trees of poplar, willow, aspen

Black bear
- Breeding season: June–July
- Mating behavior: Probably monogamous
- Breeding age: 3
- Young per season: 2 every other year
- General habitat: Coniferous and deciduous forests
- Food habits: Omnivorous—fruits, berries, small mammals, deer, cattle

Bobcat
- Breeding season: Year-long
- Mating behavior: Monogamous
- Breeding age: 1
- Young per season: 2–4
- General habitat: Coniferous and deciduous forests
- Food habits: Small mammals

Coyote
- Mating behavior: Monogamous
- Breeding age: 1
- Young per season: 5–7
- General habitat: Open coniferous and deciduous forests
- Food habits: Small mammals, deer, fawns

Deer mouse
- Young per season: 4–5/litter and 3–4 litters
- General habitat: Coniferous and deciduous forests
- Food habits: Insects, seeds, fruits, nuts
- Comments: Prey species for many birds and mammals

Eastern cottontail
- Breeding season: February–September
- Mating behavior: Promiscuous
- Young per season: 4–5/litter with 3–4 litters
- General habitat: Brush understory in forests and woodlands
- Food habits: Forbs, grasses, fruits, twigs

Elk
- Breeding season: September–October
- Mating behavior: Polygamous
- Breeding age: 3
- Young per season: 1
- General habitat: Coniferous forests, woodlands, meadows
- Food habits: Grass, forbs, browse

Gray fox
- Breeding season: January–February
- Mating behavior: Monogamous
- Breeding age: 1
- Young per season: 3–5
- General habitat: Coniferous and deciduous forests
- Food habits: Small rodents, fruits, berries

Gray squirrel
- Breeding season: December–January
- Mating behavior: Promiscuous
- Breeding age: 10–11 months
- Young per season: 2–4/litter, may have 2 litters
- General habitat: Hardwood forests
- Food habits: Hard mast, fruits, fungi

Grizzly bear
Breeding season: June–July
Mating behavior: Monogamous
Breeding age: 3–5
Young per season: 2
General habitat: Mountain, coniferous forests
Food habits: Small and large mammals, fish

Hoary bat
General habitat: Coniferous forests
Food habits: Flying insects

Long-tailed vole
General habitat: Grassy areas away from water
Food habits: Roots and bark

Long-tailed weasel
Breeding season: July–August, delayed implantation
Mating behavior: Polygamous
Breeding age: 1 for males, 3–4 months for females
Young per season: 6–9
General habitat: Varied forest conditions
Food habits: Small mammals

Lynx
Breeding season: January–February
Mating behavior: Monogamous
Breeding age: 1
Young per season: 1–4
General habitat: Dense coniferous forest
Food habits: Small mammals, rabbits preferred

Marten
Breeding season: July–August
Mating behavior: Promiscuous
Breeding age: 1
Young per season: 3–4
General habitat: Dense coniferous forests
Food habits: Small mammals especially red squirrels

Mink
Breeding season: January–March
Mating behavior: Polygamous
Breeding age: 1
Young per season: 5–6
General habitat: Lakes and streams in coniferous and deciduous forests
Food habits: Small mammals, fish

Moose
Breeding season: September–November
Mating behavior: Polygamous
Breeding age: 2
Young per season: 1–2
General habitat: Coniferous and deciduous forests with willow and aspen near water
Food habits: Woody plants, aspen and willow

Mountain lion
Breeding season: January–February
Mating behavior: Monogamous
Breeding age: 2–3
Young per season: 1–4
General habitat: Coniferous forests and woodlands
Food habits: Small rodents, deer

Mule deer
Breeding season: September–February
Mating behavior: Polygamous
Breeding age: 2
Young per season: 1–2
General habitat: Coniferous forests
Food habits: Browse, forbs, grasses

Northern flying squirrel
Breeding season: February–March
Young per season: 3–6/litter, may have 2 litters
General habitat: Dense coniferous and deciduous forest
Food habits: Seeds, fruits, mast

Porcupine
Breeding season: September–November
Young per season: 1
General habitat: Coniferous and hardwood forests
Food habits: Bark and twigs

Pygmy shrew
General habitat: Dense forests with moist areas
Food habits: Insects
Comments: Smallest mammal in North America, 2–4grams

Raccoon
Breeding season: February
Mating behavior: Monogamous
Breeding age: 1
Young per season: 3–6
General habitat: Hardwood forests near water
Food habits: Crayfish, fish, small rodents, eggs

Red fox

Breeding season:	January–February
Mating behavior:	Monogamous
Breeding age:	1
Young per season:	4–10
General habitat:	Coniferous and deciduous forests with brushland
Food habits:	Small mammals, fruits, nuts

Red squirrel

Breeding season:	February–March
Breeding age:	1
Young per season:	4–5
General habitat:	Coniferous forests
Food habits:	Seeds of conifers, mushrooms
Comments:	Stores cones in "middens"

River otter

Breeding season:	January–April
Mating behavior:	Monogamous
Breeding age:	1
Young per season:	2–3
General habitat:	Streams and lakes in coniferous and deciduous forests
Food habits:	Fish, small mammals, invertebrates

Snowshoe hare

Breeding season:	February–August
Breeding age:	1
Young per season:	3–4/litter, 2–3 litters
General habitat:	Coniferous and hardwood forests
Food habits:	Bark, buds, twigs
Comments:	Population rises and falls in a cycle

Striped skunk

Breeding season:	February–March
Mating behavior:	Polygamous
Breeding age:	1
Young per season:	4–7
General habitat:	Coniferous and deciduous forests with brush
Food habits:	Small rodents, insects, fruits

Virginia opossum

Mating behavior:	Promiscuous
Breeding age:	1
Young per season:	5–7
General habitat:	Coniferous and deciduous forests near water
Food habits:	Will eat anything available (dead or alive)

White-footed mouse

Young per season:	4–5/litter, 3–4 litters
General habitat:	Brushy areas in coniferous and deciduous forests
Food habits:	Seeds, nuts, insects

White-tailed deer

Breeding season:	September–February
Mating behavior:	Polygamous
Breeding age:	1
Young per season:	1–2
General habitat:	Coniferous and deciduous forest with brush understory
Food habits:	Browse, forbs, grasses

Woodchuck

Breeding season:	March–April
Young per season:	4–5
General habitat:	Borders of deciduous forests and meadows
Food habits:	Forbs, grasses

Index

Photo credits
Cover: lalcreative/Shutterstock.com
Figures 1-3, 1-4, 2-5: oleggano/Shutterstock.com; rudvi/Shutterstock.com; eva_mask/Shutterstock.com; stasia_ch/Shutterstock.com; lhnatovich maryia/Shutterstock.com; alfaori/Shutterstock.com; ylk2007/Shutterstock.com; Potapov Alexander/Shutterstock.com
Figure 1-5: Yuliya Koldovska/Shutterstock.com; rraya/Shutterstock.com
Figure 1-8: Bullion Photos/Shutterstock.com